Biomass, energy, and environment

Biomass, energy, and environment

A developing country perspective from India

N. H. RAVINDRANATH

Centre for ASTRA and Centre for Ecological Sciences, Indian Institute of Sciences, Bangalore

and

D. O. HALL

Division of Life Sciences, King's College, London

OXFORD NEW YORK TOKYO
OXFORD UNIVERSITY PRESS
1995

Oxford University Press, Walton Street, Oxford OX2 6DP

Oxford New York
Athens Auckland Bangkok Bombay
Calcutta Cape Town Dar es Salaam Delhi
Florence Hong Kong Istanbul Karachi
Kuala Lumpur Madras Madrid Melbourne
Mexico City Nairobi Paris Singapore
Taipei Tokyo Toronto

and associated companies in
Berlin Ibadan

Oxford is a trade mark of Oxford University Press

Published in the United States
by Oxford University Press Inc., New York

A catalogue record for this book is available from the British Library

Library of Congress Cataloging in Publication Data
Ravindranath, N. H.
Biomass, energy, and environment: a developing country perspective from India / N.H. Ravindranath and D.O. Hall.
Includes bibliographical references and index.
1. Renewable energy sources–Environmental aspects–Developing countries. 2. Biomass energy–Developing countries. 3. Renewable energy sources–Environmental aspects–India. 4. Biomass energy–India. I. Hall, D. O. (David Oakley) II. Title.
TD195.E49R38 1995 33.95'39'0954–dc20 95–10584
ISBN 0 19 856436 8

Typeset by Footnote Graphics, Warminster, Wilts

Printed in Great Britain by Bookcraft (Bath) Ltd
Midsomer Norton (Avon)

Preface

There have been significant improvements in renewable energy technologies in recent times along with declines in cost. A number of success stories globally and the growing concern for the environment (land degradation, deforestation, climate change) and sustainable development, have led to world-wide interest in renewable energies and bioenergy in particular. Biomass can be converted into modern energy forms such as liquid and gaseous fuels, electricity, and process heat to provide energy services needed by rural and urban populations and also by industry. These features of modernized biomass are becoming widely recognized and this is reflected by the predominant role projected for bioenergy, in parallel with conventional commercial energy, in the renewable energy-intensive scenarios of the UNCED-sponsored studies (Johansson *et al.* 1993) and the World Energy Council (WEC 1993).

The authors have been working on various facets of biomass energy for over 15 years at the micro, national, and global level. Out of this experience grew the firm belief, as will be expanded in this book, that in a world dominated by concerns for energy requirements, for sustainable and equitable development and for environmental conservation, bioenergy will have a significant role in both developing and industrialized countries. It is this resolve and practical understanding that prompted us to attempt to write a book on bioenergy and environmental issues. We hope that the present analysis of a large developing country such as India will be a precursor for detailed biomass energy analyses in many more developing (and industrialized) countries in the decades to come.

In the literature, which is generally poor on bioenergy in developing countries, there is a lack of detailed and country-specific studies on the potential, prospects, and problems of bioenergy. It should be recognized, however, that any analysis of bioenergy must start at the local or country level. This approach reflects the variations in socio-economic and biomass resource situations, especially because policy decisions are made at the country or regional level within a country. India offers a good opportunity to start such an analysis owing to its relatively good database, the diversity of existing biomass resources and opportunities, and a number of innovative programmes which have been implemented with varying degrees of success.

In the past 10 years or so, considerable practical experience has accumulated in India, as well as in other developing and industrialized countries, on biomass energy production and conversion. India is a pioneer among developing countries, with significant indigenous efforts in promoting renewable energy technologies, with the setting up of a Department of Non-Conventional

Energy Sources in 1982, later upgraded to a Ministry in 1992. In fact, some of the bioenergy programmes, e.g. biogas production, were launched even prior to the first oil crisis in 1973, although they achieved focus and significant funding only with the formation of the Department.

In this book we have attempted to put together the relevant experiences on bioenergy in India. We analyse the current situation, compare bioenergy and other options for promoting development, explore the potential for bioenergy, endeavour to understand the socio-economic and environmental impacts, and suggest policy decisions and guidelines required to promote bioenergy options.

Our case study shows that the energy needs of the rural population (accounting for 75 per cent of India's 844 million population) could be met using modern bioenergy options. Additionally, this would make a significant contribution to alleviating the stress on forests and the national electricity grid, thus paving the way for sustainable development and reduced dependence on oil imports. The basic premise of the analysis is to provide quality fuels to meet the energy services required to promote the quality of life of the rural population. This could be achieved through technologies that are sustainable and which promote local control, self-reliance, and equity. The new bioenergy technologies satisfy these criteria and further provide other tangible benefits such as local employment, land reclamation, and reduction in emissions of greenhouse gases. Land availability is not a constraint even in a densely populated country such as India.

The majority of developing countries share common problems with India, such as high population growth rates, rural populations with little access to modern fuels, high dependence on traditional biomass energy, low commercial energy use, large and non-sustainable dependence on oil imports, and ultimately low quality of life, particularly for rural populations. Our analysis is thus relevant to the majority of developing countries. If bioenergy options are feasible for India, which is a large, densely populated country with poor biomass resources, then, technically, bioenergy should be more feasible for most other developing countries in sub-Saharan Africa, South-East Asia, and Latin America; these areas are relatively rich in biomass resources, have lower population densities, and have higher forest and pasture area per capita.

We are conscious of the fact that even though the potential for bioenergy exists there are financial and institutional barriers, although none are insurmountable given public acceptability and political commitment. Bioenergy as an energy source is also certainly not without its problems, like all other energy sources, but biomass has several significant advantages compared to other energy sources. Undoubtedly the energy plans of any country will consist of a diversity of energy sources and technologies, including renewable as well as non-renewable sources, and, consequently, bioenergy will only be a part of the overall supply mix. The present study is a modest attempt to focus on the large bioenergy potential which exists in any country but at present is very poorly utilized in supplying energy services in parallel with conventional energy systems. We also show the unique opportunity that bioenergy provides for local, national, and global benefits without any of the serious conflicts that often characterize conventional energy options.

It is hoped that the present analysis of bioenergy options for India will assist energy planners and researchers in other developing countries, thereby encouraging widespread analysis of the bioenergy potential. The approach adopted in this study may be a useful starting point to allow more rigorous assessments of, and planning for, bioenergy programmes. One of the major problems faced by researchers on issues related to biomass is the poor data availability. We would feel that the book has served a purpose if it leads to initiation of studies to improve the analysis of biomass for practical use in energy planning and implementation. This would lead to a better understanding of the crucial role bioenergy can play in developing countries in promoting sustainable and equitable development.

Bangalore and London N.H.R.
March 1995 D.O.H.

Acknowledgements

This book resulted from nearly 20 years' work in rural South India, which included 10 years of continuous stay, sustained interaction with communities, and research and development work in Ungra and neighbouring villages. This is also a product of interaction spanning over a decade with David Hall. My association with the two centres of excellence in India, the Centre for ASTRA and the Centre for Ecological Sciences of the Indian Institute of Science, and the dedication of my colleagues to high-quality scientific work as well as their social commitment have contributed a great deal to my efforts. Particularly, I am grateful to Madhav Gadgil and K. S. Jagadish for their sustained encouragement and support over the years (N.H.R.).

We have benefited enormously from the suggestions and comments from Gerald Leach, Niranjan Joshi, and Veena Joshi. We also remember the many useful discussions with Frank Calle, Jeremy Woods, and Jo House at King's College, London. We acknowledge the support provided by H. I. Somashekar, D. M. Bhat, P. R. Bhat, C. M. Shastry, and others at Ungra and Sirsi over the years (N.H.R., D.O.H.).

We thank Sarah and Shashikala for reading through the manuscript and for many useful editorial comments and L. S. Laxmi for so patiently and efficiently helping us with the word processing. P. Sudha and P. V. Rekha also helped us in the final stages (N.H.R., D.O.H.).

We acknowledge the financial assistance from the Stockholm Environment Institute, the Commonwealth Science Council, the Hall Foundation, and King's College London. Support from the Ministry of Environment and Forests, New Delhi to the Centre for Ecological Sciences helped us to undertake many small studies (N.H.R., D.O.H.).

I am grateful to Shailaja for her enormous moral support and thank her and Darshini for tolerating my frequent and long absences. Krishna Rao and Srinath helped me in so many ways to facilitate my stay and work during all my trips to London. I thank U. Shrinivas for solving many of my problems and making life a bit easier for me so I could complete the book (N.H.R.).

Contents

1
Introduction

1.1 ENERGY AND DEVELOPMENT

Developing countries firmly believe that economic growth and development are necessary to reduce poverty and human deprivation. Such growth and development also offer a way of limiting unsustainable population increases; any constraints on growth could be counterproductive (WEC 1993). Thus developing countries are keen on achieving higher growth and exports. Inherent in their plans for economic expansion is the positive relationship between energy and GDP (gross domestic product). For example, the commercial energy growth rate in India during the 1980s was around 4.5 per cent (electricity at 9.2 per cent) per annum, comparable with a GDP growth rate of about 5 per cent (PC 1992). Similarly, in low-income countries generally, commercial energy consumption is growing at 3.6 per cent annually compared to a GDP growth rate of 2.5 per cent (WB 1992). What is disturbing is that the oil and electricity intensities of GDP for a country such as India are increasing even after past experiences of oil supply and price crises (TERI 1992*a*; OECD 1993).

Commercial energy consumption is expected to continue to rise more rapidly than the growth of national economies of developing countries (OTA 1992). Future projections, assuming a business-as-usual scenario under a GDP growth rate of 5 per cent, show that the commercial energy demand for India for the year 2009–10 is projected to be 600 Mtoe (million tonnes of oil equivalent) compared to the 1990 consumption level of 200 Mtoe (TERI 1992*a*). The proportion of developing countries' national budgets dedicated to promoting commercial energy is very high. In the case of India, 28 per cent of the five-year development plan outlay for the period 1992–97 will be spent on commercial energy (PC 1992). Such a large allocation for energy is at the cost of many other aspects of development.

A majority of developing countries in South and South-East Asia, sub-Saharan (SS) Africa and South America import oil and are acutely aware of the problems attached to such a dependency. Such oil-dependent growth is highly vulnerable to fluctuations in oil and foreign exchange markets. Currently, most of the emphasis of energy development is exclusively on electricity (centralized grid) and oil. What is often neglected is the high dependence of developing countries on traditional biomass fuels. In developing countries 38 per cent of total energy is provided by traditional biomass fuels (Hall *et al.* 1992), and in countries such as Bangladesh, Kenya, and Paraguay the share of traditional biomass fuels can be as high as 75–90 per cent (UNEP 1994). Such neglect of biomass fuels is obvious from the low or insignificant allocation of funding to bioenergy development in India compared with commercial energy (PC 1992).

The current pattern of commercial energy-oriented development has led to very large increases in the consumption of electricity and oil, resulting in inequities, external debt, and environmental degradation, as has been experienced by many developing countries. For example, large proportions of the rural population and urban poor (as will be seen in detail later) continue to depend on low-quality energy sources such as fuelwood and inefficient devices such as low-efficiency stoves for cooking and kerosene lamps for lighting. The question arises as to how long this dependency and deprivation can continue. There is a need to question the desirability and sustainability of dependence on commercial fuels, particularly for the oil-importing countries. Many conclude that the conventional paradigm of development characterized by commercial energy-supply-oriented policies is non-sustainable (Reddy *et al.* 1991). Thus the search is on for alternative approaches to energy and development which include the efficient use of energy and alternative sources of energy, particularly renewables.

An energy strategy for a country should be coupled with the goals of development so as to provide the energy services that people require. If the goals include the criteria of sustainability, promotion of equity, and self-reliance, the energy strategies of many countries may need to be redefined.

1.2 ENERGY AND ENVIRONMENT

Energy consumption patterns have two critical components, namely, socio-economic and environmental. Only the environmental aspects are considered here while socio-economic issues are discussed in later chapters. Major environmental issues confronting the world today are considered to be deforestation, decline in biodiversity, pollution of air and water, global warming leading to climate change, and the degradation of soil and water resources. One of the major contributors to these environmental problems is energy extraction and use. The industrialized countries' energy use is considered to be the major contributor to the degradation of the global environment. But what is probably more important is the degradation of the local environment, for which individual countries bear a larger responsibility, particularly in the developing world. The fuels consumed by developing countries include biomass, oil, and coal, which account for 38, 25, and 23 per cent, respectively, of the total energy use (Hall *et al.* 1992). Fuelwood and charcoal are the dominant fuels in developing countries in the domestic sector and to some extent in certain industries (such as charcoal-based pig-iron manufacture in Brazil, and wood use for brick and tile manufacture in many countries).

It has been suggested that domestic fuelwood use in India is the single most important cause contributing to the depletion of forests (FSI 1988). This contention is now being questioned; as will be seen later, the fuelwood requirement of rural communities may not be a factor contributing to 'deforestation' but surely contributes to forest 'degradation' due to the non-sustainable removal of woody biomass. In Africa, fuelwood consumption is already contributing to deforestation and soil erosion in highly populated areas, and many regions in SS Africa are facing an acute scarcity of fuelwood (Davidson 1992). Non-sustainable use of forest biomass also leads to the loss of a range of forest products and plant diversity. A further serious environmental problem is that combustion of fuelwood for cooking leads to indoor air pollution with serious health consequences (Smith 1993), while the emission of CO_2 from fuelwood used for cooking is estimated to contribute to about 2 per cent of global warming (Ahuja 1990). The use of coal in developing countries is associated with environmental degradation in mining, transportation, and, more importantly, during combustion, leading to local air pollution and global warming (ADB 1991). Hydroelectric power projects, although a renewable energy option, are not environmentally benign as they may lead to the submergence of forests and the displacement of communities.

Environmental concerns have generally received a low priority in developing countries so far, but are now increasingly becoming an important factor in the making of decisions on energy. For example, India's development plan (1992–97) states that efficient use of energy resources and long-term sustainability are the two key factors in energy planning (PC 1992). Developing countries cannot ignore environmental issues while striving for economic growth since, in the final analysis, the poor are harmed more by pollution and resource degradation than the rich (Siddayao 1992). Thus energy planners are increasingly looking towards renewable sources of energy.

1.3 ROLE OF RENEWABLE ENERGIES

Renewable energy sources such as wind, solar, hydro (particularly small hydro), and biomass are receiving increased attention in developed as well as developing countries. Such renewable energies offer developing countries the prospect of increasing their energy supplies in a self-reliant way at national and local levels along with the attendant economic, social, and security benefits (WRI 1994). Long-term sustainable development in all countries, and particularly in the developing world, requires a gradual shift towards renewable sources of energy that are more equitably distributed and less environmentally destructive than fossil fuel sources (Goldemberg *et al.* 1988). Renewable energy technologies have evolved markedly over the past two decades while the costs of all renewables have declined significantly (WB 1992) and are projected to decline even further in the next 5–10 years

(Ahmed 1994). Thus the commercial development of renewables may be justified even on non-environmental grounds. In India, for example, the goals of energy planning include the promotion of decentralized energy technologies based on renewable resources in the medium term, and the promotion of energy supply systems based on renewable sources of energy in the long term (PC 1992). Nevertheless, other than hydroelectricity, renewables have not made much impact in developing countries as suppliers of modern energy services despite numerous and varied plans and projects over the past 20 years. This is due to various reasons, such as a lack of commitment on the part of energy planners, technical failures, and higher costs as compared to petroleum (Foley 1992*b*). Renewables continue to face these and other barriers although enhanced global awareness of environmental degradation due to dependence on conventional energy sources, along with technological developments of renewables, has given new impetus to renewable sources of energy.

New projections show that by 2025, given adequate support, renewables could contribute nearly 30 per cent of direct fuel use and 60 per cent of global electricity supplies. Further projections for 2050 show that much of the world's growing energy needs could be met by renewables at prices lower than those forecast for conventional energy (Johansson *et al.* 1993). The recent World Energy Council report (WEC 1993) makes a conservative projection (Table 1.1) for renewables and, according to the minimum possible scenario for 2020, 'new' renewables would meet 3–4 per cent of total energy, amounting to 539 Mtoe (compared to the 1990 level of 164 Mtoe). Under the maximum possible scenario with major policy initiatives, renew-

Table 1.1 World Energy Council projections of 'minimum'/'maximum' contribution from 'new' renewable energy (WEC 1993)[a]

	In 2020 'minimum'		In 2020 'maximum' with major policy support	
	Mtoe[b]	% of total	Mtoe	% of total
'Modern' biomass	243	45	561	42
Solar	109	20	355	26
Wind	85	16	215	16
Geothermal	40	7	91	7
'Small' hydro	48	9	69	5
Oceanic	14	3	54	4
Total	539	100	1345	100
% of total energy demand		3–4		8–12

[a] In 1990 'new' renewables contributed 164 Mtoe (1.9%) to total energy demand.
[b] Mtoe = million tonnes of oil equivalent.

ables could provide 8–12 per cent of total energy by 2020 (1345 Mtoe). The recent *Energy Policy* review concluded that renewables offer considerable potential for displacing conventional energy sources, and in some cases are already competitive with them, in addition to offering environmental advantages (Jackson 1992). This emphasizes the need to make detailed studies at the country level, both to estimate the potential of each renewable energy source to meet future needs, and to highlight environmental and socio-economic benefits. Among the renewables, the focus of this book is on biomass energy, and the rationale for this is presented in the next section and in detail in the later chapters.

1.4 WHY BIOMASS ENERGY?

Biomass is a renewable fuel used in nearly every corner of the developing world as a source of heat, particularly in the domestic sector. Biomass energy includes energy from all plant matter (tree, shrub, crop) and animal dung. Currently biomass energy is characterized by a low efficiency of use (Dutt and Ravindranath 1993) and a low quality of life due to the drudgery associated with its gathering and use. However, biomass, unlike other renewables, is a versatile source of energy which can be converted to 'modern' forms such as liquid and gaseous fuels, electricity, and process heat. Bioenergy also permits operation at varying scales. For example, small-scale (5–100 kW), medium-scale (1–10 MW), and large-scale (about 50 MW) electricity generation systems or biogas plants of a few cubic metres (Indian and Chinese family plants for cooking) to several thousand cubic metres (Danish systems for heat and electricity). This variety of scales is useful for power generation for decentralized applications at the village level as well as for supply to the national grids. Bioenergy is particularly attractive for decentralized applications for producing gaseous fuels or electricity. Unlike solar, wind or micro-hydroelectric systems, modern biomass energy systems could be set up in virtually any location where plants can be grown or domestic animals are reared. Renewables such as solar, wind, and micro-hydro require 'spare' or additional capacity to produce adequate energy when the conditions are right, such as water flow or wind speed. This intermittent feature of such renewable energy sources necessitates electricity storage facilities, especially with small and local systems, if sustained demand requirements are to be met (WEC 1993). Bioenergy sources such as producer gas systems do not require electricity storage.

In short, **biomass energy** systems offer an opportunity for **sustainable** (as biomass can be grown sustainably), **self-reliant** (biomass is available in all countries and can be converted to gaseous or liquid fuels and electricity, leading to a decrease in imports of oil), and **equitable development** (between and within countries due to universal availability of biomass and the fact that decentralized bioenergy systems lead to local control and employment).

According to WEC (1993), under the minimum case scenario, modern biomass is the most important of the renewables and is projected to account for 45 per cent of the new contribution by renewables to world energy by 2020. In the maximum scenario projections, modern biomass will account for 42 per cent of the total renewable energy contribution by 2020. The US Department of Energy suggests that biomass power will be the most important renewable energy option for the next quarter of a century and has projected a possible 25 GW in installed power in the US alone by 2010 (USDOE 1993). This clearly establishes a dominant role for biomass energy among all the renewables.

A study on the potential role of biomass gasifiers in India has shown that the electricity needs for lighting and stationary shaft power of half a million villages could be met sustainably through decentralized biomass electricity while requiring only a part of the existing degraded lands for biomass production (Ravindranath 1993). Biomass-based village energy systems, where all the energy services of a village ecosystem are met by bioenergy technologies, have been shown to be technically feasible (Ravindranath and Chanakya 1986).

In addition to meeting the heat and electricity requirements at a micro level, biomass has the potential to supply electricity to the national grid (e.g. the existing 8500 MW in USA) and liquid fuels for transportation (12 billion litres of ethanol used annually in Brazil). Biomass energy options are feasible with currently available off-the-shelf technologies (e.g. biogas, producer gas, and alcohol) and with those that will become available soon (e.g. advanced gasifier/gas turbines). Additionally, the large-scale utilization of biomass energy can provide a basis for rural development and employment in both developed and developing countries. A further significant benefit is that, if used in a sustainable way, bioenergy could lead to no net emission of CO_2 and, in fact, would lead to a reduction of greenhouse gas emissions when used as a substitute for fossil fuels (Hall *et al.* 1990). Land reclamation is also possible if woody biomass is grown on degraded lands.

In spite of all its attributes, biomass is not the complete answer to the energy problems of developing countries, since it has its own limitations. It requires land and water, has a low energy density, and can lead to loss of biodiversity. However, biomass has many advantages and can overcome the perceived problems. One may rightly ask why, with so many advantages, biomass energy has not commanded the attention of energy planners in developing countries and why it has not made any significant impact on commercial energy supplies? It is a fact that in many countries, such as India and China, biomass energy technologies, in spite of several subsidies and incentives over the past 10–15 years, have not achieved any serious market penetration. Sceptics may, for example, point to the failures of biomass gasifier projects in developing countries as in South-East Asia (BTG 1993) or the problems with the small gasifier programme in India (Subramanya *et al.* 1993), or the poor performance of the community biogas programme in India (Venkataramana 1991). Some may raise the old and often repeated question of food-versus-fuel and land availability in developing countries for biomass production in the face of growing populations. Large biomass energy systems may also result in loss of biodiversity, unacceptable levels of effluents and emissions, and loss of alternative outputs (WEC 1993). Biomass energy production and utilization is also site specific, and still faces many economic, social, and institutional barriers that need to be resolved at both local and national levels (Hall 1991). All these issues will be considered in some detail in this book. However, it is worth making a few points on why biomass merits a new look.

First, technological developments in the recent past and the possibility of efficient and reliable bioenergy systems before the turn of the century are very promising while their relative costs are declining (Johansson *et al.* 1993; Ahmed 1994). Secondly, a number of examples exist today of functioning small, as well as large, biomass energy systems in industrialized, as well as developing, countries. Thirdly, there is greater awareness of the environmental impacts of conventional energy sources at both the local level (land submergence or local air pollution) and the global level (global warming or loss of biodiversity); governments are now more receptive to environmentally sound and sustainable options such as bioenergy. Fourthly, conventional energy-based options have not resulted in equitable and sustainable development, nor have they led to provision of reliable, affordable, and quality energy services, particularly to rural populations in developing countries. Finally, bioenergy provides an opportunity to convert the universal biomass fuel which is currently used for low-grade heat in an inefficient and polluting way, to more acceptable high-quality fuel forms, such as liquid or gaseous fuels and electricity, for rural as well as urban applications. Bioenergy systems will only involve shifts from the current traditional forms of use of biomass to conversion to modern forms through the involvement of new technology and investment. Biomass production programmes for energy could even become part of land reclamation and rural development programmes.

A review of renewable energy technologies reveals that even technologies that are competitive, under common assumptions of energy economics, are unable to compete equitably against conventional fuel sources within the existing infrastructure and policy framework (Jackson 1992). This review further concludes that, provided that the appropriate institutional steps are taken and there is sufficient will and institutional flexibility, renewable energy offers considerable scope for a sustainable, affordable, and equitable energy future. At the same time, one must be realistic about the feasibility of significant shifts in energy supply systems, given that in many developing countries subsidy levels on conventional fuels are too high (Foley 1992*a*). For example, in India electricity and kerosene are acknowledged to be highly subsidized (CMIE 1994). Energy pricing is more a political decision than an economic decision to permit any rational choice of options. Furthermore, energy utilities are largely state controlled in developing countries, although some changes are taking place.

Assessing the potential of biomass energy in developing countries and arriving at effective implementation policies is a complex task. It involves consideration of social and economic features and technical capabilities at both country and local levels. At the country level, there is a gradient of development among developing countries, such as from Brazil to Burundi and Malaysia to Mali. Country-level studies alone can reveal the overall bioenergy potential, prospects, and problems.

1.5 A CASE STUDY OF INDIA

India, with a population of 844 million in 1991, is the world's second most populous country after China. Among developing countries, India offers an excellent opportunity to assess the potential and performance of bioenergy in contributing to national development and environmental conservation. India recognized the potential of bioenergy early and had a clear policy to promote it through support for research, development, and demonstration programmes (MNES 1993*a*). In 1982 it set up a Department of Non-Conventional Energy Sources, which was upgraded to an independent ministry in 1992. The ministry has well-defined programmes on biomass production, conservation, and conversion to energy. Currently bioenergy is a dominant component among all the renewable energy programmes of the ministry (MNES 1993*a*). Most of the bioenergy technologies being implemented are indigenously developed and thus also provide possibilities for South–South technology transfer. Fortunately, India has a reasonably well-documented set of field projects on the performance of bioenergy technologies, which have been disseminated through commercial as well as government subsidized and managed programmes. India thus provides a wide spectrum for analysis and learning. One problem a researcher has to contend with in dealing with the biomass situation in many developing countries is the lack of data on various aspects of biomass (Hall and Mao 1994). India has a relatively good database compared to many developing countries which enables detailed analysis of the bioenergy potential.

In India, energy planning is considered at three time levels; short, medium, and long term. The country has clearly set out to promote renewable energy programmes in the medium and long term, and bioenergy is an important component of the development plan programme. A detailed analysis of biomass energy programmes in India is necessary to understand the features of current biomass energy use patterns, learn from the programme experiences so far, estimate the potential, and formulate policy measures that could lead to the realization of the full potential of biomass energy. There is also an important need to analyse specifically the social, economic, and environmental implications of such bioenergy options.

This book sets out to examine bioenergy options as follows: the Indian experience is considered first, by analysing the current energy consumption patterns and their social, economic, and environmental implications (Chapters 2 and 3); secondly, by analysing the past and current attempts to promote bioenergy programmes (Chapter 4) with examples from some case studies from India as well as other countries (Chapter 5); thirdly, by analysing the potential of bioenergy as a modern fuel, the options and strategies for utilizing the potential of bioenergy to provide energy services to promote development (Chapter 6), analysis of land availability and sustainability issues (Chapter 7), and the implications for local and global environments (Chapter 8); fourthly, by the economic analysis of bioenergy options (Chapter 9); then by considering an approach to implementation of bioenergy options (Chapter 10); by comparing the bioenergy potential and feasibility in India with other developing countries (Chapter 11); and, finally, by presenting a policy framework and a set of guidelines to overcome the barriers for ensuring an acceptable place for bioenergy in national development (Chapter 12). The role of local, national, and international agencies, and the justification for promoting bioenergy, will also be highlighted.

India is a large country with diverse socio-economic situations, resource endowments, and levels of development. Thus we are acutely aware of the limitations and pitfalls of making generalizations based on a few case studies or examples. Extrapolations are a means of exploring the upper limits of the contribution of biomass to energy supply and of placing bioenergy on a comparable basis to other energy-supply strategies. Such extrapolations are facilitated by a certain element of commonality in large parts of India, as will be seen in later chapters. Biomass energy consumption patterns, in particular the dominance of biomass energy in the domestic sector, the dependence on low-quality kerosene for lighting, and the gathering of fuelwood and fetching of water with considerable human effort and drudgery, contribute to the low quality of life. No matter what the social or cultural diversity or the resource situation, every rural settlement (there could be some exceptions which at this moment are difficult for us to visualize), wherever situated, is likely to require or aspire to piped potable water, electricity for lighting, quality fuel for cooking, and a mechanized flour mill. This is important as no one would like to walk miles to fetch water or gather thorny twigs or spend hours preparing animal-dung cakes when this precious time could be used for other more productive or leisure activities. Since an energy-use ladder exists in urban areas in a situation of increased incomes and access to better fuels, why not facilitate an energy ladder for rural areas? We readily recognize that the ultimate choice of fuels or devices must be left to the individual farmer, housewife, or entrepreneur in rural areas, but it is the responsibility of the

Table 1.2 Land, population, forest area, energy use, and oil imports of selected developing countries for 1990

Countries	Population 1990 (million)	Geographical area (Mha)	Rural population (% of total)	Population density (no./km^2)	Forest area per 1000 population (ha)	Population growth rate (1990–95)	Per capita annual energy use (GJ)			% of total oil use imported
							Commercial	Biomass	Total	
Asia										
India	844	328	74	257	76	2.20	9.0	4.0	13.0	40
Indonesia	184	181	72	107	595	1.78	9.0	8.1	17.1	21
Philippines	62	29	58	223	126	2.07	9.0	5.6	14.6	99
Sri Lanka	17	6	78	261	102	1.33	4.0	5.0	9.0	100
Africa										
Ghana	15	22	67	72	637	3.00	3.0	10.5	13.5	100
Kenya	24	56	77	46	50	3.35	3.0	14.6	17.6	100
Senegal	7	19	62	41	1078	2.70	4.0	0.6	4.6	100
Tanzania	26	88	69	32	1290	3.36	1.0	13.8	14.8	100
Zaire	37	226	61	18	3061	3.17	2.0	10.1	12.1	100
Zambia	8	74	45	12	4037	2.84	6.0	16.2	22.2	100
Zimbabwe	10	38	73	28	890	2.97	19.0	6.5	25.5	100
Latin America										
Brazil	149	845	28	185	3766	1.59	22.0	12.8	34.8	48
Developing world	3828	7645	–	50	700	–	18.3	12.5	30.8	–

Sources: Population, geographical area, and population density, WRI (1992); rural population, UN (1991); forest area, FAO (1993*a*); population growth rates, UNEP (1994); commercial energy and % of oil use imported, UN (1992); biomass, FAO (1993*b*).

development planner to create affordable and accessible improved energy alternatives to improve the quality of life.

It would be wrong to say that a certain programme (such as a community biogas or producer gas system or community forestry programme) may work in one district and may not work in the next due to social problems (or conflicts), absence of skills, or a low level of development. The approach to a programme may vary, for example the mode of implementation, whether it be by government agency, private enterprise, through non-governmental organizations (NGOs), or by the local community itself. There could be questions as who should manage the systems (government, private enterprise, or local community); whether to use biogas for cooking or electricity generation; whether to promote a single megawatt system or ten 100 kW systems, or a system for every village or a cluster of villages; and the *modus operandi* of programmes, such as raising a community energy forest (species choice, mode of harvest, management). Such problems are complex but resolvable, as will be shown in later chapters. Thus, through appropriate institutional and financial arrangements, nearly every location has the potential to succeed in organizing and managing a decentralized development plan based on bioenergy.

We have attempted to explore the potential of biomass energy in meeting the goals of development and to highlight the technical potentiality and broad policy initiatives for India as a whole, which could be a starting point for more detailed location-specific studies in India. We also consider that it may be feasible to draw some lessons from this exercise for other developing countries.

1.6 BIOMASS ENERGY IN INDIA AND DEVELOPING COUNTRIES

It is useful to compare the energy situation in India with that in other developing countries and to look for commonalities, if any. Thus we can explore the need for an alternative approach to energy planning, the potential role of biomass energy, or at least the technical potentiality of biomass in meeting the energy needs of development in other developing countries. We fully realize at the very outset the limitations of such comparisons, let alone suggestions for any policy initiatives. Developing countries are all different from one another, so what is appropriate for Brazil may not be so for Bangladesh, given the wide variation in resource base besides the differences in cultural, social, and economic situations. Keeping these constraints in mind, we try to analyse the essential features of the energy situation amongst the developing countries, selecting a set of countries to represent some major regions but making no claim to having covered all regions or of the representative nature of these countries. Some of the features seen in Table 1.2 are as follows.

Dominance of the rural population People in most developing countries live in rural areas, except for countries such as Brazil and Zambia. Rural predominance varies from 58 per cent in the Philippines to 78 per cent in Sri Lanka (and 74 per cent in India). The predominance of rural populations is important since energy consumption patterns and requirements vary for rural and urban areas.

Population density Asian countries have higher population densities than SS Africa or Brazil. India has a higher population density than other major Asian countries. Thus the human pressure on land in India is higher than in most African and Asian countries, or in Brazil.

Forest resource When the per capita area under forests is assessed, India is seen to be poorly endowed with forest resources (FAO 1993*a*). Indonesia, Brazil, and SS Africa have between 8 and 50 times more area under forest per capita than India. Thus any biomass-based energy option which is feasible for India with its relatively poor biomass resource, should theoretically be more feasible for the majority of developing countries.

Commercial energy consumption Commercial energy use (1990) is less than 10 GJ/capita/year in most developing countries (except Brazil and Zimbabwe; UN 1992). For most of SS Africa it is less than 6 GJ/capita/year. India (9 GJ/capita/year) is in the same league as Indonesia and the Philippines. India is therefore a mid-level country compared to, say, Brazil on the one hand and SS Africa on the other. This masks the rural and urban inequities in commercial energy use. The crucial issue here is the extremely low commercial energy use among most developing countries compared with industrialized countries, where levels are over 200 GJ/capita/year (OECD 1993).

Biomass energy use Table 1.2 includes only fuelwood (plus charcoal) data and does not include crop residues and cattle dung, and hence gives an underestimate of total biomass energy use. Fuelwood use in India is lower than in Indonesia and SS Africa (FAO 1993*b*). The share of fuelwood in India is 30 per cent of the total energy use, whereas it is higher in all other developing countries, particularly in SS Africa. In some countries fuelwood alone accounts for over 80 per cent of total energy (e.g. Tanzania, Kenya, and Zaire). Thus fuelwood is an important and dominant energy in developing countries.

Dependence on oil imports The majority of developing countries in Asia and SS Africa, and even Brazil, are oil-importing countries. India imports about 40 per cent of its oil consumption, whereas the Philippines, Sri Lanka, and most SS African countries depend wholly on imported oil. Like India, other developing countries would be keen on reducing their dependence on imported oil if that were at all possible.

Population growth rates Population growth in Asian countries including India is lower than in SS Africa, where it is around 3 per cent for most countries. This will lead to an escalation in demand for both biomass fuels and commercial energy, which could lead to increased pressure on forests, and to external debt and large outflows of foreign exchange—a situation that all countries would want to avoid.

We conclude from this comparison with a sample of developing countries in Asia and SS Africa that there are common features encompassing a dominance of rural populations, high population growth rates, high levels of dependence on biomass fuels, low commercial energy consumption, and high dependence on imported oil. Other developing countries also face problems similar to those of India but in some respects their problems are more severe because of higher population growth rates and higher dependence on imported oil. Interestingly, most other developing countries are better endowed with land and biomass resources than India. Thus while it looks as though biomass energy has a high potential to provide modern fuels in India, and to contribute to an improved quality of life, theoretically at least, most other developing countries could also usefully explore their own biomass energy options since they have larger biomass resources than India and lower population densities. Lower population densities entail extensions of the grid that would prove not only expensive for consumers but also for the utilities, as shown for a number of African countries such as Zambia, Ethiopia, Lesotho, and Somalia (Ranganathan 1992). In such countries, decentralized biomass-based systems should be even more attractive as populations are thinly spread over a larger geographical area compared to India.

Land-based programmes, such as bioenergy, are location-specific, and technical potential is only one of the many conditions required for modern bioenergy to make a realistic beginning or future impact. We fully recognize that the analysis of biomass resources, technological options, potential and feasibility of bioenergy systems and their economics for India allows only some broad lessons and generalizations to be drawn for other developing countries. Thus we do not make any policy recommendations for any country. For any specific recommendations, country studies are required and, if possible, region-specific studies within a country.

2
Energy and biomass consumption patterns in India

The main focus of this book is to explore the potential and feasibility of bioenergy options in meeting the energy needs of a developing country such as India and to draw some lessons from this for other such countries. The first step in such an exercise is to understand the current energy scene, particularly that of biomass energy. Detailed information is required on the sources and end uses of biomass fuel, sustainability of its supply, features of use such as efficiency and socio-economic aspects, and production and conservation potential. Unfortunately, the data available on biomass energy are very poor compared to commercial energy data in most countries. Among developing countries, India was a pioneer in rural energy studies in the aftermath of the first oil crisis, with detailed micro-level (Ravindranath *et al.* 1981; Reddy 1982) and macro-level (NCAER 1981, 1985) studies. The Indian energy research scene has been active and several studies are available on rural energy use (ABE 1985; Leach 1987; Veena 1988; Bose *et al.* 1991; Bhide 1992; Joshi *et al.* 1992*b*). This does provide an opportunity to analyse the bioenergy situation in India. But as seen later in this chapter, the database is rather weak on certain aspects of biomass energy, for example, on the sources of the fuelwood used in rural India. In this chapter, an attempt is made to consolidate the existing information on various aspects of energy use in India with particular reference to biomass energy.

2.1 PRIMARY ENERGY USE IN INDIA

Primary energy consumption in India was in the range 356–425 Mtoe (or 14.9–17.8 EJ/year) in 1991, depending on the extent of biomass use considered for the domestic sector (discussed in later sections). The share of biomass energy ranged from 36 to 46 per cent (Table 2.1). Among the commercial sources, coal was the dominant source of energy followed by oil, and among biomass fuels, the most dominant was fuelwood. The total primary energy use per capita was in the range 17.7–21.1 GJ/yr, which is very low compared to the OECD countries (200 GJ/capita/yr). This tenfold difference prompts energy planners in countries such as India to promote energy-supply-enhancing policies.

Table 2.1 Primary energy use in India (1991)[a]

Fuels	Quantity[b]	Using 'mean estimate' for biomass[c] Total energy use (PJ/yr)	Per capita energy (GJ/yr)	Using 'low estimate' for biomass[c] Total energy (PJ/yr)	Per capita energy (GJ/yr)
Coal	217 Mt	6325 (35.4)	7.49	6325 (42.2)	7.49
Oil	53.7 Mt	2293 (12.8)	2.71	2293 (15.3)	2.71
Natural gas	17 890 Mm^3	665 (3.7)	0.79	665 (4.4)	0.79
Electricity[d]	77 779 GWh	280 (1.6)	0.33	280 (1.9)	0.33
Total commercial		9563 (53.5)	11.32	9563 (63.8)	11.32
Fuelwood	298/[227][e] Mt	4470 (25.0)	5.29	3405 (22.7)	4.03
Crop residues	156/[97][e] Mt	2267 (12.8)	2.68	1500 (10.0)	1.77
Cattle dung	114/[37][e] Mt	1562 (8.7)	1.85	507 (3.5)	0.60
Total biomass energy		8299 (46.5)	9.82	5412 (36.2)	6.40
Total energy		17 862 (100)	21.14	14 975 (100)	17.72
Total energy		425 Mtoe[f]	0.50 toe	356 Mtoe	0.42 toe

[a] Sources: Commercial energy, TERI (1992*b*); TSL (1992); CMIE (1994). Biomass energy; refer to Table 2.2. Estimates for 1991, population 844 million.

[b] Energy conversion: coal = 1 Mt (million tonnes) = 29.15 PJ; Oil = 1 Mt = 42.0 PJ; electricity = 1 TWh = 3.6 PJ; natural gas = 1 billion m^3 = 37.19 PJ; fuelwood (air dry) = 1 t = 15 GJ; crop residue (air dry) = 1 t = 13 GJ; dung cake = 1 t = 13.7 GJ; sugarcane bagasse (dry) = 1 t = 19 GJ; rice husks = 1 t = 12.6 GJ.

[c] Joshi *et al.* (1992*b*) made three estimates for rural biomass use; low, mean, and high estimates. Section 2.5.1 explains why low and medium estimates for biomass use are employed.

[d] Includes hydro and nuclear sources of electricity.

[e] Values within [ ] are low-end estimates of fuelwood, crop residues, and cattle dung cake. Figures in ( ) are percentages of the column total.

[f] Mtoe = million tonnes of oil equivalent.

However, the limitation in using the per capita primary energy data for international comparisons needs to be recognized since there are large variations in the efficiency of energy end use and there are differences in the services requiring energy. For example, India and most other tropical countries do not require energy for heating homes in winter. Furthermore, in the case of cooking, which is one of the chief energy-using activities, it can be shown (Fig. 2.1) that to cook a given quantity of food, the primary energy required declines significantly with a shift from traditional to modern fuels; for example, by a factor of five on changing from traditional wood stoves to liquid petroleum gas (LPG), due to differences in the efficiencies of the devices. Thus not only is the energy use level low in India, but even the efficiency of use of a dominant fuel such as biomass is low.

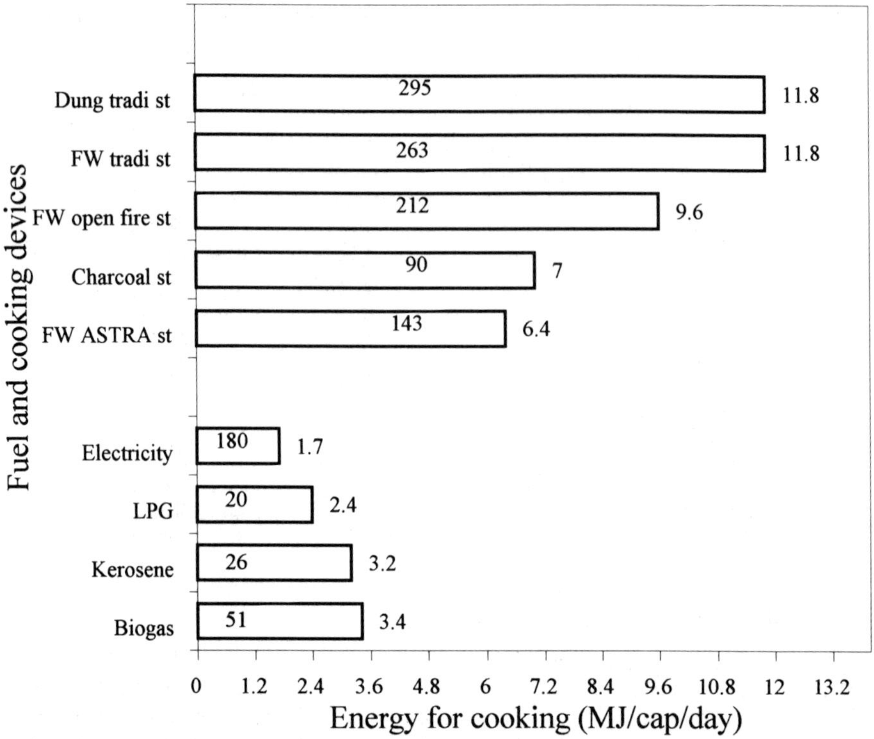

Fig. 2.1 Energy required for cooking a typical meal in the Ungra area of South India using different fuel–device combinations. Values outside the bar are in MJ/capita/day. Values inside the bar are annual per capita energy required (in kg) for fuelwood (FW), dung, charcoal, LPG (liquid petroleum gas), kerosene; in kWh/capita/yr for electricity; and in m^3/capita/yr for biogas. st, Stove; ASTRA, an improved stove design. (From Dutt and Ravindranath 1993.)

2.2 GROWTH OF ENERGY CONSUMPTION

Total energy

Figure 2.2 shows clearly that the use of all fuels in India is increasing. This could be due to increasing population, economic growth rate, and energy intensity per unit of GDP. In developing countries as a whole, energy intensity is increasing, whereas it has been declining in developed countries since 1970 (OECD 1993).

Between 1981 and 1991, the Indian use of commercial energy increased by 4.5 per cent annually, and projections for energy use reveal that the growth rates are likely to continue (TERI 1992*a*). To give an example, coal use is expected to reach twice the 1991 level of 217 Mt by the year 2000 (ADB 1991) and to treble by 2010 (TERI 1992*a*). Electricity use in India has been growing

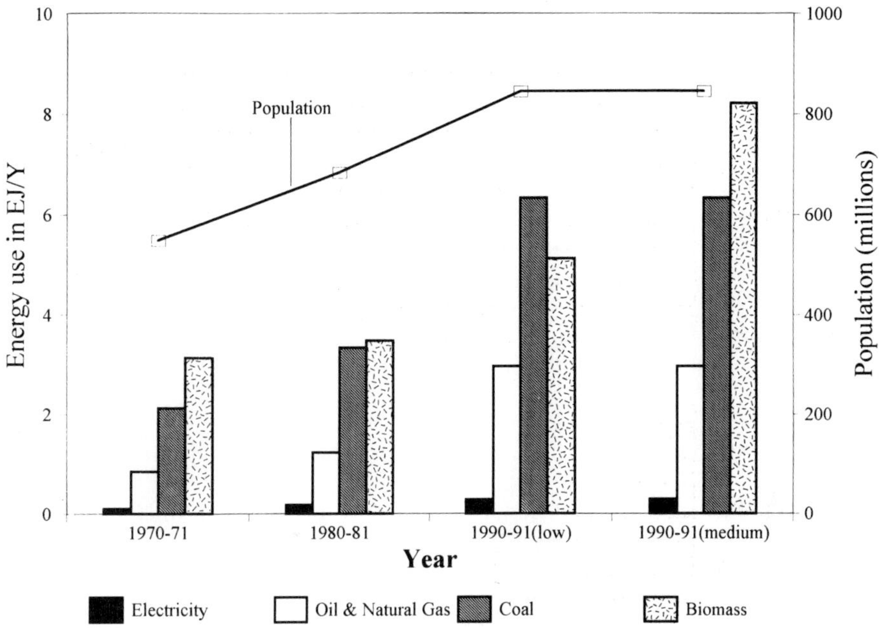

Fig. 2.2 Primary energy use and population growth in India from 1970–71 to 1990–91 (from Ministry of the Environment 1986; PC 1992; TERI 1992*b*; CMIE 1994). 1990–91 (low) corresponds to a 'low estimate' of fuelwood use during 1991, and 1990–91 (medium) refers to a 'medium estimate' of fuelwood use for 1991 (from Joshi *et al.* 1992*b*).

at over 8 per cent annually since 1985 (TSL 1992). The questions that arise from such projections are: first, the desirability of a supply-oriented approach to development since alternative approaches are available where, without increasing energy use, an improvement in quality of life is possible (Reddy *et al.* 1991); secondly, whether India can afford quantitatively to import oil, generate electricity from coal, and grow trees (to supply fuelwood) to meet the projected energy needs; and, finally, whether the mere increase in energy supply at the national level will lead to improved quality of life, particularly in rural areas.

Dynamics of biomass energy use over the years

Historically, as economies developed there was a shift from biomass-based fuels to fossil fuels and electricity. Thus the role and dependence on biomass energy in India is expected to decline with economic development. However, comparison over a period of nearly 20 years has shown that biomass energy use (fuelwood, crop residues, and dung) has increased continuously (Fig. 2.3).

Over the period 1970–90, the rural population increased by 43 per cent

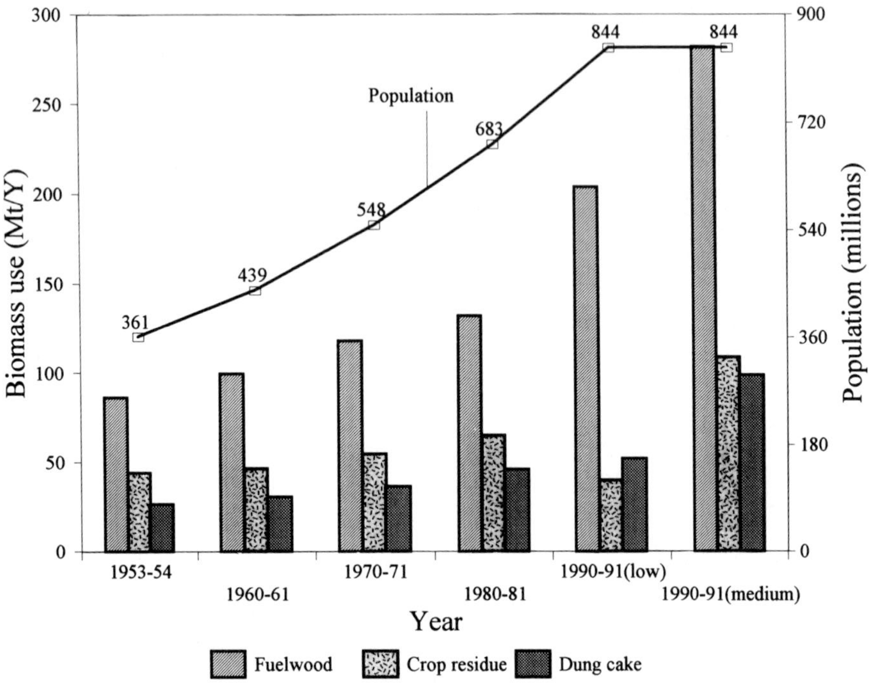

Fig. 2.3 Growth in use of fuelwood, crop residues, and dung cakes in the domestic sector in India along with the growth in population (Ministry of the Environment 1986; Joshi *et al.* 1992*b*). Low and medium estimates as in Fig. 2.2.

and, in comparison, biomass energy use increased at nearly the same rate when the 'low estimate' of biomass energy use in the domestic sector is considered (Joshi *et al.* 1992*b*). Thus, in absolute terms, bioenergy use has continuously increased and the per capita level of bioenergy use has remained nearly constant (Joshi *et al.* 1992*b*). It is important to note that the large increase in commercial energy use has not decreased the total use of biomass energy (Fig. 2.2). Population growth, especially in rural areas, coupled with the absence of alternate fuel options or efficient devices, has contributed to the growth in biomass use. Projections on biomass use as a fuel for 2004–5 show that 316 Mt of fuelwood may be required annually, with possibly all the dung produced in India being burnt as fuel (ABE 1985). Fuelwood use in other sectors (such as industries and establishments) may also increase.

The crucial implication of increasing biomass energy demand is that the pressure on biomass resources, such as forests, village trees, crop residues, and cattle dung, will increase rather than decrease, even though the use of fossil fuels in the country as a whole is increasing. We conclude that the absolute role of biomass energy has not diminished in India and is likely to continue to play an important role in the years to come. The current energy-related development programmes in India, as will be seen in detail later (Chapter 4), are aimed at conserving biomass (efficient devices) and examining how biomass may be grown to meet the increasing needs of fuelwood for the growing population (MNES 1993*a*). There are no programmes specifically aimed at replacing biomass in any significant way in the foreseeable future.

2.3 SECTORAL USE OF BIOMASS: DOMESTIC AND INDUSTRIAL

The domestic sector dominates biomass energy use, and fuelwood is the dominant source of energy in this sector in rural areas (Table 2.2). Fuelwood is also used by urban households. Dung cake use is wholly in the domestic sector while crop residue use is prevalent in both the domestic and the industrial sectors. Presently development efforts are mainly concentrated on fuelwood conservation and production, mainly for the domestic sector.

Table 2.2 Sectoral and per capita use of biomass energy in India

Type of biomass	Domestic sector (Mt/yr)[a]		Industry and establishments (Mt/yr)		Total use (Mt/yr)	Per capita energy (GJ/yr)
	Rural	Urban	Rural	Urban		
Mean estimate[b]						
Fuelwood	252	30[c]	6[d]	10[e]	298	5.29
Crop residue	99			57[f]	156[g]	2.68
Dung cake	109	5			114	1.85
Total						9.82
Low estimate[b]						
Fuelwood	181	30	6	10	227	4.03
Crop residue	40	–	–	57	97	1.77
Dung	32	5	–	–	37	0.60
Total	–	–	–	–	–	6.40

[a] Domestic sector involves mainly cooking.
[b] Mean and low estimates are from Joshi *et al.* (1992*b*), see section 2.5.1 for explanation.
[c] Urban fuelwood use in domestic sector, assuming per capita estimates of Ranganathan *et al.* (1993) and the 1991 urban population.
[d] 6 Mt for brick industry in rural areas (see section 2.5.4).
[e] Urban fuelwood use 10 Mt (see section 2.5.4).
[f] Sugarcane bagasse production in 1991–92 of 82 Mt fresh = 779 PJ (at 19 GJ/t). Rice husk production in 1991 was 16 Mt = 201 PJ (12.6 GJ/t). Total energy value (of 57 Mt) = 980 PJ.
[g] Crop residue energy value = 779 PJ of cane bagasse + 201 PJ of rice husks + 1287 PJ of crop residue as domestic fuel (99 Mt at 13 GJ/t) = 2267 PJ for 'mean estimate'.
Similarly, for 'low biomass estimate' = 779 PJ of cane bagasse + 201 PJ of rice husks + 520 PJ of crop residue (40 Mt at 13 GJ/t) = 1500 PJ.
Per capita fuelwood use/year in rural areas = (181 Mt/yr)/627 million population = 0.288 t.

Biomass energy use on a country-wide basis is estimated taking the lower and mean consumption estimates of Joshi *et al.* (1992*b*). The 'lower estimates' are considered because they seem to be consistent with the past trends projected for fuelwood in Fig. 2.3 (section 2.5.1).

In India fuelwood is also used in rural industries such as brick and tile making, lime manufacture, and tobacco curing. Crop residue use is widespread, e.g. sugarcane bagasse is used as a fuel in sugar mills. Unfortunately, data on fuelwood use in industry are not available at the national level. However, its use in major industries such as brick and tile manufacture is slowly being replaced by coal and rice husks. These rural industry sectors, although not the dominant fuelwood users, are important in any effort for conserving wood as it is much easier to transfer technology to large establishments which operate on commercial lines as compared to domestic users. Thus it is necessary to initiate studies to estimate the fuelwood use and explore the potential for efficiency improvements in industrial and commercial establishments.

2.4 REGIONAL VARIATION IN BIOMASS USE

Agro-climatic zone

India is a large country with a geographical area of 328 Mha, with vast climatic variations and vegetation types ranging from tropical moist evergreen forests to dry deciduous forests and temperate montane forests. We can thus expect variations in the type and extent of biomass use according to agro-climatic zones, since biomass use is likely to be determined by local climate (temperature and rainfall), vegetation type, fuelwood availability, and access to fuelwood (especially in the case of forests). India has been classified into 15 agro-climatic zones (PC 1989). The variation in type and extent of biomass use according to agro-climatic zones can be seen in Table 2.3.

Fuelwood use ranges from 3.36 kg/capita/day in the eastern plateau and hills to 0.29 kg/capita/day in the upper Gangetic zone. The lowest use of fuelwood in the upper Gangetic zone seems to be offset by the highest use of cattle dung. Cattle dung use is insignificant in zones such as the west coast plains and hills and eastern plateau and hills. The type of biomass used depends on the access to woody biomass or forests, as is evident from the high dependence on fuelwood in forested areas of the Himalayan zones, east and west coast plains and hills. A large study in the South Indian state of Karnataka showed that fuelwood use in the hilly and forested districts is nearly double that in the plains, which have lower areas under forests (Ranganathan *et al.* 1993). The high rate of dung use in some regions with poor access to fuelwood has implications for the environment and quality of life; these aspects are discussed in later chapters.

In order to obtain a better micro-level understanding of the biomass energy situation, a set of village case studies has been selected.

Table 2.3 Variations in type of biomass fuel used according to different agro-climatic zones in India (kg/capita/day) based on 'mean estimates' (Joshi *et al.* 1992*b*)

Agro-climatic zone	Fuelwood (kg/capita/day)	Cattle-dung cake (kg/capita/day)	Crop residue (kg/capita/day)	Energy value of biomass fuel use (GJ/capita/yr)
Western Himalayan zone	1.09	0.17	NA	6.82
Eastern Himalayan zone	1.90	0.20	0.37	13.16
Lower Gangetic zone	1.93	0.56	0.62	10.83
Middle Gangetic zone	0.45	0.41	0.30	5.94
Upper Gangetic zone	0.29	1.29	0.04	8.23
Trans Gangetic zone	0.38	0.57	0.34	6.54
Eastern plateau and hills	3.36	0.01	0.24	19.58
Central plateau and hills	0.95	0.65	0.26	9.69
Western plateau and hills	0.94	0.18	0.50	8.42
Southern plateau and hills	0.94	0.25	0.82	10.29
East coast plains and hills	1.78	0.91	1.35	20.70
West coast plains and hills	0.94	0.16	0.12	6.52
Gujarat plains and hills	0.87	0.23	0.31	7.38
Western dry regions	0.57	0.61	0.13	6.79

Case studies of villages in different regions

Given the large size of the country and the diversity of resource base, patterns of energy use, and levels of development, six villages have been selected to represent major regions (Table 2.4).

Table 2.4 Selected villages

Village	State	Region
Maulaganj	Uttar Pradesh (Gorakhpur District)	North India
Bishnapur	Uttar Pradesh (Bahraich District)	North India
Dhanwas	Haryana (Gurgaon District)	North-west India
Nanuda	Gujarat (Banaskanta District)	Western India
Bandinarayanpura	Orissa (Ganjam District)	Eastern India
Ungra	Karnataka (Tumkur District)	South India

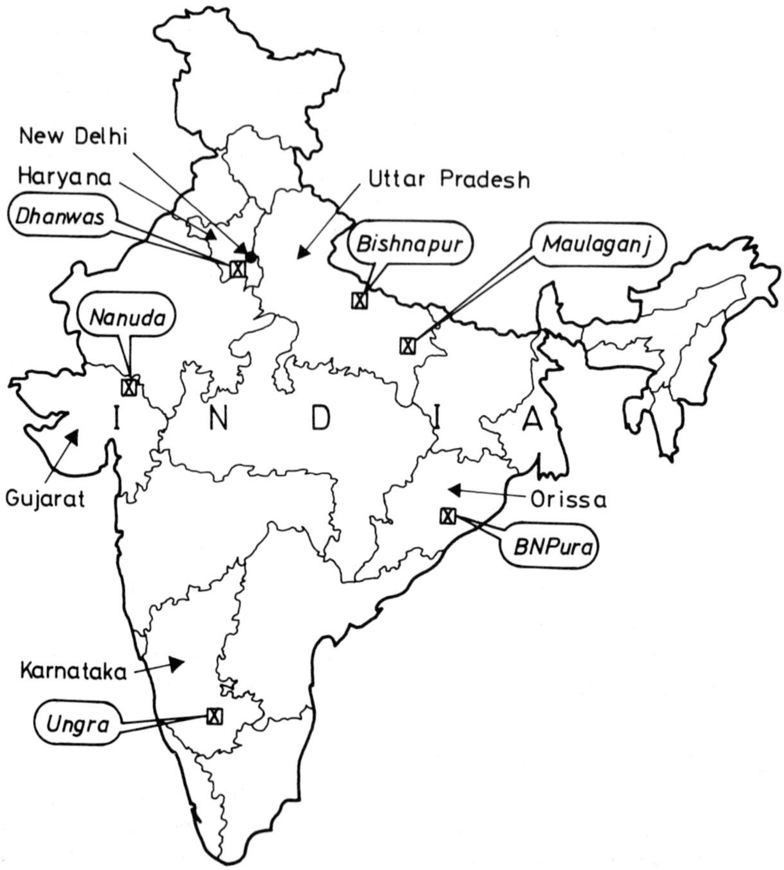

Fig. 2.4 Map of India showing the locations of villages selected for detailed analysis. BNPura, Bandinarayanpura.

The selection of villages was governed by the availability of detailed published data and their location in order to represent the major geographical regions (Fig. 2.4). Two villages were selected in Uttar Pradesh state as it is a large state (with a population of 139 million) and the two villages are from different districts; Bishnapur is near the forest and Maulaganj away from the forest, with both lacking an electricity supply. The same set of villages is used in later chapters for analysing the potential for bioenergy. Different features of the selected villages are considered in the various sections of this chapter.

Per capita biomass use (Table 2.5)

The annual biomass energy use per capita varied from location to location but four of the six villages had a per capita biomass use of 8.3–10.8 GJ/yr.

Table 2.5 Biomass fuel use in the domestic sector in the selected villages in India[a]

Village	Population	Family size	Fuelwood use (t/yr)	Crop residues use (t/yr)	Dung cake use (t/yr)	Total biomass energy (GJ/yr)	Biomass energy (GJ/capita/yr)
BNPura	593	6.0	351	18	304	9635	16.3
Maulaganj	1143	6.0	333	238	305	12 267	10.7
Bishnapur	931	10.0	645	30	–	10 065	10.8
Dhanwas	833	7.1	42	115	375	7262	8.7
Ungra	1047	5.1	295	329	–	8702	8.3
Nanuda[b]	838	6.2	121	173	50	4749	5.6

[a] BNPura (Bandinarayanpura) (Nishanka and Misra 1990); Maulaganj and Bishnapur (Bose *et al.* 1991); Ungra (Ravindranath and Chanakya 1986); Dhanwas (Puri 1988).
[b] Nanuda: in addition to biomass, 6 l/capita/yr of kerosene are used (Veena 1988).

In the Gujarat village, biomass use was low because of the use of kerosene for cooking at 6 l/capita/yr (Veena 1988). The use of kerosene in one of the selected villages is a pointer to the fact that even in rural areas kerosene is presently accepted as fuel for cooking. It is not clear why biomass energy use is high in Bandinarayanpura (BNPura) (Nishanka and Misra 1990).

Share of different biomass fuels (Table 2.6)

Biomass is used in all the villages and its share in total energy ranges from 99 per cent in Bishnapur to 79 per cent in Nanuda, indicating its predominance in the total village energy use. In the two villages of the southern and western regions, dung was not used. In North India dung cakes are a major fuel source, as is seen in Dhanwas (in Haryana) and Maulaganj (Uttar Pradesh (UP)). One of the reasons for this could be that these states have smaller areas under forest per unit of population as compared to the southern states (0.04 ha/capita in Uttar Pradesh, 0.038 ha/capita in Bihar and 0.003 ha/capita in Haryana in the north, compared to 0.1 ha/capita in states such as Andhra Pradesh and Karnataka in the south; FSI 1992). Bishnapur village in Uttar Pradesh is exceptional as it has access to fuelwood from a nearby forest and thus shows the highest dependence on fuelwood (95 per cent; Bose *et al.* 1991). In Bishnapur village the availability of crop residues with fuel potential was high, but despite such large availability it was not used as fuel (section 2.8.1). This clearly demonstrates that if fuelwood is available, households prefer it to cattle dung or crop residues. The Ungra village example also demonstrates this. Commercial energy accounted for less than 20 per cent of the total energy apart from lighting in all villages except Nanuda where kerosene is also used for cooking. For Ungra (Ravindranath *et al.* 1981) in the south and BNPura in the east (Nishanka and Misra 1990) even

Table 2.6 Biomass energy use in relation to total energy use in selected villages (GJ/yr)[a]

Energy source	Maulaganj	Bishnapur	Dhanwas	Nanuda	Ungra	BNPura
Wood	4995	9675	630	1815	4425	5704
	(38.9)	(95.0)	(8.0)	(30.2)	(43.5)	(49.9)
Crop residues	3094	390	1495	2249	4277	253
	(24.1)	(3.8)	(19.0)	(37.4)	(42.0)	(2.2)
Dung	4178	–	5137	685	–	3678
	(32.6)		(65.2)	(11.4)		(32.1)
Biomass total	12267 (95.6)	10065 (98.8)	7262 (92.2)	4749 (79.0)	8702 (85.5)	9635 (84.2)
Kerosene	291	118	72	526	217	129
	(2.3)	(1.2)	(0.9)	(8.7)	(2.1)	(1.1)
Diesel	269	–	346	57	–	57
	(2.1)		(4.4)	(0.9)		(0.5)
Electricity	–	–	198	682	278	31
			(2.5)	(11.4)	(2.7)	(0.3)
Commercial energy total	560 (4.4)	118 (1.2)	616 (7.8)	1265 (21.0)	495 (4.8)	217 (1.9)
Animal[b]					129	579
					(1.3)	(5.1)
Human[b]					848	1012
					(8.4)	(8.8)
Total energy	12827	10183	7878	601	10174	11443
	(100)	(100)	(100)	(100)	(100)	(100)

[a] Figures in brackets indicate percentage of total energy.
[b] Animal and human energy use data are available for Ungra and BNPura only.

when animate energy (e.g. draught animal power) is included biomass continues to dominate.

These village examples clearly demonstrate that in different regions of India, biomass energy dominates the village energy matrix, and furthermore that there are no major differences in the share of biomass and commercial fuels in different geographical locations. This infuses some confidence into our attempts to formulate a national-level analysis of the bioenergy potential, taking these sample villages as indicators.

2.5 MAJOR END USES OF BIOMASS ENERGY

Biomass energy is used in both rural and urban areas in the domestic sector (cooking and heating bath water), in industries, and in establishments. These end uses are discussed in this section.

2.5.1 Cooking

A national-level study during 1978–79 (NCAER 1985) showed the following proportion of households using biomass in the domestic sector:

(1) rural: fuelwood, 51 per cent; crop residues, 16 per cent; dung, 21 per cent; others, including commercial fuels, 12 per cent;
(2) urban: fuelwood, 35 per cent; dung cake, 4 per cent; others, 61 per cent.

According to a recent large survey in Karnataka (covering 300 villages and 17 towns and cities and a household sample of 2000), 99 per cent of rural households and 78 per cent of urban households use fuelwood (Ranganathan 1993). The migration of poor people to urban areas contributes greatly to the fraction of urban households depending on fuelwood.

It is evident that fuelwood is used by the majority of India's households. The authors' field observations have shown that a household is more likely to use a combination of biomass fuels on a typical day, and it is common to observe a mixed pile of fuelwood and crop residues stored in front of households or in the kitchen in rural India; exceptions occur in villages such as Bishnapur, where fuelwood is available from forests and only fuelwood is used. Thus it may be inaccurate to say that only 51 per cent of the households use fuelwood (based on the national survey; NCAER 1985). Even households reporting crop residue and dung as their fuel probably also use fuelwood. It is likely that the figure of 51 per cent is for households using fuelwood as their dominant fuel type, since most studies do not give the mix of fuels used by the households.

Biomass use in selected villages

Biomass use for cooking in the selected villages as well as in a sample of other villages is listed in Table 2.7. Some studies give data for cooking and bath-water heating separately but most do not make this distinction and give data only for cooking. Based on observations from four village case studies and two Karnataka state-level studies, it is possible to conclude that the majority of villages fall within a range of biomass use for cooking of 13–23 MJ/capita/ day, barring exceptions such as BNPura. The majority of the values are closer to the 'lower estimate' of 16 MJ/capita/day at the national level (Joshi *et al.* 1992*b*).

National-level rural fuelwood use estimates for cooking

Various estimates ranging from 93 Mt (110 kg/capita/yr) to 309 Mt (492 kg/capita/yr) of annual (1991) wood use are given in Fig. 2.5. This excludes crop residues and dung use for cooking. The village case studies (Table 2.7) showed that per capita fuelwood use in most villages ranged between the two extremes. According to the three estimates for India given by Joshi *et al.* (1992*b*), based on a large recent study, the 'mean estimate' is 252 Mt/yr. The estimates of fuelwood use by FSI made for 1986 (FSI 1988), by NCAER for

Table 2.7 Biomass energy use for cooking and bath-water heating in villages in Karnataka and India (per capita/day)

End use	Village	Fuelwood (kg)	Crop residues (kg)	Dung cake (kg)	Total biomass energy (MJ)	Total biomass energy (GJ/capita/yr)
Cooking	Ungra	0.59	0.57	–	16.2	5.9
	Nanuda	0.34	0.73	0.14	16.5	6.0
	Gopularam[a]	0.24	0.97	0.12	17.9	6.5
	Dhanwas	0.13	0.35	1.00	20.2	7.4
Cooking + bath-water heating	Maulaganj	0.49	0.19	0.62	18.3	6.7
	Bishnapur	1.34	–	–	20.1	7.3
	BNPura	0.83	1.00	1.40	44.5	16.2
	Arro[b]	0.32	0.06	1.31	23.5	8.6
Bath-water heating	Ungra	–	0.28	–	3.9	1.5
	Dhanwas	3.00	0.03	0.24	3.7	1.3
Karnataka[c] cooking (rural)	North Karnataka	0.72	–	–	10.8	3.9
	South Karnataka	0.89	–	–	13.3	4.8
Karnataka[d] domestic (rural)	State mean	1.50	–	0.03	22.9	8.3
Cooking[e] (rural) India	Low	0.79	0.17	0.14	16.0	5.8
	Mean	1.10	0.43	0.47	28.5	10.4

[a] Gopularam (Bowonder *et al.* 1985).
[b] Bose *et al.* (1991).
[c] Jagadish and Ravindranath (1990).
[d] Ranganathan *et al.* (1993).
[e] Joshi *et al.* (1992*b*). The national means are different from Table 2.2, as that table includes industrial use of fuelwood and crop residue.

1978–79 (NCAER 1985), and FAO (1993*b*) calculated for the 1991 population are lower than this 'mean estimate'.

The question arises as to which estimate to assume, given the wide variation. The fuelwood use trends shown in Fig. 2.3 and the majority of the per capita values given in Table 2.7 and the national total estimates of IREP, FAO, and NCAER (Fig. 2.5) are all closer to the 'lower end estimate' of 181 Mt of Joshi *et al.* (1992*b*). Thus, of the three estimates made by Joshi *et al.* (1992*b*), the 'lower estimate' of 181 Mt of fuelwood is used for further calculations for rural India in this book.

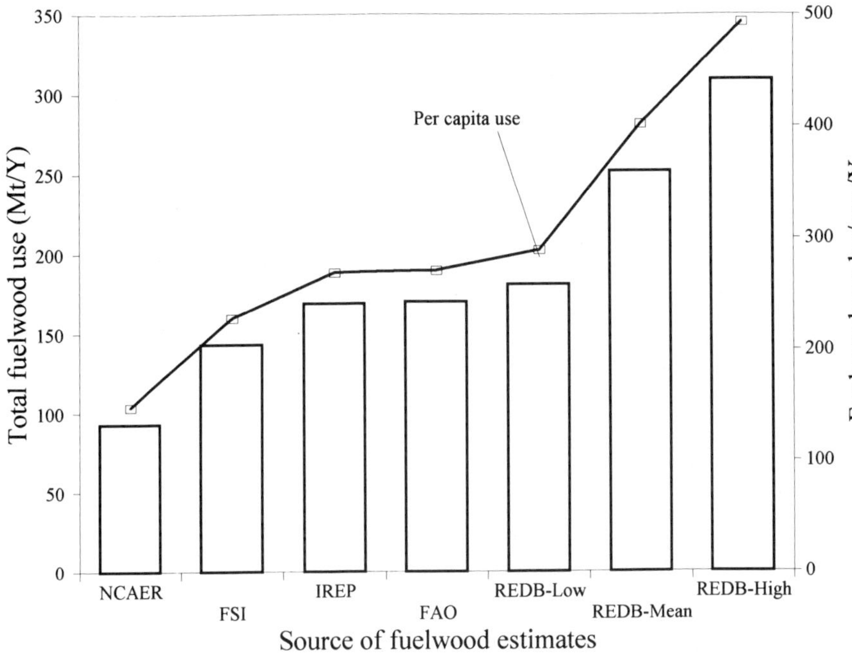

Fig. 2.5 Total and per capita annual rural fuelwood use for cooking in India from various studies estimated for the 1991 population. Sources: NCAER (1985), FSI (1988), and FAO (1993*b*). IREP (integrated rural energy programme); low, medium, and high REDB (rural energy database) estimates from Joshi *et al.* (1992*b*).

2.5.2 Heating bath-water

Very often the analysis of biomass use in the domestic sector is restricted to cooking alone. The possible reason could be that according to estimates, heating bath-water accounted for only 3.6 per cent of the household energy consumption, as compared to cooking which used 88 per cent (NCAER 1985). However, it can be estimated from Table 2.7 that in two villages biomass use for heating bath-water ranged from 19 per cent (Ungra) to 15 per cent (Dhanwas) of total biomass use. Currently most people in rural areas bathe only once or twice in a week, possibly more due to the effort involved in fetching water than in collecting fuelwood. Our personal observations in South India have shown that if households have to depend on fuelwood and crop residues they prefer to reserve fuelwood for cooking and use residues for heating bath-water. This is because the density of fuelwood is higher than that of crop residues or dung cakes, thus requiring less fire tending and attention while cooking.

2.5.3 Biomass use in urban areas

The importance of biomass fuels in urban areas has been recognized, and a number of studies are available for Bangalore (Reddy and Reddy 1983), Hyderabad (Alam *et al.* 1985), Raipur (Macauley 1989), and 17 urban centres in Karnataka (Ranganathan *et al.* 1993). Fuelwood and dung cake use is considered from the Karnataka study since it is a recent study and included urban centres of size less than 0.1 million as well as Bangalore city (4 million population). Surprisingly, according to the Karnataka study, 78 per cent of the urban households use fuelwood, compared to only 35 per cent according to the 1978–79 national study (NCAER 1985). Another recent study (Kulakarni *et al.* 1994) in three Indian cities in Maharashtra state showed that fuelwood use for cooking was low, in the range 4–7 per cent even among lower-income households. However, fuelwood use for water heating for the low income households ranged from 19 to 41 per cent. Thus urban households seem to be shifting away from biomass fuels to liquid and gaseous fuels. The Karnataka and Maharashtra studies show large variation. If the Karnataka level of fuelwood use in urban areas is considered, the national fuelwood use in the urban domestic sector would be 30 Mt. The urban fuel use is of consequence since 85 per cent of the fuelwood used there is in the form of logs (wood cut from tree stems and large branches), which is likely to come from the felling of trees. The Karnataka study further showed that 29 per cent of the fuelwood sold in the urban areas came from forests. This will have implications for forest degradation and deforestation.

2.5.4 Biomass use in industries and establishments

Fuelwood is used in the manufacture of housing materials such as bricks, tiles and lime, and in agro-processing, such as jaggery (crude sugar) preparation, the curing of tobacco, and the preparation of some spices such as cardamom. In rural areas nearly all building construction materials are still fired using fuelwood. However, manufacture of bricks for the urban market has slowly shifted from fuelwood to coal and rice husks, as the price of wood has increased. A survey of brick kilns around Delhi showed that 307 out of 312 used coal as the main fuel (FAO 1993*c*). Our observations around Bangalore showed that here fuelwood is no longer used for brick manufacture. However, in villages and around small towns fuelwood is still used for firing bricks.

Bricks

According to an estimate for the development plan period (1992–97), the annual brick requirement is 32 billion in urban areas and 25 billion in rural areas. At 0.25 kg of fuelwood per brick, and assuming all the bricks for rural needs are fired in wood kilns, the wood required is estimated as 6 Mt annually

(BMTC 1992). Fuelwood use for tiles is restricted mainly to some parts of India. Thus when compared to the 211 Mt of fuelwood used in the rural plus urban domestic sector (Table 2.2), its use in the housing industry is not significant. Even though less than 5 per cent of fuelwood is used in making building materials, it is important to note that nearly all of it is derived from the felling of trees in villages, tree plantations, and even forests. Brick kilns are operated on a commercial basis and so would be candidates for conserving fuelwood through efficiency improvements.

Biomass use in establishments (hotels, bakeries, hostels, marriage halls, etc.) in urban areas is estimated to account for 22 per cent of total fuelwood use in a large city such as Hyderabad (Alam *et al.* 1985) to 45 per cent in Raipur, a moderate-sized town (Macauley 1989). The details of fuelwood use in Bangalore and Hyderabad given in Table 2.8 show that industries, bakeries, and hotels dominate such wood use. Assuming the Raipur rate of fuelwood use for all towns with less than 1 million people, and the Hyderabad rate for cities with >1 million people, national fuelwood use in establishments in urban areas is estimated to be 10 Mt annually (Table 2.2). These commercially operated institutions which purchase cut wood originating from felled trees in villages or forests are also prime candidates for efficiency improvements for conserving trees or forests.

Crop residues

The two major crop residues used in industries are rice husks and sugarcane bagasse. Rice husk production in India is used in rice mills for parboiling rice,

Table 2.8 Fuelwood use in urban industrial and institutional sectors

Industry/establishment	Fuelwood use in Bangalore city, % of total non-household sector[a]	Fuelwood use in Hyderabad city, % of total non-household sector[b]
Hotels	18	21
Bakeries	19	22
Hostels	9	4
Crematoria	4	2
Marriage halls	8	4
Industries	41	47
Others	1	–
Total (%)	100	100
Total fuelwood use in non-household sector (10^3 t/yr)	102	116

[a] Bangalore data from Reddy and Reddy (1983).
[b] Hyderabad data from Alam *et al.* (1985).

in cement factories, and in brick kilns. The density of rice husks is low and thus the cost of long-distance transport from dispersed rice mills in rural areas to urban centres, the main users, merits attention. Rice husks could be used as a fuel in biomass gasifiers for power generation or thermal applications, near to the site of production. What is already evident, however, is that rice husk is not wasted and it has an opportunity cost. The use of sugarcane bagasse as fuel is discussed in Chapter 6.

2.6 EFFICIENCY AND COMPARATIVE REQUIREMENT OF BIOMASS FUELS

Thermal efficiency

Given the large quantities of biomass energy involved, the efficiency of its use becomes very important. Traditional wood stoves were condemned as inefficient and blamed as a cause of deforestation. Earlier studies assumed efficiencies as low as 8 per cent for cooking stoves in India (ABE 1985) and concluded that such traditional stoves are very inefficient.

Several studies on thermal efficiency conducted using standard tests have, however, indicated higher values for traditional stoves (Fig. 2.6). A study in

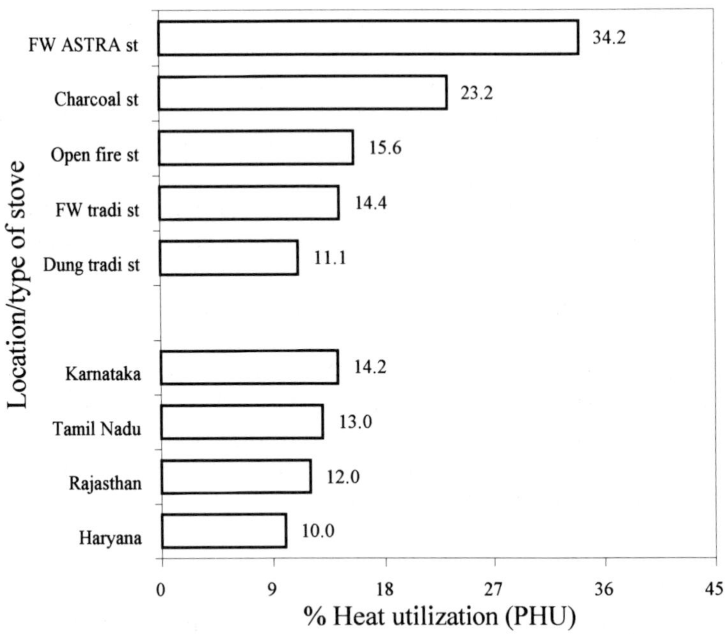

Fig. 2.6 Thermal efficiency of traditional and improved ASTRA stoves in different regions. FW, fuelwood; ASTRA stove, an efficient stove in South India; st, stove. (From Joshi *et al.* 1989; Ravindranath *et al.* 1989*a*; Dutt and Ravindranath 1993.)

South India showed traditional stoves using fuelwood having a PHU (percentage heat utilization) of 15.7 per cent (Ravindranath *et al.* 1989*a*). It is possible to conclude that the PHU of traditional stoves is generally low but somewhat higher than that assumed earlier.

Conclusions from a recent study (Dutt and Ravindranath 1993), conducted in South India using the standard PHU tests for different biomass fuels in traditional and improved stoves, are given in Fig. 2.6. The traditional wood stove in South India (Ungra area) is a three-pan stove, with a fuelwood feeding port at the centre. It is a versatile stove where fuelwood, crop residues, and dung cakes are used and vessels of varying dimensions can be used. The water boiling tests showed a lower PHU for dung cakes compared to fuelwood. Surprisingly, the much condemned and out-moded three-stone or open-fire stove had a higher PHU (15.6 per cent) than the currently used three-vessel traditional stove. The three-stone fire has a relatively higher PHU since the cooking vessel is immersed into the fire and a large vessel surface area is exposed to the flame and hot flue gases, leading to greater heat transfer. If this is the case, what is the rationale for shifting from a three-stone fire to a traditional three-pan stove? Traditional stoves in South India had a relatively higher thermal efficiency than stoves in other parts of the country because they are multiple-vessel stoves and simultaneous cooking is possible. This is discussed later. Dung cakes used in traditional stoves have the lowest PHU, probably because the low flame generated does not reach the vessel and only hot flue gases are able to heat the contents of the vessel.

Charcoal stoves give a PHU of 23.2 per cent. However, one should consider the low efficiencies of charcoal production (about 20 per cent, Dutt and Ravindranath 1993). In such a case the efficiency of charcoal stoves, starting from the wood used to produce charcoal and the useful heat obtained in the stove, would be 4.6 per cent. This explains the rationale, especially in rural areas, for using fuelwood directly rather than as charcoal. Many states in India have banned charcoal production. The scarcity of fuelwood has led to escalation of charcoal prices and in any town in South India its price is four times that of fuelwood.

This concern about the low efficiency of use of fuelwood led to large national and international stove programmes, with the intention that any small improvement in efficiency could bring large benefits in terms of less deforestation, reduced drudgery in fuelwood gathering, improved health, and savings in cash to fuel-buying households.

Cooking zones

A large variety of cooking stoves is used in India, with a varying number of pans (1–3), shapes, and sizes. The stove type used depends on the type of food cooked (or the major food crop grown) and the local climate. In Karnataka (population 44 million), a study has identified three food zones:

Table 2.9 Cooking zones[a] and features of cooking in Karnataka

Food zone	Main dishes cooked	Features of cooking	Typical cooking time (min)
Ragi[b]	3 dishes ragi flour rice curry	Three-pan stoves used, all simultaneously cooked by interchange of vessels between pans—central pan receives maximum heat Cooking involves bringing water to boil and adding rice, ragi, and pulses, allow to boil and simmer Sequence of cooking: rice—curry—ragi	55
Sorghum	2 dishes sorghum bread vegetable + pulses curry occasionally rice	Mainly two-pan stoves Boiling water, adding pulses, vegetables on pan 1, shift to pan 2 Baking sorghum flour to make flat roti or bread on pan 1	172
Rice	2 dishes rice fish/vegetable curry	Mainly two-pan stoves Cooking rice and fish or vegetable + grams	103

[a] A cooking zone consists of thousands of villages with similar cereal/millet-dominated dishes and nearly identical cooking practices.
[b] Ragi = finger millet (*Eleusine coracana*).

a rice zone (mainly coastal districts where rice is the dominant crop), a sorghum zone (mainly northern districts), and a ragi or finger millet zone (southern districts) (Ravindranath *et al.* 1989*b*). Cooking practices varied from zone to zone (Table 2.9).

For example, in the ragi zone the cooking involved boiling and simmering of rice, vegetable curry and ragi dishes simultaneously on three-pan stoves. In the sorghum zone, cooking involved mainly baking of bread using a low heat on a flat pan; this bread baking requires longer cooking times than in the ragi zone. In the rice zone mainly two dishes, rice and a curry were prepared using two-pan stoves. There was also a large difference in cooking duration between zones, depending on the type of food cooked. For example, in the ragi zone the cooking duration was shortest (55 minutes) since cooking involved only boiling and simmering on low heat. In the sorghum zone the cooking duration was longest (172 minutes) as cooking involved baking bread on a low heat for a long time. Designers of improved stoves must keep in mind the variations in cooking practices in different food zones and allow for flexibility in choice of design.

2.7 SOURCES OF FUELWOOD IN INDIA

One of the intriguing questions for rural energy researchers in India is to explain the sources of fuelwood and the gap between the demand and supply. Many reports and studies have estimated the fuelwood use or requirements over the years, but none of them has assessed the issue of the origin of the cooking fuel (ABE 1985; FSI 1988; Joshi *et al.* 1992*b*). Based on a survey of households (during 1978–79) an attempt was made to understand the sources of fuelwood using the data from the survey of users. According to the study, the dominant sources are 26 per cent each from forests and own farms, 17 per cent from roadside collections and other sources, including neighbours' farms, and the rest obtained as a part of wages and not gathered (NCAER 1985). It is difficult to draw any conclusions as to the sources of fuelwood for the present period from these studies. In an attempt to trace the sources of fuelwood, FSI (1988) explained the source of only 19 Mt, as supplied by the Forest Department, out of a total of 157 Mt for 1985–86, and finally concluded that overexploitation of forests for meeting cooking-fuel needs was the main reason for forest degradation.

Analysing the supply side of fuelwood, Leach (1987) suggested that only 12 per cent of the fuelwood was from logs involving the felling of trees, and the other 88 per cent was from the cutting of branches. The study did not explain the sources of fuelwood.

The Karnataka state study (Ranganathan *et al.* 1993), which attempted to explain the sources of fuelwood, could not conclusively establish them, as the annual woody biomass productivity from forests was estimated to account for only 18 per cent of the total supply of 20 Mt/yr. It is not clear where the remaining fuelwood requirement is obtained from and how the rate of extraction is sustained. Similarly, a biomass budget for Karnataka state shows that fuelwood accounts for nearly 50 per cent of the total biomass used in the domestic sector (11 Mt/yr) and, furthermore, that the forest department supplied only 4 per cent of the fuel demand while the unreserved forests and other sources such as village trees were the dominant sources of fuelwood (Gadgil and Sinha 1986).

It is clear from all these studies that there are several fuelwood sources and that the contribution of different sources is not known. Furthermore, the sustainability of the extraction rates is not known as even twigs and branches can be collected in a non-sustainable mode. We need to understand the sources of fuelwood and the rates of extraction at national level in order to assess the consequences of fuelwood use on deforestation and to plan for meeting both short- and long-term fuelwood needs. Currently fuelwood conservation and production programmes are based on the assumption that fuelwood demand contributes to deforestation.

Fuelwood is harvested from tree and shrub species:

(1) trees: from clear felling of forests or plantations, selective felling of trees, sustainable removal of twigs and, branches, e.g. lopping, and non-sustainable removal of twigs, branches, and stem;
(2) shrub species such as *Prosopis juliflora, Lantana camara, Cassia auriculata,* and *Eupatorium* are harvested as fuel.

Sources of fuelwood can be forests, degraded forests, tree plantations, strips or rows of trees along roads/streams/railway lines/canals/cropland boundary and bunds (separating small plots), common lands, and homestead gardens. The annual use of fuelwood is estimated to be 227 Mt/yr, out of which 181 Mt is used in the domestic sector for cooking in rural areas. We attempt to explain the sources of such a large quantity of fuelwood from the supply side.

2.7.1 Forests as a source of fuelwood

Information on the area under forests, the standing biomass, the productivity of forests, and extraction rates is necessary for estimating the contribution of forests to providing fuelwood in the absence of direct estimates of the contribution of forests.

Forest area

The area under forests has been estimated to be 63.5 Mha according to FSI (1992). This area has been categorized on the basis of crown cover: 37.8 Mha with greater than 40 per cent tree crown cover and 25.7 Mha with 10–40 per cent cover. Standing stock and productivity depend on forest type, nature of vegetation, tree crown cover, rainfall, and altitude. Indian forests have been classified into 14 types (Table 2.10). Tropical moist deciduous and tropical dry deciduous forests are the dominant types, accounting for 64 per cent of the total forest area.

Standing biomass in forests

Using data on standing biomass (from the literature) and tree crown cover (FSI 1992) for different forest types, the estimates of standing biomass with five different crown cover levels for the major forest types (Tiwari and Singh 1987) and by extrapolation to other forest types, Ravindranath *et al.* (1992*b*) have estimated the standing biomass for the Indian forests (Table 2.10). The tropical moist deciduous and tropical wet evergreen forests are the dominant types, accounting for 66 per cent of the total standing biomass, with an overall mean biomass of 126 t/ha of forest.

Table 2.10 Area and standing biomass of forest types in India (during 1986) (Ravindranath *et al.* 1992*b*)

Forest type	Area (10^3 ha)	Standing biomass (t/ha)	Average crown cover fraction	Total standing biomass (Mt)
Tropical wet evergreen	5289.3	607.7	0.4700	1496.5
Tropical semi-evergreen	2575.7	468.0	0.4782	576.4
Tropical moist deciduous	23 054.7	409.3	0.4259	4018.5
Littoral and swamp	383.4	213.8	0.4426	36.3
Tropical dry deciduous	18 083.0	93.8	0.4313	731.5
Tropical thorn	1573.0	40.0	0.3989	25.1
Tropical dry evergreen	134.2	40.0	0.4156	2.2
Subtropical broad-leaved	267.8	108.7	0.3818	11.1
Subtropical pine	4567.5	210.8	0.4581	141.1
Subtropical dry evergreen	1201.0	159.7	0.2901	55.6
Montane wet temperate	2581.9	237.6	0.4875	299.1
Himalayan moist temperate	2242.8	562.2	0.4263	537.5
Himalayan dry temperate	30.5	169.1	0.4079	2.1
Subalpine and alpine	2027.7	127.4	0.4822	124.5
Total	64 012.5	–	–	8 057.5[a]

[a] Mean standing tree biomass/ha = 8057 Mt/64.01 Mha = 125.87 t/ha ~ 126 t/ha.

Annual increment of woody biomass

An estimate of the annual increment is necessary to determine the potential extent of the annual supply of fuelwood in the form of both branches and twigs. Using data on the standing biomass and annual increment available for some forest types from the literature, Ravindranath *et al.* (1992*b*) have estimated the total annual increment for the forest to be 138 Mt, of which branches and twigs account for 42 Mt (31 per cent of standing biomass, Seebauer 1992).

Sources of fuelwood, mode of removal of fuelwood, and contribution of forests is considered next.

Collection of twigs and branches

In India rural and tribal households are allowed to carry headloads of fallen twigs and small branches as fuel for family consumption and often for sale in nearby markets. It is often alleged that rural and tribal communities remove fuelwood in a non-sustainable way, leading to forest degradation. Most forests in India are undoubtedly subjected to use due to the high population density and dependence on biomass. Thus it could be assumed that a minimum of

42 Mt of fuelwood potentially available from the annual increment of forests is removed.

Felling of trees

In many states, Forest Departments provide fuelwood through fuelwood depots to local communities. This mostly comes from the felling of trees, since only cut logs and branches are sold in the depots. According to an estimate of FSI (1988), 19 Mt of fuelwood is supplied annually from the forests.

Logging wastes

Timber consumption (soft- and hardwood) was estimated to be 21 Mt for the year 1991 (using 1986 data from FSI (1988)) and, assuming 31 per cent of the above-ground biomass as branches and twigs, 9.5 Mt of branches and twigs would be available when trees are felled to produce 21 Mt of timber. Generally, whenever trees are felled the twigs and small branches are sold to local communities or left to be collected freely by them.

Thus the total contribution of forests to meeting the fuelwood demand is estimated to be 70.5 Mt, out of which only 27 per cent is expected to come from the felling of trees, possibly leading to deforestation (Table 2.11). It is also clear that logging wastes do not presently contribute significantly to fuelwood supplies in India.

2.7.2 Tree plantations as a source of fuelwood

Since 1950 India has had a tree plantation or afforestation programme, which was accelerated after 1980 under various social forestry programmes. One of the main goals of social forestry is to meet the fuelwood requirement of local communities. It includes farm forestry on farmers' lands, community forestry on village commons, FD (Forest Department) plantations on degraded forest lands, and strip plantations along roads and streams. The total area established under such tree plantations under various afforestation programmes between 1975 and 1990 was estimated to be 15 Mha (PC 1992), and 2 Mha were estimated to be added during both 1991 and 1992. Most of the tree plantations on village commons, farmers' lands, and trees planted along side roads, canals, and streams are not included in the national forest area of 63.5 Mha (Ravindranath *et al.* 1992*b*). The contribution from afforested plantations to fuelwood supplies is estimated to be 40 Mt/year (see footnote [b] of Table 2.11). The rest of the plantation wood production is used as industrial and structural wood (details of productivity and contribution are discussed in Chapter 7).

Table 2.11 Summary table of sources of fuelwood used in the domestic sector, mainly for cooking

Sources	Details	Total fuelwood contribution in Mt/yr
Forests (> 10% tree crown) (FSI 1988)	1. Felling of trees[a]	19.0
Forests = 64 Mha (Table 2.10)	2. Lopping of twigs and branches in existing forests[a]	42.0
	3. Logging wastes[a]	9.5
	Total	70.5
Tree plantation[b] (17 Mha)	Tops + twigs + small branches + poles	40.0
Farm tree[c] (Table 2.13)	Twigs + branches	46.0
Homestead garden (Table 2.14)[d]	Twigs + branches	16.0
Degraded land[e] (52.8 Mha)	52.8 Mha at 0.5 t wood/ha/yr	26.4
Land for shrub estimation[f] (78.5 Mha)	78.5 Mha at 0.25 t/ha/yr (shrubs, climbers)	19.6
Total wood supply		218.5

[a] See text for explanation (section 2.7.1).
[b] Tree plantations include farm forestry (on farm land) + Forest Department (FD) forestry (on degraded forest land) + community forestry (on village commons). Tree plantation = farm forestry (8.5 Mha) at 4.2 t/ha/yr + community forestry and FD forestry (8.5 Mha) at 2.6 t/ha/yr (from Table 7.4).
Of the total annual productivity of 57.8 Mt, a maximum of 17.8 Mt is assumed to be used as industrial and structural wood (section 7.4) and the rest is assumed to be used as fuel.
[c] Farm trees: rows of trees along cropland boundary and bunds.
[d] Homestead gardens: trees around homes.
[e] Degraded lands = degraded forest < 10% tree crown (11.2 Mha) + permanent pastures (11.8 Mha) + miscellaneous traditional tree groves (3.5 Mha) + culturable waste land or fallow (15.2 Mha) + other fallow (11.1 Mha) = 52.8 Mha (see section 7.4 and Table 7.1 for data and explanation of land categories).
[f] Land for shrub production = 52.8 Mha of degraded land + 25.7 Mha of open forests (with 10–40% tree crown, Table 3.1).
For detailed explanation for degraded lands see section 7.4.

2.7.3 Trees on farms as a source of fuelwood

The contribution of trees on farms to meeting the fuelwood requirement was rarely recognized (Leach and Mearns 1988). This was probably due to the difficulties in estimating their contribution, the absence of studies and information on trees on farms and the level of productivity of such trees. For example, a recent study in Pakistan has shown that of the estimated number of 1434 million trees in the country, trees on farmland accounted for 50 per cent of the total, while forests and other sources accounted for the rest (Leach 1993), thereby clearly showing the importance of farm trees in southern Asia.

Three studies are available on the trees on farms from three regions: South India (Ungra village), North Indian plains (villages in Uttar Pradesh state) and north-west India (Haryana, two villages). In India it is common to find that the land holding of a farmer is divided into a number of smaller fragments and plots, and these are separated by a small bund with a mound of soil about 30–60 cm wide. This bund area accounts for 3–10 per cent of the cropland area and, traditionally, farmers plant trees on these bunds (Ravindranath *et al.* 1991; Saxena 1991). Based on these and other studies, an attempt is made here to estimate the contribution of trees on farms to fuelwood supplies.

*Ungra village in South India (*Ravindranath *et al.* 1991)

Ungra is a semi-arid village (population 1047) in the Tumkur district in Karnataka, with a village ecosystem area of 360 ha incorporating a cropland area of 240 ha. The distribution of trees in the village (excluding coconuts and block plantations of *Eucalyptus* and *Casuarina*) is given in Table 2.12. Local tree species such as *Pongamia pinnata, Acacia nilotica*, and *Azadirachta*

Table 2.12 Tree population distribution of Ungra village ecosystem (semi-arid) in South India.[a] (Source: Ravindranath *et al.* 1991)

Species	Total number of trees in village	%	Standing biomass (dry t)	%	Mean dry weight per tree (kg)
Eucalyptus hybrid	2666	23.6	54.4	6.3	20.4
Pongamia pinnata	2265	20.1	176.1	20.3	77.7
Syzygium cuminii	1043	9.2	96.6	11.1	92.6
Azadirachta indica	835	7.4	59.3	6.8	71.0
Acacia nilotica	827	7.3	69.7	8.0	84.2
Salix tetrasperma	490	4.3	55.6	6.4	113.4
Casuarina equisetifolia	426	3.8	75.9	8.8	178.1
Alangium salvifolium	413	3.7	NA	NA	NA
Ficus glomerata	255	2.3	90.5	10.4	354.9
Ficus bengalensis	238	2.1	74.5	8.6	313.0
Others[b]	1829	16.2	114.6	13.3	62.6
Total	11 287	100.0	867.2	100.0	79.7

[a] Includes farm trees along cropland bunds and boundaries.
[b] 'Others' includes a further 47 tree species.
NA, data not estimated.
Coconut palms present in the village ecosystem were not classified as trees but were considered as plantation crops.

indica dominate the tree distribution. The density of trees on dry and irrigated cropland bunds was 51 and 13 trees/ha, respectively, and 63 trees/farm. Using the standing biomass data for the trees in the village ecosystem, an annual increment in woody biomass was estimated to be 53 t. Thus Ungra village can sustainably harvest 50 kg/capita or 147 kg/ha of fuelwood annually from trees on cropland bunds and boundaries. Coconut is considered a plantation crop and thus its contribution is included as a crop residue.

Rural Uttar Pradesh (North India)

The tree distribution in two villages each from Allahabad, Muzaffurnagar, and Nainital districts, representing different regions, were studied (Saxena 1991). The number of trees per farm in four of the six villages in three districts ranged from 253 to 787 trees (Table 2.13).

Rural Haryana in north-west India

The number of trees has been estimated for two villages, namely Barkha Alimuddin and Dhanwas (Table 2.13) (Chaturvedi 1990). Using the limited available data, the tree population, standing biomass, and annual increment of woody biomass on farms is projected for India (Table 2.13). Taking a conservative annual increment rate of 2.84 per cent of standing biomass, an aggregate annual increment of 46 Mt of woody biomass is estimated to be the significant contribution of trees on farms (see footnotes of Table 2.13 for details).

2.7.4 Homestead gardens as a source of fuel

In many places, such as West Bengal, the hilly regions of Uttar Pradesh, the western Ghat regions of Karnataka, and Kerala, there is a tradition of growing trees around the house, where the houses are situated in the midst of tree gardens. These trees provide a range of biomass products including fuelwood for domestic use. Data from a number of studies including the three studies on the presence of trees on homestead gardens in Kerala, Karnataka in the south, and the hilly region of Uttar Pradesh, and in two villages from Haryana in the north are presented in Table 2.14. The annual contribution of homestead gardens in Kerala state is estimated to be 8 Mt (80 per cent of 9.9 Mt) (Kutty 1992).

According to the National Sample Survey (as quoted in NWDB 1986), private house gardens and trees around houses are estimated to provide 30 Mt of fuelwood annually in India. However, based on the data from five studies in four states, a conservative estimate of 16 Mt is taken by us as the annual contribution of homestead gardens to fuelwood use in India (see footnotes to Table 2.14 for explanation).

Table 2.13 Farm tree distribution in selected locations and extrapolation to national level in order to obtain total number of trees and standing biomass

Location of study site (state)[a]	Mean no. of trees/farm (range)	Mean trees/farm selected for estimation	States grouping[b]	No. of farms[c] (million)	No. of trees[d] (million)	Standing biomass above ground (Mt)[e]
Uttar Pradesh	From 6 villages in 3 districts: 17, 21, 253, 377, 443, 787	253	UP, Bihar, HP, JK, MP, Punjab, WB	47.5	12 017	1382
Haryana	7	7	Haryana	1.3	9	1
Karnataka	63	63	Andhra Pradesh, Maharashtra, Orissa, Tamil Nadu	32.5	2048	235
Gujarat	16	16	Gujarat	3.1	50	16
Rajastan	21–347	21	Rajastan	4.7	100	11
			Total	89.1	14 224	1635

[a] Sources: Uttar Pradesh, Saxena (1991); Karnataka, Ravindranath *et al.* (1991); Gujarat and Rajastan, Saxena (1991).
[b] Due to lack of studies, states are grouped considering broad features. This is a crude classification. UP, Uttar Pradesh: HP, Himachal Pradesh; JK, Jammu and Kashmir; MP, Madhya Pradesh; WB, West Bengal.
[c] Number of farms in states from NIRD (1991).
[d] Number of trees estimated by multiplying mean number of trees/farm by number of farms.
[e] Standing biomass estimated by multiplying number of trees by mean weight/tree. Mean weight of village tree: 79.7 kg/tree (mean of 11 287 trees in Ungra) (Table 2.12); 115 kg/tree (1987 trees in Barkha Alimuddin); 146 kg/tree (mean of 1295 trees in Dhanwas (Chaturvedi 1990)); 217 kg/tree (mean of 347 million trees in Kerala (Kutty 1992)). Weight of a mean tree assumed to be 115 kg/tree.

Net annual increments as % of standing biomass = 6.14% (tropical dry deciduous forest) (Cannel 1982); 2.84% (tropical dry deciduous forest) (Sharma *et al.* 1991).

A conservative value of 2.84% is taken to estimate the net annual increment of woody biomass of trees on farms from standing biomass, as farms in most parts of India can be classified as tropical dry deciduous forest.

Total annual increment = 1635 Mt × 0.0284 = 46 Mt. This is the woody biomass contribution for cooking.

Table 2.14 Trees on homestead gardens and above-ground tree biomass in selected locations, and projections to national level

State	Location	Units	No. of trees	Above-ground standing biomass (t/HH)
Karnataka	Sirsimakki village western Ghat (heavy rainfall)	No. of trees/HH	18	2.07 (at 115 kg/tree)
	Ungra (semi-arid)	No. of trees/HH	1	0.08 (at 77 kg/tree)
Kerala	State-level data	No. of trees/HH	97	21.00 (at 217 kg/tree)
		Total no. of trees in the state	347 million	
Haryana	B. Alimuddin	Trees/HH	3	0.34 (at 115 kg/tree)
	Dhanwas	Trees/HH	3	0.44 (at 146 kg/tree)
UP	Garhwal region data from 17 villages	Standing biomass up to 100 t/village = 12 villages Standing biomass 200–300 t/village = 3 villages Standing biomass > 300 t/village = 2 villages Total = 17 villages		

HH households.

Contribution of homestead gardens as follows:

1. Kerala = 8 Mt/yr was estimated as the contribution of homestead gardens by Kutty (1992).
2. West Bengal and Assam = 21 t tree biomass/HH (same as Kerala) × 11.3 million HH × 0.0284 t annual increment/t standing biomass (from Table 2.13) = 6.7 Mt/yr. West Bengal and Assam have homestead gardens similar to Kerala.
3. Rest of India = 0.34 t tree biomass/HH (same as Haryana village) × 86 million rural households × 0.0284 t annual increment/t standing tree biomass = 0.83 Mt/yr.
4. Total fuelwood contribution of homestead gardens (1+2+3) = 15.53 Mt/yr.
5. References: Sirsimakki (based on a survey for this book), Ungra (Ravindranath *et al.* 1991), Kerala (Kutty 1992), Haryana (Chaturvedi 1990), and Uttar Pradesh (Sharma *et al.* 1991).

2.7.5 Village commons and degraded forest lands

Village commons are lands traditionally allocated for use by local communities for meeting their biomass needs. No studies are presently available on the vegetation status and contribution of village commons. In India, degraded forests or village commons, and degraded cropland left fallow for long periods also provide fuelwood to rural communities (in addition to forest, cropland, settlement, water body, and roads). Total land under these categories is 52.8 Mha (Table 2.11). A large study in 12 semi-arid districts of India during 1984 showed that 66–84 per cent of the total domestic fuel needs of marginal farmers (with <2 ha of land) and the landless poor were derived from the village commons; furthermore, 8–32 per cent of large farmers also obtained their domestic fuel from village commons, again showing the significant contribution of village commons (Jodha 1990). Village commons may have degraded since 1984. In the absence of any data on the vegetation status of degraded lands, a very conservative estimate of annual increment in woody biomass of 0.5 t/ha/yr is assumed and the total fuelwood contribution of degraded lands is estimated to be 26 Mt annually (Table 2.11).

2.7.6 Non-tree biomass

It is common to observe rural people collecting twigs, branches, and roots of shrubs and climbers as fuel. Some dominant examples are *Prosopis juliflora, Lantana camara, Cassia auriculata* and *Tecoma stans*, where all parts of such shrubs (excluding leaves) are collected as fuel. Unfortunately no studies are available on the distribution and growth of such shrubs or their contribution to fuel supplies. However, it is common knowledge that in many districts in South India, such as Ramanathapuram and Madurai (Tamil Nadu), Anantapur and Cudappa (Andhra Pradesh), and Bellary and Raichur (Karnataka), *Prosopis* is already meeting the cooking fuel requirements of a significant proportion of rural households, and that the surplus is sold as fuelwood or converted into charcoal for urban markets. According to a field study in five villages of Anantapur district in Andhra Pradesh state (Ravindranath *et al.* 1994*b*), 68 per cent of village households met all their cooking energy requirements from *Prosopis* and 18 per cent of households met three-quarters of their fuel requirement from *Prosopis*.

Lantana is also present in most regions, particularly in open and degraded forests, and the dried *Lantana* is used extensively as fuel. From these observations it is possible to conclude that a significant proportion of the villages obtain their domestic fuel requirements from species such as *Prosopis* and *Lantana*. These are not necessarily favoured species, because of the presence of thorns in the case of *Prosopis* and low density in the case of *Lantana*. Also, for the same reason, large-scale commercial exploitation does not yet appear

Table 2.15 Biomass fuel potential of shrubs in North India in their natural state, and of *Prosopis juliflora* in experimental plots

Shrub species[a]	Fuel potential (t/ha/yr)
Tamarix dioca	1.1
Sesbania aegyptica	21.3
Adathoda vasica	0.5
Vitex negundo	2.8
Nyctanthus arbortristis	5.7
Prosopis juliflora[b]	12.0
Prosopis juliflora[c] (Phaltan, Maharashtra) (semi-arid)	22.0

Sources:
[a] Shrub species other than *Prosopis*, Singh (1988).
[b] *Prosopis juliflora*, Gurumurthy *et al.* (1984).
[c] *Prosopis juliflora*, Chaturvedi (1993).

possible, particularly for *Prosopis*. However, some research projects have been initiated in India to breed thornless, hardy *Prosopis* varieties.

Results from a study of productivity of some dominant shrubs in North India in their natural habitat are given in Table 2.15. Production of biomass with potential as fuel (woody portion) for the three shrub species was >42 GJ/ha/yr (about 3 t/ha/yr). Unfortunately no studies on the productivity of *Prosopis juliflora* and *Lantana camara* in their natural habitats in different regions of India are available. In experimental plots, 12–22 t/ha/yr of woody biomass has been reported for *Prosopis* (Table 2.15).

It should be noted here that all the degraded lands may not have shrubs or may have shrubs at varying density. So we have assumed a conservative woody shrub biomass productivity of 0.25 t/ha/yr for all the degraded non-forest land and open forests (78.5 Mha), with a total fuel contribution of 19.6 Mt/yr (Table 2.11).

2.7.7 Contributions of different fuelwood sources

The contributions of different sources of fuelwood supply are given in Table 2.11. Forests are estimated to contribute 32 per cent of the total wood coming from various sources. The crucial few points to be noted are: first, various sources contribute significantly to meet fuelwood demands; secondly, the contribution of forests is only slightly higher than that of other sources; thirdly, the contribution of plantation residues is significant even though tree plantations are harvested mainly for urban and non-fuel uses; and, finally, clearfelling of trees contributes only about 9 per cent of the total fuelwood obtained from trees.

The above analysis shows the difficulties involved in estimating the contribution of various sources of woody biomass and the importance of a range of land categories as sources of fuelwood. Our present work is one of the few attempts to explain the sources of fuelwood used in India from the supply side. We have explained nearly all the rural fuelwood use if demand is assumed to be 181 Mt annually. However, the 40 Mt of urban fuelwood demand, which is mainly in the form of logs, has been left out. Forests, farm forestry plantations, trees on farms, and trees in urban areas probably all contribute to the urban supply. Such an urban analysis would assist in evaluating the impact of fuelwood demand on forests, which is discussed in the next chapter. A recent survey of urban fuelwood dealers in Karnataka reported 29 per cent of them procuring fuelwood from forests (Ranganathan *et al.* 1993) with most of it in the form of cut logs coming from felled trees.

2.8 CROP RESIDUES

The importance and contribution of crop residues as a source of fuel for domestic use has been recognized in several studies (NCAER 1985; Leach 1987; Barnard 1990). The general conclusion of such studies is that once fuelwood resources are depleted, households will be forced to shift to crop residues. This section presents data on crop residue production and use in the selected villages as well as at the national level in India.

2.8.1 Crop residue production and use in selected villages

Among the end uses of crop residues, the dominant use in the six villages was for fodder and fuel (Table 2.16). Only in Bishnapur were crop residues not used as fuel, as in this case there is a forest nearby to provide wood (Bose *et al.* 1991). In the other villages, where data are available, the extent of residue used as fuel ranged from 28 per cent (BNPura) to 71 per cent (Ungra) of the total available residues with fuel potential. In Maulaganj the use of residues was more than the production, and they were probably obtained from nearby villages. The contribution of crop residues is significant.

2.8.2 Crop residue production and use at the national level

Crop residue production and end uses in India for fuel and fodder are given in Table 2.17 for major crops, including plantation crops such as coconut, tea, and coffee. Crop residues from cereals, which are low in lignin, are used as fodder for cattle. However, this excludes maize cobs and rice husks. Rice husks contain silica and mature maize cobs are hard for cattle to consume.

Table 2.16 Crop residues production and use as fuel in selected villages

Village	Human population	Area cropped (ha)	Residue production (t/yr)	Fuel potential (t/yr)	Residue use as fuel (t/yr)	Fodder use (t/yr)
Maulaganj	1143	220	1361	152	238 (156)[a]	428
Bishnapur[b]	931	185	642	245	0	142
Dhanwas	833	201	616	211	115 (55)	195
Ungra	1047	236	2593	462	329 (71)	428
Nanuda[c]	838	637	NA	NA	173	NA
BNPura	593	94	290	63	18 (28)	NA

[a] Figures in brackets indicate percentage of crop residue production with fuel potential used as fuel.
[b] Bishnapur: crop residue not used as fuel.
[c] Nanuda: production data not available (NA).

Table 2.17 Crop residue production and use as a source of fuel in India[a] (Mt dry/yr for 1985–86)

	Fodder for livestock	Domestic fuel	Fuel in industry and establishments	Other uses or unused[b]
Cereals + millets	254	2	16[c]	5.5
Pulses	19	10	–	–
Oil seeds	12	16	–	0.8
Fibre crops (cotton, jute)	–	12	–	–
Mulberry	–	2	–	–
Sugarcane	–		28[d]	19.3
Coconut + areca nut	–	7	–	2.2
Coffee + tea	–	3	–	–
Total	285	52	44	27.8

[a] Estimates made by author using data on area under crops, share of residues, and use of residue as fuel, based on a number of field studies and observations. Minor crops, such as vegetables, fruits, spices, and tobacco, are not included. Leaf biomass of all other annual crops (other than sugarcane) and plantations are not considered.
[b] Other uses include: thatch material (rice straw and coconut leaf) and unused components such as sugarcane leaves and tops.
[c] Rice husk use in small restaurants, brick firing, and rice parboiling.
[d] Sugarcane bagasse (dry) use as fuel in sugar mills and crude sugar manufacturing units.

Among the pulses (legumes), only pigeon pea (*Cajanus cajan*) stalks are woody and thus are used as fuel (10 Mt annually).

The residues of oil seed crops, such as groundnut (husks), niger, castor, and the stalks of fibre crops such as cotton and jute are all used as fuel. Mulberry stalks provide a continuous supply of fuel to farmers growing mulberry for its leaves (to feed silkworms). The productivity of mulberry stalks is estimated to be about 10 t/ha/yr (Ravindranath and Somashekar 1991).

The area under coconuts in India was 1.5 Mha in 1991. In Ungra village in South India, the annual contribution of coconut as fuel was estimated to be 272 t in 1990 and it accounted for 40 per cent of the biomass fuel used for cooking (Ravindranath and Somashekar 1991). Coconut produces several types of residues and their potential fuel productivity is 5.4 t/ha/yr (for young plantations) to 11.7 t/ha/yr (for mature plantations) at a density of 100 trees/ha (Ravindranath *et al.* 1991). Thus crops such as mulberry and coconut have a large potential for providing fuel for domestic use for the farmers growing them.

The fuel potential of crop residues in India is estimated to be about 52 Mt annually. However, Joshi *et al.* (1992*b*) estimated that crop residue use for cooking was 40 Mt/yr (Table 2.2). The point to be noted is that the residues of all crops unsuitable for use as fodder and having significant lignin, and which can be harvested and transported within the village, are all used as fuel. Crop residues are usually not wasted. The use of rice husks and sugarcane bagasse as energy in industry has already been discussed.

2.9 CATTLE DUNG PRODUCTION AND FUEL USE

Cattle dung use is a dominant feature only in South Asia, and, within India, only in certain parts (Table 2.3). Traditional stoves using dung cakes have lower thermal efficiency compared to those using wood (Fig. 2.6) and there is also significant drudgery in preparing dung cakes. The question is why dung cakes are still used in many parts of India as fuel.

Dung has the following advantages for household energy supply:

(1) It is available within the house, in the case of cattle-owning households. In major dung-using states, such as Uttar Pradesh, Bihar, and Madhya Pradesh (MP), 80, 59 and 70 per cent of rural households, respectively, own cattle (ABE 1985).
(2) It is available throughout the year as a by-product with no additional effort to procure, although it takes effort to prepare dung cakes.
(3) It can be stored easily for the rainy season, as is common in North India.
(4) It provides slow and continuous heat.
(5) It does not involve any cash expenditure.

Dung is thus a convenient fuel from the supply point of view for the rural housewife in the absence of fuelwood. These factors are also applicable to urban areas.

Dung yield

Yields vary from location to location and very few studies report data on dung yields from field measurements. Unfortunately, many estimates used in

Table 2.18 Dung yield: (cattle shed stall collection of dung) in selected locations (fresh weight in kg/day)

Village	Bullocks	Buffaloes	Cows	Calves
BNPura (East)	3.65	4.57	2.37	1.46
Sirsi (south-west)[a]	3.90	10.40	5.21	Cross-bred cows: 9.80 kg
Ungra (south)[b]	5.78	5.78	6.89	NA
UP villages[c] (north)	5.5–6.4	8.1–8.9	NA	NA
Data used at national level according to ABE[d] (kg/bovine/day)				
UP	13.2	10.2		
Orissa	10.6	12.7		
Karnataka	11.6	10.1		

[a,b] Measurements made for the study.
[c] Bose *et al.* (1991).
[d] ABE (1985). The values may include dung production during a full day and not merely stall collection.

the past were based on high yields obtained from animals with high feeding levels (ABE 1985). Field studies on dung yields in the selected villages (Table 2.18) show lower yields compared to the values used in the national estimates by agencies such as ABE (1985) and PC (1992). There are also seasonal variations as animals have access to more grass during the rainy season compared to the summer, thus resulting in higher yields in the rainy season. What is important from a dung cake or biogas point of view is not the total dung yield per day but the collection rate during the period of the day when the cattle are tied in the cattle shed. When cattle are grazing or working in the field, the dung produced cannot easily be collected and transported, due to the human effort involved.

Dung production

Livestock population and dung production in India is shown in Table 2.19. The total cattle population is 257 million, giving a cattle to rural human population ratio of 0.41. The dung of sheep and goats is not used as fuel as their rate of dung production is low and the dung cannot easily be made into cakes, due to the lower moisture content; but such dung can be used as feedstock for biogas production. The total collectable fresh dung production nationally averages about 1.2 Mt/day. The solids content of dung is 18 per cent (82 per cent moisture) and if a 10 per cent moisture is assumed (for dung cakes ready for burning), the dung cake potential on an annual basis is 123 Mt of air-dry weight. Currently just over a quarter of this is burnt annually in India (Table 2.2, 'low estimate'). After preparation of the dung cakes they are stored in heaps for use in the rainy season, thus theoretically

Table 2.19 Livestock population (for 1987) and dung production (in fresh weight) at national level

State	Cattle population (million)	Sheep + goats population (million)	Total dung production (million kg/day)
Andhra Pradesh	21.91	13.08	103.8
Assam	7.31	1.77	33.6
Bihar	20.85	13.53	99.2
Gujarat	11.43	5.65	53.7
Haryana	5.69	1.36	26.1
HP	2.79	2.75	13.7
JK	2.88	2.91	14.1
Karnataka	14.95	9.34	71.0
Kerala	3.5	2.01	16.6
MP	33.67	8.52	154.9
Maharashtra	20.13	10.37	94.7
Orissa	14.26	6.92	66.9
Punjab	7.85	0.61	35.6
Rajastan	15.14	28.91	79.7
Tamil Nadu	13.57	10.78	65.4
UP	41.96	11.99	193.6
WB	16.64	12.27	79.8
Other states	2.48	0.77	11.5
Total	257.01	143.54	1213.9

Dung yield data used in the estimation of dung production: 4.5 kg/bovine/day and 0.4 kg/day/sheep or goat.

all the available dung could be used if necessary. In villages such as Ungra and Bishnapur where dung cake is not used as fuel, it is instead used as manure for crop production.

2.10 SOCIO-ECONOMIC ASPECTS OF BIOMASS ENERGY USE

Socio-economic aspects of biomass use as fuel are very crucial to the goals of development and better quality of life. Access to fuels, drudgery for women, and the quality of different biomass fuels used may depend on the socio-economic status of the household.

2.10.1 Quality of biomass fuels

The quality of different types of biomass used as fuels differs with respect to density, calorific value, moisture content, and emission of pollutants.

Calorific value

A study of 22 commonly used tree species showed that three-quarters had calorific values of 14–19 MJ/kg oven-dry wood (Jain 1991). A study in the Ungra area showed a mean calorific value of 15.9 MJ/kg oven-dry biomass for five shrubs (Reddy 1982). There are no large differences between different species nor between trees and shrubs. The moisture content of fuelwood and shrubs used as fuel was 8.4 per cent. The calorific values of crop residues such as maize cobs, cotton stalks, and gram legume stalks are 17.4 MJ/kg, 13.7 MJ/kg, 16.5 MJ/kg, respectively (TERI 1992*b*). Crop residues and dung cakes have a lower calorific value compared to fuelwood, indicating a larger volume of fuel to be used for cooking.

Density

Crop residues and dung cakes have lower densities than that of wood (0.67). Among tree components, stem wood has a higher density than twigs (which depends on lignin content). There are also differences among tree species. For the housewife, lower density means handling more volume and continuous feeding and tending of the fire; thus ideally she prefers fuelwood from trees to crop residues or dung, as is evident from observations in Bishnapur and Ungra.

Moisture content

The presence of moisture leads to extra smoke and increased tending of the fire. Moisture also results in waste of heat to evaporate the water. Good air-dried wood contains about 10 per cent moisture and the housewives would prefer such properly dried wood, but often they have to use wood with higher moisture contents, especially during the rainy season, thereby increasing drudgery.

Emissions characteristics

The emission levels of pollutants are higher for dung cakes as compared to fuelwood (NIOH 1980), indicating the hazards of cooking with dung cakes.

2.10.2 Socio-economic status and type of fuel used in rural areas

It is important to understand the relationship between socio-economic status of households and the type of fuel used in the domestic sector for cooking. Urban areas provide an excellent example since a range of solid, liquid, and gaseous fuels, and electricity are available. The results of a large study in urban areas on the relationship between income levels and biomass fuel use are shown in Fig. 2.7. It is seen that at very low income levels biomass fuels,

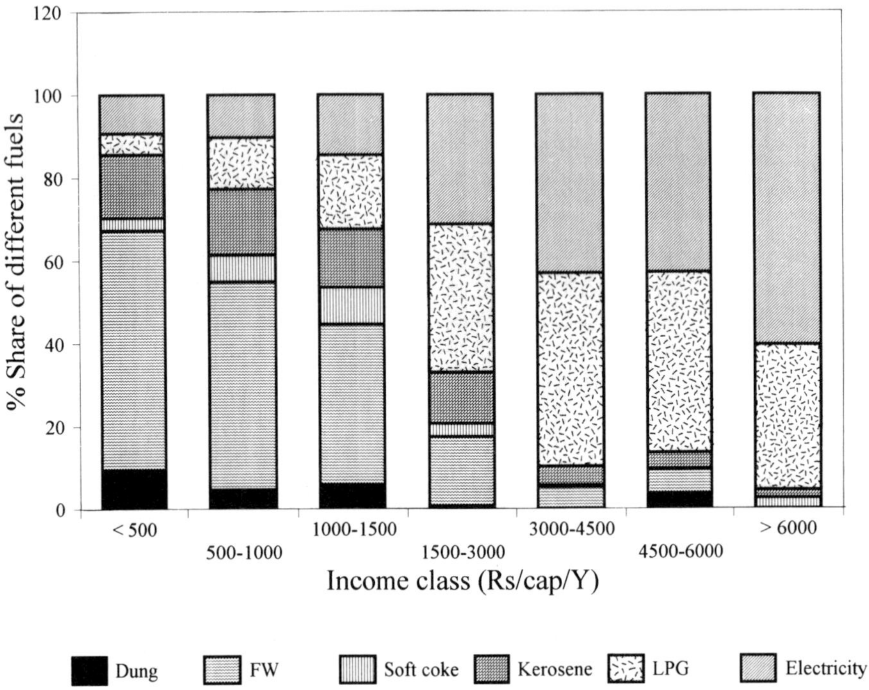

Fig. 2.7 Impact of income on fuel use mix in urban households in India (from TERI 1992*b*); FW, fuelwood; LPG, liquid petroleum gas.

especially fuelwood, dominate energy use and that biomass use declines as income levels increase. At mid-income levels biomass use is not significant and at higher income levels it is nearly absent (TERI 1992*b*; Reddy and Reddy 1994). As the income levels increase there is a shift away from biomass to liquid fuels and liquid petroleum gas (LPG) at mid incomes, and to gaseous fuels and electricity at higher income levels. In urban areas fuelwood is usually the only serious solid fuel option available.

Among the solid fuels there is also a hierarchy from dung cake to crop residues to tree twigs to cut wood, depending on the density and amount of fire tending required during cooking. The type of biomass used for two villages is given in Fig. 2.8, according to land-holding size, taken as an indicator of economic status; landless are the poorest and those having large farms the most affluent.

In Nanuda village, crop residues (and twigs together) are the dominant types of fuel, followed by fuelwood, except that large farms seem to depend a little more on fuelwood. Dung cake use was low in all land-holding groups. In Dhanwas village dung cakes were the dominant fuel followed by crop residues for all income groups. Similar results were obtained in villages of

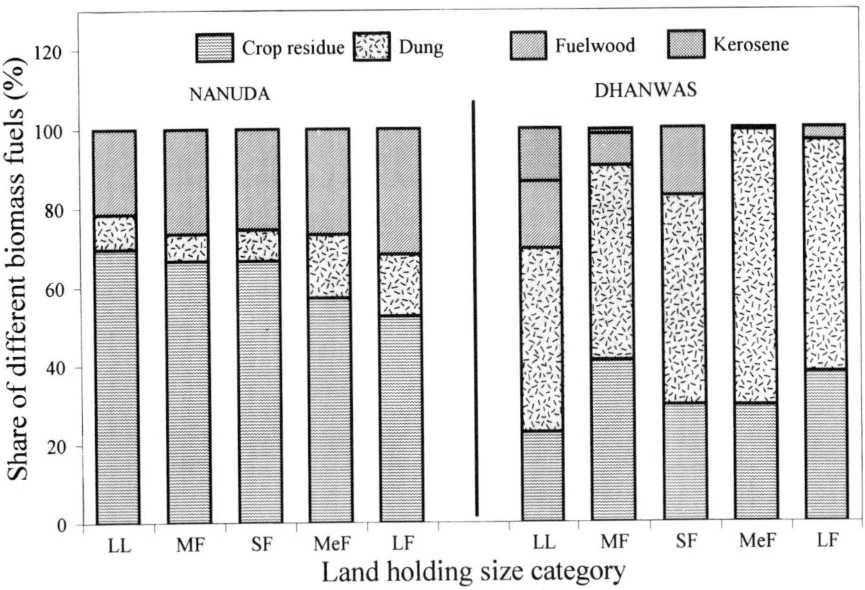

Fig. 2.8 Impact of size of farm (proxy for income) on type of biomass used in two villages in India. LL, landless; MF, marginal farms; SF, small farms; MeF, medium-sized farms and LF, large farms.

Himachal Pradesh (Singh 1993) and Haryana (Bhide 1992). In Bishnapur village all households, irrespective of socio-economic status, wholly depended on fuelwood from forests. Thus among rural households of a given area it appears that the types of fuels used were nearly uniform among all income groups.

There are, however, differences in type of biomass fuel used in different agro-climatic regions, but within a given region access to different fuels, particularly fuelwood, seems to be generally the same (Table 2.3). The landless fall into a different category as they do not own any land or trees. Nevertheless, in Nanuda and Dhanwas the type of fuel used by the landless seems to follow the same general trend as that of the landed, since they may gather dung or obtain crop residues from farmers as wages.

2.10.3 Socio-economic status and quantity of fuel use

As countries develop and income levels improve, fuel type use may change (Fig. 2.7). It is important to understand the impact of increased incomes on the quantity of fuels used. A large study in Haryana state has shown that domestic fuel consumption in rural as well as urban areas generally increases with increased incomes (Bhide 1992). It should, however, be noted that in rural areas the type of biomass used changes little with income. Thus one could expect increased per capita use of biomass energy in rural areas since, with increased incomes, households probably cook snacks, tea, and meat more frequently, or may entertain more guests, than the landless, marginal farmers, or poorer households. Thus the implications of such a pattern is that, with development, the demand for biomass energy per capita might increase if higher-quality fuel is not made available.

2.10.4 Mode of procurement of biomass

Biomass fuels are obtained from several sources: gathered from own tree resource or local areas or purchased.

Karnataka A six-village study showed that households depended on gathering biomass from public land or their own sources, and only 20 per cent of households purchased fuelwood (Reddy 1982).

Gujarat In selected villages fuelwood and crop residues were obtained free from their own sources or gathered from common land; the proportion of households gathering biomass was higher for the landless and smaller landholding households (Veena 1988).

Haryana In Dhanwas village, the forest was the dominant source of fuelwood. Among crop-residue-using households, 58 per cent obtained residues from their own source and 25 per cent (mainly landless) obtained them as a part of their wages, while dung was wholly obtained from their own cattle (Puri 1988).

Recent state-level studies in Haryana and Karnataka showed that for over three-quarters of households, gathering continues to be the major mode of obtaining fuelwood (Bhide 1992; Ranganathan *et al.* 1993). However, in urban areas fuelwood-using households largely purchased the fuel.

2.10.5 Human effort for gathering fuel

The human effort involved in gathering fuelwood has received significant attention as it leads to drudgery for the gatherers who often walk in the hot sun or rain on roads and fields often littered with stones and thorns, with a load on their head. Several studies are available showing the human effort required in a number of locations (Table 2.20).

At the national level one can conclude that the average number of hours spent on gathering biomass (including fuelwood, crop residues, and dung) is about 2 hours/day/household. Such human effort for gathering biomass may increase with declining supply due to degradation of fuelwood sources. An

Table 2.20 Human effort in gathering fuelwood for domestic use

Location	Human hours spent/week/household
6 villages (Karnataka) (1977)[a]	15.1
Dhanera, 10 villages (Gujarat) (1988)	11.2
Dhanwas (Haryana) (1984)	14.3
BNPura (Orissa), summer months for 6 months (1986)	35.0

[a] The year given in parentheses refers to the period of study.
Sources: Karnataka, Reddy (1982); Dhanera, Veena (1988); Dhanwas, Puri (1988); BNPura, Nishanka and Misra (1990).

important aspect of gathering fuelwood is the dominance of woman and child labour, which has important social and economic consequences for the family and the country as a whole (Reddy 1982). Thus shortages of cooking fuels would put increasing pressure on women and children.

Type of fuelwood gathered

A six-village study (Reddy 1982) as well as a state-level study in Karnataka (Ranganathan *et al.* 1993) show that twigs in the form of thin tree branches are the main type of wood gathered and that logs or main stems are not gathered at all as these usually have a market value and are thus sold.

2.10.6 Access to biomass energy

It is necessary to appreciate the access problem to different sources of biomass fuels for households with differing socio-economic status in rural areas, as it determines the extent of drudgery in gathering biomass.

Fuelwood Rural communities are allowed to collect headloads of dry or fallen twigs from forests that are protected by the Forest Department. Thus all categories of households have equal access, though sometimes influential families with the connivance of forest officials have greater access to fuelwood. The village commons in India have become 'no-man's-lands' with equal and uncontrolled access to all households for biomass. With the decline of local management systems for the village commons, leading to their degradation, poorer households such as the landless and marginal farmers have lost their main source of fuelwood (Jodha 1990). Trees on farmlands are usually used by the land owners themselves. In South India, there are no restrictions on any person collecting fallen twigs from trees, even on private farm lands. The uniform trend in type of biomass used by different land-holding groups shows (Fig. 2.8) that access to biomass sources is generally the same for all in a given location.

Crop residues In the past, crop residues that did not have fodder value were freely distributed or burnt in the field. Now all categories of land-owning households use crop residues (Fig. 2.8). The changed situation is evident in some locations, such as Dhanwas, where poorer households obtained crop residues as a part of their wages.

Dung cakes These are usually prepared from the dung of household cattle. The proportion of households owning cattle was over 70 per cent in all the dominant dung-using states of North India (ABE 1985). However, even the landless use dung cakes at the same rate as the landed (Fig. 2.8). They either own cattle or gather dung from fields and village streets.

2.10.7 Gender aspect of biofuel use

Rural energy problems are often equated with the rural fuelwood problem in developing countries. This could be further extended to infer that the rural fuelwood problem is actually a crisis for women. Biomass is largely gathered and prepared by women, who also cook using it and are thus exposed to drudgery in gathering as well as exposure to smoke during use. Any scarcity of fuelwood or shift to lower-quality fuels (such as from fuelwood to dung cakes) will thus affect the quality of life of women most adversely.

2.11 COMMERCIAL ENERGY USE IN RURAL INDIA

Commercial energy use in India accounted for 63 per cent of the total primary energy (Table 2.1). Out of the total commercial energy use, industry dominates (accounting for 50.5 per cent) followed by transport (24.5 per cent), household (13.7 per cent), and others (2.3 per cent), while agriculture accounted for only 9 per cent (PC 1992). This clearly shows that urban and industrial sectors dominate commercial energy use. Thus it is necessary to understand the extent of use of different commercial energy sources and their features in rural areas.

All the well-known energy databases or reports such as CMIE (1994), PC (1992), TSL (1992), and TERI (1992*b*) provide very little information on commercial energy use in rural areas. For example, none of these reports give any data on kerosene or electricity use for lighting in rural India. Thus, based on case studies, projections are made below on commercial energy use in rural areas.

Table 2.21 Electricity use in the four selected villages (kWh/yr) (these villages are selected for detailed bioenergy analysis in Chapter 6)

End use	Ungra[a]	Nanuda[b]	BNPura[c]	Maulaganj[d]
1. Residential (lighting + appliances)	21 900	30 780	8760	0
2. Flour milling	10 950	0	0	0
3. Pumping irrigation water	39 960	154 775	0	9 diesel pumps
4. Agro-processing	4500	0	0	8 diesel pumps
Total electricity use	77 310	185 555	8760	
Per capita electricity use	74	221	15	0

[a] Ungra: 100 out of 177 houses electrified + 7.5 kW flour mill + 36 irrigation pumps (5–7 hp).
[b] Nanuda: 73 out of 135 houses electrified + 12 irrigation pumps + other uses. Detailed breakdown of end uses not given.
[c] BNPura: 15 out of 98 houses electrified.
[d] Maulaganj: diesel-operated systems (non-electrified village); 9 irrigation pumps and 8 crop threshing machines.

Electricity use

Electricity use in four of the six villages (see Chapter 6 for a detailed bioenergy analysis) is given in Table 2.21. Electricity use ranged from a low of 15 kWh/capita/yr in BNPura to 221 kWh/capita/yr in Nanuda. The percentage of households electrified also shows a wide range. Pumping irrigation water is the dominant electricity-using activity in Ungra and Nanuda. In Maulaganj, a non-electrified village, diesel-operated shaft power systems are in use. Electricity and diesel use may thus depend on water availability and the resulting demand for energy for pumping irrigation water.

Electricity use at the national level is given in Table 2.22. Electricity use in rural areas is estimated to be 67.7 TWh, which accounts for only 23 per cent of national total (although 75 per cent of the population live in rural areas) with a per capita energy use of 108 kWh/capita. However, the achievement of rural electrification in India is truly commendable, with 84 per cent of the 0.55 million villages already electrified (TSL 1992). But only a quarter of the rural houses are electrified. Rural electrification seems to be largely aimed at providing energy for pumping irrigation water (PC 1992). The electricity use for pumping water is estimated assuming that all the 9.1 million pumps are in use and the rate of consumption is 5362 kWh/pump (according to CEA 1991).

It is also interesting to note that the share of agriculture in total electricity use has increased from 10 per cent in 1970–71 to 28 per cent in 1991–92 (CMIE 1994). This shows the growing importance of the power needs of the agricultural sector.

Table 2.22 Estimated electricity, diesel, and kerosene use in rural India (for 1991)

Sectors/activity	Number of households or units or villages	Electricity, diesel, and kerosene use (units)	Total electricity/kerosene /diesel use per year
Lighting	28 million HH[a]	408 kWh/HH/yr[b]	11.4 TWh
Agriculture	9.1 million pumps[c]	5362 kWh/pump/yr[d]	48.8 TWh
Agro-industry, processing	0.484 million electrified villages	15 450 kWh/village/yr[e]	7.5 TWh
Total	–	–	67.7 TWh
Diesel pumps[f]	5.5 million pumps	854 kg diesel pump/yr	4.7 Mt
Lighting, kerosene	627 million	6.6 litres/capita/yr[g]	3.35 Mt

[a] HH, households.
No. of rural HH in 1971, 79.6 million (CMIE 1992); in 1981, 90.8 million (CMIE 1992); in 1991, 104 million assumed, taking a family size of six.
Extent of HH electrification = 27% (NSS 1992).
No. of HH electrified in rural area = 104 million rural HH × 0.27 = 28 million.
[b] Mean electricity use/HH/yr = mean of electricity use in (Ungra, BNPura, and Nanuda) = 408 kWh/HH/yr.
[c] Source: CEA (1991).
[d] Electricity use/pump during 1991 from CEA (1991).
[e] Electricity use in agro-industry: Ungra level of one flour mill and one small processing unit assumed at 15 450 kWh/village/yr for all the electrified villages.
[f] Diesel use data for pumping from PC (1992).
[g] Kerosene use/capita, refer to text (section 2.11).

Diesel use

Diesel use in rural areas is largely for pumping irrigation water and, nationally, agriculture accounts for 9 per cent of total oil use. Apart from the use of oil in pumping, diesel is also used to fuel tractors (for draught power in agriculture and rural transport) and agro-processing. It is surprising that even though 84 per cent of villages are electrified, there are 5.5 million diesel pumps; more than one for every two electric pumps. This has to be viewed against the fact that electric pumps are cheaper to operate than diesel pumps (Rajagopalan and Demaine 1994). The use of diesel pumps could be due to non-availability of electrical connection from the grid, low reliability of electricity supply, or the advantage of mobility of diesel engines.

Kerosene

Kerosene is the fuel used by most households in India, with 97 per cent of them using it in rural areas (NCAER 1985). Kerosene is used chiefly for lighting in rural areas in both electrified and non-electrified households.

The use of kerosene in rural areas is: according to a national survey (during 1978–79), 5.1 l/capita/yr (NCAER 1985); according to a recent study in Karnataka, 6.6 l/capita/yr (Ranganathan *et al.* 1993); and in a study of three non-electrified villages in North India it ranged from 6.9 to 8.5 l/capita/yr (Bose *et al.* 1991). The rural kerosene use would be around 3.4 Mt annually (at 6.6 l/capita/yr). Kerosene use for lighting is characterized by low conversion and luminous efficiency (ADB 1993). Information on kerosene use in rural areas for cooking is limited. Kerosene was used for cooking only in Nanuda (Veena 1988). In urban areas kerosene is the preferred cooking fuel, next only to LPG (Fig. 2.7).

2.12 CONCLUSIONS

Energy use pattern, particularly biomass energy use, was analysed in this chapter. Although the share of biomass in national total primary energy use is low (36 per cent of the total), it is the dominant source in rural areas. Fuelwood is the dominant biomass energy used, followed by crop residues and cattle dung. The sources of 181 Mt of fuelwood used for cooking in rural areas were analysed and the findings show that forests, tree plantations, farm trees, homestead gardens, and village commons are the diverse sources, and that each source contributes significantly. All the crop residues having potential for use as fuel are used as fuel in domestic and industrial sectors. Biomass energy use in the domestic sector is characterized by low efficiency of use, drudgery in gathering and use, emission of smoke in the kitchen, and ultimately a lower quality of life. Kerosene is the fuel used by the majority of rural households, mainly for lighting. It has poor conversion efficiency and provides poor-quality lighting. Electricity is used only by a quarter of rural households for lighting. Agricultural water pumping is the dominant activity using electricity, as well as diesel, in rural areas. Even commercial energy use is characterized by low efficiency (irrigation pumps and kerosene wick lamps) and poor quality of life.

There is a need to explore the potential for providing quality fuels and efficient devices to promote development in rural areas. Before that, it is necessary to understand the environmental impact of the current patterns of biomass energy use, to enable comparison with alternative options.

Photo 2.1 Traditional three-pan mud stove for cooking in a South Indian village (Ungra).

Photo 2.2 Industrial use of fuelwood—a brick kiln near Bangalore.

Photo 2.3 Homstead gardens providing fuelwood, along with coconuts and tree products in Western Ghat mountain region.

Photo 2.4 *Prosopis juliflora* is increasingly an important source of fuelwood in rural India.

Photo 2.5 Fuelwood depot in an urban centre marketing supplies from the forest.

Photo 2.6 Women with fuelwood headloads for household use.

3

Environmental impacts of biomass energy

The extraction, conversion, and utilization of various forms of energy is recognized as one of the major contributors to environmental degradation at the global as well as the local level; be it greenhouse gas emissions and local air pollution due to combustion of fossil fuels, submergence of forests by hydroelectric power projects, or forest degradation due to fuelwood use. It is necessary to understand the environmental implications of biomass use as an energy source via:

(1) contributions to deforestation and forest degradation, leading to loss of biodiversity and soil degradation;
(2) atmospheric pollution from emissions of greenhouse gases during combustion of wood, with its implications for climate change;
(3) release of pollutants (indoor air pollution) leading to domestic health hazards (particularly for women during cooking); and
(4) loss of nutrients due to combustion of cattle dung and crop residues.

Some of the impacts analysed in this chapter are based on the biomass energy consumption patterns in India described in Chapter 2.

3.1 DEFORESTATION AND FOREST DEGRADATION

A common perception among many is to link deforestation to fuelwood use. This has been the subject of intense debate and controversy in India between the Forest Department and researchers. The Forest Department links fuelwood requirements to deforestation and forest degradation (FSI 1988; PC 1992). The FSI report (1988) justified this by concluding that only 19 Mt of fuelwood came from recorded harvests in forests and another 30 Mt from forest degradation. This left an unexplained gap of over 100 Mt of fuelwood supply for the year 1986. It concluded that the imbalance between demand and production of fuelwood is the one single cause that has contributed the most to forest depletion (FSI 1988).

Rural domestic sector

Total rural fuelwood use in the domestic sector was estimated to be 181 Mt annually (Table 2.2). Based on discussions in earlier sections we attempt here to understand the impact of the use of fuelwood in rural areas on deforestation, using the following arguments:

1. Gathering by women and children is the dominant mode of procurement of fuelwood (as head loads) and they are unlikely to fell trees.
2. Mostly twigs and small branches are gathered, which do not require tree felling.
3. Fuelwood in the form of twigs is likely to come from sustainable harvesting using mainly the current year's biomass increment (Table 2.11).
4. Fuelwood gatherers (mainly women and children and those from poorer households) are not likely to cut trees in forests, especially as it is banned by the Forest Department. Anyone caught cutting a tree would face severe punishment as local forest guards can easily locate such infringements, particularly by the poor.
5. Monitoring of tree felling in South India at village level (Ravindranath *et al.* 1991) and district level (DES 1985) showed that it was largely aimed at urban and industrial markets.
6. Annual woody biomass productivity of forests is estimated to be 137 Mt (Ravindranath *et al.* 1992*b*) which, if sustainably harvested, could meet 75 per cent of the rural domestic fuelwood needs.
7. A study in Kerala state (population 29 million) showed that only 5 per cent of the fuelwood used came from forests (Kutty 1992). Similarly, a study in Haryana state in the north showed that only 9 per cent of the domestic energy use came from wood cut from trees (Bhide 1992). A large study in Karnataka showed that only 6 per cent of fuelwood used in rural areas came from logs of felled trees (Ranganathan *et al.* 1993).
8. Confirmatory evidence also comes from investigation of the sources of fuelwood (Table 2.11). Twigs and branches obtained in a sustainable way from non-forest tree and shrub sources, such as social forestry plantations, village commons, croplands, and homestead gardens, dominate the fuelwood supply scene.
9. A recent national-level study of the vegetation status in forests and tree plantations has concluded that 'illegal cuttings were not so rampant in primary forests or in plantations, and social forestry measures in India may have arrested further degradation of forests or even that some of the forests have been rehabilitated' (Seebauer 1992).

So far no study has provided any significant evidence to prove that rural or tribal needs for fuelwood are met by clearfelling forests, by cutting large trees, or that they contribute to deforestation. All the available evidence shows that the rural fuelwood requirement does not seem to lead to deforestation. It is apt to quote Leach and Mearns (1988), 'if all fuelwood and charcoal use is stopped tomorrow, deforestation rates would hardly be altered'.

Urban domestic sector

The fuelwood used in urban areas (40 Mt annually) is generally a high-density-type cut wood, since logs or stems are transported using diesel at high cost. In Haryana state about half of the fuelwood used is accounted for by wood cut from felled trees (Bhide 1992). A recent survey of fuelwood dealers in the urban areas of Karnataka showed that 29 per cent of wood sold through traders originated from forests (Ranganathan *et al.* 1993). Earlier studies in Bangalore (Reddy and Reddy 1983) and Hyderabad (Alam *et al.* 1985) have shown a much smaller percentage as coming from forests.

A study of fuelwood use in urban centres and deforestation, using satellite data, for the periods 1972–75 and 1980–82 concluded that open forests have increased and closed forests are being converted to open forests in the vicinity of urban centres, and that this decline is attributed to fuelwood extraction (Bowonder *et al.* 1988). It must be noted that this study refers to the period mainly before 1980, and thus prior to the Forest Conservation Act 1980, and it refers only to forests in the vicinity of urban centres.

Fuelwood use in industries and establishments

Fuelwood use in industries and establishments accounts for less than 5 per cent of the total use. The conclusion drawn above for the urban domestic sector, that a part of the fuelwood used comes from private sources and village trees and a part from forests, may be true for fuelwood use in industrial units and establishments, although unfortunately no specific data are available.

Overall contribution to forest degradation and deforestation

If we assume that 29 per cent (Ranganathan *et al.* 1993) of the total urban use of fuelwood (40 Mt, Table 2.2) comes from sources termed as forests, the area that may have to be clearfelled is 92 000 ha annually (at 126 t/ha of average standing biomass of forest, from Table 2.10) to obtain 11 Mt. If we assume that 19 Mt (as reported by FSI (1988) and in Table 2.11) of fuelwood used comes from forests (including the urban use), it may be necessary to clearfell 150 790 ha of forests, or plantations on forest land, annually. Given the large quantity of fuelwood required, even if less than 10 per cent is assumed to come from clearfelling of forests, it would lead to significant deforestation. It is likely that a part of 19 Mt or 11 Mt could be obtained from tree plantations, and not necessarily all from primary forests. Thus it is also wrong to conclude that fuelwood demand does not contribute to deforestation or forest degradation in India, as every million tonne of cut wood or logs supplied requires clearfelling of nearly 8000 ha of forests or plantation annually. It should be noted that the total fuelwood used in India was 227 Mt in 1991 and is increasing along with population growth. However, urban fuelwood demand contributes relatively more to forest decline than rural demands (Ranganathan *et al.* 1993).

3.2 LOSS OF BIODIVERSITY AND LAND DEGRADATION

In dry areas, stripping land of vegetation for fuelwood leads to soil erosion due to wind and water effects, and on a world-wide basis such overexploitation is estimated to contribute to 7 per cent of soil degradation (WRI 1992). In India it is common to observe fuelwood gatherers collecting uprooted shrubs and overlopping trees on village common lands, which exposes the soil to tropical monsoon rains. Excessive removal of vegetation and damage to ground vegetation during removal of fuelwood could affect plant diversity. It should again be noted, however, that fuelwood collection may not be a dominant factor contributing to land degradation or decline in biodiversity.

3.3 DOMESTIC AIR POLLUTION FROM BIOMASS COMBUSTION

In most locations in India cooking is done mainly indoors and the majority of rural houses usually have no chimney system to remove smoke (Ramakrishna *et al.* 1989; Ranganathan *et al.* 1993). A study (Reddy 1982) in South India showed that the number of hours spent cooking ranged from 3.3 to 4.6 hours/household/day, and another study (Veena 1988) in 10 villages in Gujarat state showed that 3.1–3.4 hours are spent on cooking/household/day. Women are therefore exposed to smoke from biomass combustion for long periods. When burnt in stoves, biomass fuels emit pollutants such as respirable particulate matter, carbon monoxide, nitrogen oxides, methane, and non-methane organic compounds. When compared with the combustion of fossil fuels, the emission of pollutants from wood combustion is higher. A comparative study of daily exposure during cooking in Pune (India) showed that the exposure to particulate matter was 17–26 mg.h/m^3 for biomass, compared to 0.4 mg.h/m^3 for LPG or 2.4–3.6 mg.h/m^3 for kerosene (Smith *et al.* 1994). Emission levels are shown to be high for dung cakes compared to firewood (NIOH 1980). A review of the health effects of exposure to smoke concluded that there is growing scientific evidence to support the many anecdotal accounts relating high biomass smoke levels to significant adverse health effects (Smith 1991). Cooking in smoke-filled kitchens is very inconvenient and women certainly do not like it. A large study (Ravindranath *et al.* 1989*c*) in Karnataka state on monitoring fuel-efficient and smokeless stoves showed that households reported higher fuel consumption in the improved stoves (compared to traditional stoves) and smokelessness. It was concluded that even if an improved stove design consumes more fuel than a traditional stove, women seem to prefer it because it removes smoke, and great importance is attached to this. The burning rate (kg of fuel burnt/hour) of the biomass also affects emissions of smoke and pollutants as well as thermal efficiency. Women seem to adopt higher burning rates since it decreases the cooking duration and also improves combustion with reduced smoke (Ravindranath *et al.* 1989*a*). Higher burning rates also lead to higher specific fuel consumption, so that attempts to reduce pollution from smoke would have implications for total biomass fuel use.

3.4 CARBON EMISSIONS FROM BIOMASS COMBUSTION AND DEFORESTATION

Tropical deforestation is currently a significant environmental and developmental issue and the subject of much research. Loss of tropical forests diminishes biodiversity, contributes to climate change by releasing stored carbon into the atmosphere, and often results in serious soil degradation (WRI 1992). At the global level, according to recent estimates by FAO (1993*a*), tropical deforestation for the decade 1981–90 was estimated to be about 15.4 Mha/yr. The contribution of tropical deforestation to global carbon emissions has been variously estimated as:

(1) 1.6 ± 1.0 Gt C (carbon) annually, accounting for 21 per cent of global C (IPCC 1992);
(2) 1.1–3.6 Gt of C annually (Houghton 1991); and
(3) 0.9 Gt (Dixon *et al.* 1994).

The rates of deforestation are thus crucial to estimations of global C emissions. This will also influence projections of the potential for C sequestration in forests and for using bioenergy as a substitute for fossil fuels to reduce net C emissions.

Relevance of C emissions from forestry to bioenergy

There is a need to consider carbon emissions from forestry for the following reasons:

1. Deforestation or forest degradation could result from land clearance for agriculture, timber harvest, and fuelwood extraction.
2. C emissions from forestry take place through combustion and decomposition. Combustion could be due to use as fuel.
3. Modern bioenergy technology, the main focus of this book, has the potential to reduce C emission by substituting for fossil fuel energy. Forests and tree plantations could provide the biomass feedstock.

In India, with the Forest Conservation Act of 1980, forest conversion to agriculture or pasture land has been banned. Thus deforestation or forest degradation in India could result largely from legal as well as illegal extraction of timber (industrial and structural) and fuelwood. The C emissions from forestry are analysed, even though the contribution of fuelwood to deforestation is marginal, in order to understand the relative importance of fuel and non-fuel end uses of forest biomass. It is necessary to assess the areas under forest, rates and causes of deforestation, and the contribution of fuelwood combustion to C emissions.

Area under forests

This has been the subject of intense debate and controversy in India between the Forest Department and environmental groups, as well as researchers. Until periodical satellite data became available, the area under forests had been officially estimated by the Forest Department at 75 Mha. India made its first attempt at estimating the area under forests using satellite imagery for the period 1972–75. However, systematic and periodic monitoring of the area under forests with improved satellite imagery techniques began in the period 1981–83 and was further improved and repeated for the periods 1985–87, 1987–89, and 1989–91. Nevertheless, data on area alone are not adequate since all forests do not have the same standing stock. It is also essential to have information on crown cover as an indicator of standing stock. Based on crown cover, forests were categorized into two types to reflect the standing stock of wood (Fig. 3.1).

According to the latest data, published in 1994, for the assessment period 1989–91, the total area under forests was 64.01 Mha, accounting for 19.5 per cent of India's geographical area (FSI 1994). Of the total area under forest 60 per cent had more than 40 per cent crown cover (dense forest) while the remainder had a crown cover of between 10 and 40 per cent.

Changes in area under forest

We should consider the dynamics of the area under forest along with the changes in crown cover for India for the post-1980 period, instead of the deforestation rate alone. This approach is critical to the estimation of standing stock of carbon in forests and net emissions of carbon.

When the pre-1980 period is considered, according to an Indian government estimate, about 4.3 Mha of forest was officially diverted for non-forest uses, in other words deforested, during the period 1951–80 (FSI 1988). In reality, more forest may have been lost during that period but accurate estimates are not available. Deforestation during this period resulted mainly from conversion to agriculture (62 per cent), followed by river valley (hydroelectric and irrigation) projects (12 per cent), industry (4 per cent), with the rest being used for other purposes (Central Forestry Commission 1981).

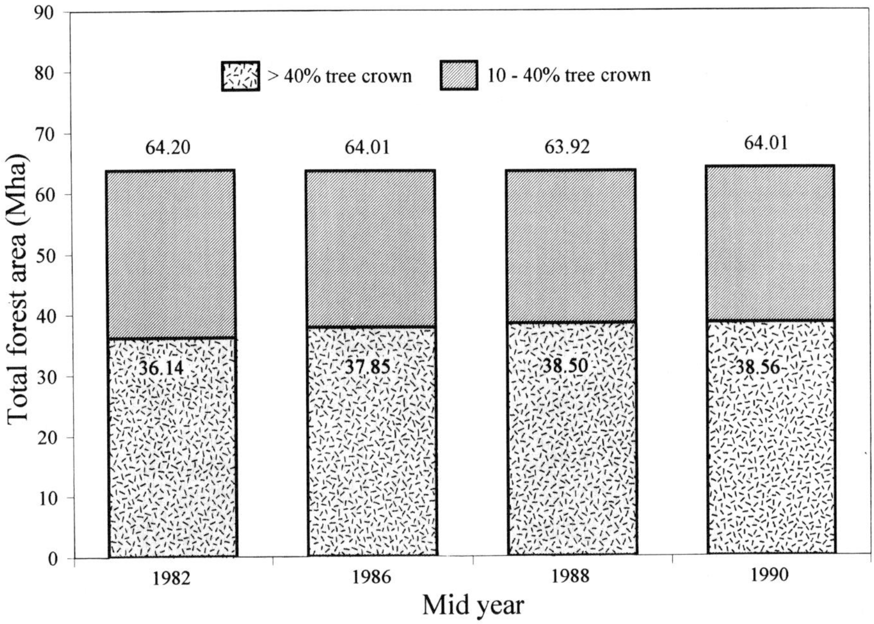

Fig. 3.1 Changes in area under forests for different tree-crown classes for four assessment periods (from 1982 to 1990), taking mid-year assessments based on data of the Indian National Remote Sensing Agency (FSI 1988, 1990, 1992, 1994).

During this period India's priority was to increase food production and to provide land for the landless, hence forests were cleared and converted to agricultural land.

Since 1982, the total area under forests has been nearly stabilized at around 64 Mha. Figure 3.2 shows that the forest area declined only by 0.07 per cent (47 500 ha) annually between 1982 and 1988 but increased between 1988 and 1990. What is more important is that the area under dense forest with > 40 per cent crown cover continuously increased from 1982 to 1990 (Fig. 3.1).

Rates of deforestation

Data on the extent of forest area clearfelled or deforested are not available to allow estimates of the rates of deforestation. Data on the area under forests at different periods are interpreted and reported at the state (province) level as the forests are managed by the state governments in India. Considering the total area under forests, 47 000 ha were lost annually during the four-year period 1982–86.

Bearing in mind the availability of data, the following procedure is adopted for estimating the area deforested for gross C emission calculations. All the

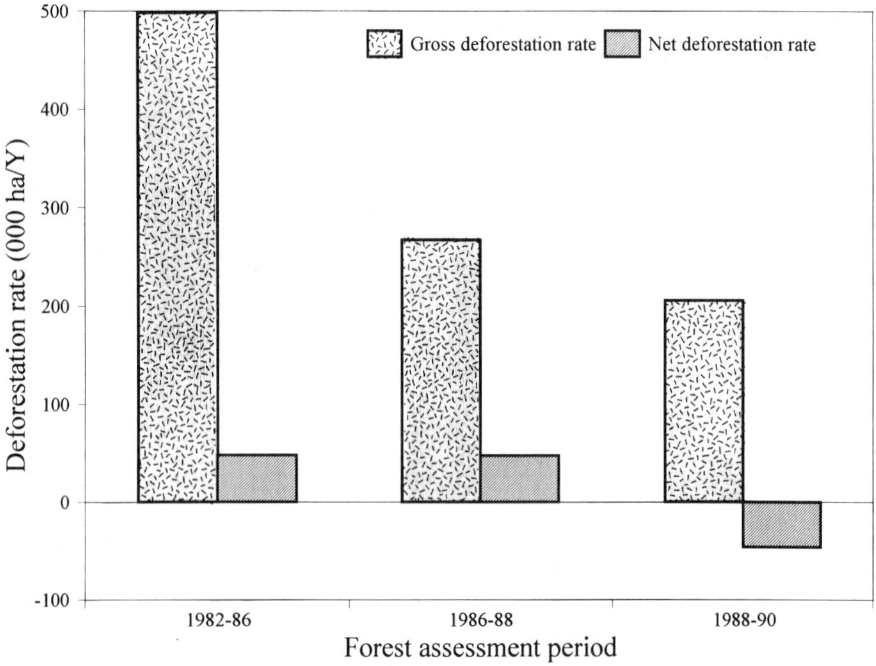

Fig. 3.2 Net and gross deforestation rates for the periods 1982–86, 1986–88, and 1988–90, based on four satellite assessments. Net deforestation rate includes part of the afforested area covered under satellite assessments (> 25 ha contiguous area) and gross deforestation rate is the difference in total area under forest between two periods (FSI 1988, 1990, 1992, 1994).

states where there was a decline in forested area, in the years 1982–86 and 1986–88, from the three satellite assessments are given in Table 3.1. The net decline in forest area of only those states where there was a reduction in area between the two periods is estimated. This is then divided by the number of years to obtain the net annual decline in area.

$$\text{Net annual decline in forest area} = \sum_{i=1}^{27} (A\,82_i - A\,86_i)$$

of only states where $A\,82_i > A\,86_i$, and:
$A\,82_i$ = area under forest for 1982 assessment in state 'i';
$A\,86_i$ = area under forest for 1986 assessment in state 'i';
i = 1–27 states and union territories.
Similarly, the decline in forest area is estimated for the period 1986 to 1988 and 1988 to 1990.

The sum of all such states is considered as the net decline in forest area for the reference period. It should be noted that part of the area brought

Table 3.1 Area under forests and changes in forest area statewise based on satellite assessment of forests (10^3 ha)

State/union territories	1982 assessment (1981–83)	1986 assessment (1985–87)	1988 assessment (1987–89)	1990 assessment (1989–91)
Andhra Pradesh	5019.4	4791.1	4279.0	4725.6
Arunachal Pradesh	6050.0	6876.3	6875.7	6866.1
Assam	2638.5	2605.8	2475.1	2450.8
Bihar	2874.8	2693.4	2666.8	2658.7
Goa (including Daman and Diu)	128.5	130.0	125.3	124.7
Gujarat	1357.0	1167.0	1190.7	1204.4
Haryana	64.4	56.3	51.3	51.3
Himachal Pradesh	1288.2	1337.7	1178.0	1250.2
Jammu and Kashmir	2088.0	2042.4	2006.4	2044.3
Karnataka	3226.4	3210.0	3219.9	3234.3
Kerala	1040.0	1014.7	1029.2	1033.6
Madhya Pradesh	12774.9	13319.1	13578.5	13539.6
Maharashtra	4741.6	4405.8	4404.4	4385.9
Manipur	1767.9	1788.5	1768.5	1762.1
Meghalaya	1651.1	1569.0	1587.5	1576.9
Mizoram	1909.2	1817.8	1885.3	1869.7
Nagaland	1435.1	1435.6	1432.1	1434.8
Orissa	5316.3	4713.6	4720.5	4714.5
Punjab	76.6	116.1	134.3	134.3
Rajastan	1247.8	1296.6	1283.5	1309.9
Sikkim	283.9	312.4	303.3	311.9
Tamil Nadu	1838.0	1771.5	1771.3	1772.6
Tripura	574.3	532.5	553.5	553.8
Uttar Pradesh	3144.3	3384.4	3360.9	3396.1
West Bengal	881.1	839.4	801.5	818.6
Other territories	786.6	785.9	1235.7	786.0
Total	64203.9	64013.0	63918.2	64010.7

Source: FSI (1988, 1990, 1992, 1994).

under tree plantations is included as forest area if it satisfies the two conditions:

(1) it has a tree crown cover of > 10 per cent; and
(2) the forest or plantation patch size is a contiguous plot larger than 25 ha.

Thus forest plantations on farmland and village commons lands are largely excluded, as they are generally small fragments < 1 ha or of a few hectares planted in different years in different locations, and thus are not likely to be in contiguous patches of > 25 ha. However, afforested patches adjacent to existing forests are likely to be included in the satellite assessments. This procedure is adopted mainly due to limitations in the available data. According to this procedure the net annual decline in forest area for the period 1982–86 is estimated to be 498 250 ha and 267 150 ha for 1986–88 (Fig. 3.2). In this study, the states where the area under forest in 1986 was higher than in 1982 are considered while estimating carbon uptake.

Using deforestation rates, carbon emissions and uptake values are estimated (Table 3.2). In India the increase in area under dense forest (>40 per cent crown cover) is a key indicator of the increase in standing stock of carbon. A recent study on carbon flows in Indian forests has taken into account the emissions from forest vegetation and soils resulting from combustion of fuelwood, decomposition, and short-term (e.g. pulp) and long-term (e.g. structural timber) use of wood, as well as C sequestration in forests due to forest succession and afforestation (Ravindranath *et al.* 1992*b*). Only the summarized findings of this study are considered here (Table 3.2).

Table 3.2 Carbon flows in India from deforestation and forest degradation for the year 1986 (Mt C)

1. Carbon stored	
forest: vegetation	4179.00
soil	5399.00
Total carbon stored	9578.00
2. Carbon emission from:	
shifting cultivation	1.56
combustion[c]	26.90
past long-term use[a]	1.38
past short-term use[b]	3.10
current short-term use	2.84
release from soil	3.91
biomass decomposition	18.92
Total carbon emissions	58.61
3. Carbon uptake (annual)	
Forest plantations and succession in forests	68.87
4. Total annual carbon uptake	68.87
5. Net annual carbon uptake	10.26

Source: Ravindranath *et al.* (1992*b*).

[a] Long-term use: timber used in furniture or housing, where carbon is stored for over 30 years. Timber used 30 years ago is expected to emit C during the reference year.

[b] Short-term use: timber used in pulp, container boxes, etc. where carbon is stored for short periods, 2–3 years. C emission during the reference year takes place from timber use during 3 years, including the reference year.

[c] Combustion: fuelwood obtained from clearfelling and forest degradation used as fuel.

Net carbon emission

When the net carbon emission in Indian forest ecosystems is considered (uptake – release), a net sequestration of 10 Mt C is estimated for the reference year (Table 3.2). This estimate differs from the annual emission estimates of 41 Mt C by Houghton (1991) and 42 Mt C by Mitra (1992). The methods used for emission estimations are different for the various studies, which generally do not consider the combined details of carbon sequestration in tree plantations and forests in succession or the delayed carbon release of wood used for structural purposes, and the different values used for rates of deforestation.

3.5 STANDING BIOMASS AND CARBON SEQUESTRATION IN FARM TREES (NON-FOREST AREA)

Trees present in farm forestry plantations, on farms, and in homestead gardens are not included in the area under forests. Such farm trees are usually not considered at all in carbon flow estimations (for example Houghton 1991; Mitra 1992; Ravindranath *et al.* 1992*b*). Village and farm trees are important in regions such as South and South-East Asia, East and southern Africa for several reasons:

(1) they contribute significantly to meeting the requirements of biomass fuel (for India see Table 2.11) as well as timber;
(2) their numbers are large and thus the standing biomass or carbon would be significant; and
(3) as an option for growing biomass for energy and/or carbon sequestration.

For example, data available for Pakistan (Leach 1993) have shown that of the total 1434 million trees in the country, trees on farmlands accounted for 50 per cent. The standing biomass of farm trees in Pakistan is estimated to be 105 Mt, compared to 81 Mt in forest lands. This clearly shows the importance of farm trees in estimating the carbon flows in southern Asia.

The standing biomass and carbon sequestered in farm trees are given in Table 3.3. The three sources of carbon storage are trees in farm forestry, along farmland boundary and bunds, and in homestead gardens. The total carbon sequestered annually in farm trees in India is estimated to be 48 Mt (refer to Table 3.3 for methods) compared to 69 Mt (Table 3.2) in forests. A part of this woody biomass productivity is harvested as fuelwood, as shown in Table 2.11.

Importance of farm trees to carbon stock

It can be seen from Table 2.11, that farm tree sources provided 82 Mt of fuelwood (farm trees + homestead gardens + 50 per cent of tree plantations

Table 3.3 Annual carbon sequestration in farm trees (farm forestry, homestead gardens, and row trees on cropland bunds/boundaries)

	Area or number of trees	Annual increment	Total annual increment in biomass (Mt/yr)	Annual carbon sequestration (Mt/yr)
Farm forestry (up to 1991)	8.2 Mha	4.2 t/ha/yr[a]	34	17
Trees in homestead gardens (Table 2.14)	–	–	16	8
Trees along cropland bunds and boundaries (Table 2.13)	14 224 million trees	2.84% of[b] standing biomass	46	23
Total				48

[a] Annual increment of woody biomass in farm forestry plantations = 4.2 t/ha/yr, based on the national-level study (Seebauer 1992) (Table 7.4).
[b] Refer to footnote of Table 2.13.

as farm forestry), equivalent to 45 per cent of total fuelwood use in India. In other words, in the absence of these farm tree sources the pressure on trees in forests would have been high, leading to degradation and loss of the standing stock of wood and carbon in forests. A large study on the impact of social forestry also supports such a contention (Seebauer 1992). In the above analysis of carbon sequestration in farm trees, perennial shrubs on village commons and tree plantations on village commons are not included as no studies are available. Thus the carbon sequestration estimated above for non-forest areas is incomplete.

3.6 NET CONTRIBUTION OF BIOMASS USE AS ENERGY FOR CARBON EMISSION

The carbon emissions from combustion resulting from clearfelling of forests and forest degradation has been estimated to be 27 Mt C, which accounts for 46 per cent of the gross emissions from forests (Table 3.2). However, it has been shown that carbon emission from biomass harvesting for fuel purposes from forests is being offset by carbon sequestration in afforested areas and natural regrowth in existing forests (forest succession). Therefore carbon emissions from the combustion of fuelwood in India do not make any net contribution to global carbon build-up.

There is now, however, a new concern about the contribution of wood combustion to greenhouse gases (GHG) due to the emission of products of incomplete combustion (PIC) resulting from the inefficient combustion of traditional stoves. These PICs are even more powerful GHGs per gram in carbon emitted than CO_2 (IPCC 1992). According to Smith (1991), an estimate of the global warming potential of non-CO_2 greenhouse gases, such as CO, CH_4, and non-methane hydrocarbons, could be in the range of 20–110 per cent as much as that of CO_2 itself, depending on the time horizon. One implication would be that even if all the 227 Mt of wood combusted comes from sustainable harvests and thus leads to no net emission of CO_2, PICs could still account for a significant contribution to the global warming potential. At the global level the overall contribution of CO_2 emissions from biomass stoves to global warming is estimated to be equivalent to about 2 per cent (Ahuja 1990). This includes an assumption that one-eighth of net deforestation results from meeting the biomass energy needs of cooking globally.

3.7 SIGNIFICANCE OF FACTORS CONTRIBUTING TO FOREST CONSERVATION

The observed trends in forest cover, particularly the stabilization of the forest area around 64 Mha in India since 1982 (Fig 3.1) could be interpreted in various ways. First, a country such as India, with a very low per capita area under forest, cannot 'afford' (due to the watershed and biodiversity roles) to lose any more forests; it is thus experiencing the lowest decline in forest area compared to tropical South American or Central African countries (FAO 1993*a*), which have large per capita forest areas. A second possible interpretation could be that India, with a high population density and rural communities dependent on forests, could have lost the maximum natural forest area that was socially tolerable. In the past, over the short term it would have been advantageous to the political leadership to condone deforestation in the guise of development. However, over the past decade both the government and local communities have resisted the temptation of indiscriminate exploitation of forest resources, and thus the conservation of forests in India has become significant. Indeed, the area under dense forest cover (>40 per cent crown cover) increased by 0.8 per cent annually between 1982 and 1990 (Fig. 3.1). Of further significance is the fact that forest conservation in India started well before the global concerns of climate warming and decline of biodiversity, and thus represents an indigenous response to environmental concerns and not a response to external pressure.

Contributing factors

The factors that appear to have contributed to forest conservation in India (Ravindranath *et al.* 1992*b*) are as follows.

The Forest Conservation Act, 1980 This bans all unnecessary and avoidable forest conversion to non-forest use. Only 46 000 ha of forest have officially (FSI 1988) been diverted to non-forest use in the 7 years after 1981, mainly for river valley projects, compared to 4.3 Mha officially diverted between 1951 and 1980.

Compensatory afforestation Any permission given for forest conversion must be accompanied by compensatory afforestation on degraded land.

Establishment of national parks and wild life sanctuaries 12.6 Mha of forest has been dedicated to nature reserves.

Afforestation programme India has launched a large afforestation programme in degraded forests, village commons, and farm land. The area afforested in India from 1980 to 1990 was 14.4 Mha; in comparison the total area afforested from 1980 to 1990 in Brazil was 2.8 Mha, 4.7 Mha in Indonesia, and 0.07 Mha in Mexico (FAO 1993*a*).

Joint forest management Under the legislation of 1988 and 1990, a joint programme involving local communities and the Forest Department has been launched with the aim of protecting and promoting natural regeneration in degraded forest lands. Under such a programme, in three districts of West Bengal, 191 750 ha of degraded forest land was revegetated between 1984 and 1990 (Poffenberger and Singh 1992)

Removal or reduction of subsidies to industry Earlier, industry enjoyed a guaranteed supply of timber at low or nominal rates from forests. With the reduction in subsidies, industry is expected to use forest resources efficiently.

Environmental movements, free press, and judiciary Environmental movements, starting with the 'Chipko movement' in the Himalayas, have spread to several states. India's free press and judiciary is generally progressive and supports forest conservation efforts and acts as a check on any indiscriminate deforestation.

India's forest conservation attempts are certainly not without failures or criticisms within the country: slow progress in afforestation, inadequate community participation in forest management and afforestation programmes (still largely Forest Department managed), insecure land or product tenure in local communities, excessive dependence on monocultures of exotic tree species, favouring industries at the cost of local communities, low productivity of forest plantations, and several institutional barriers.

3.8 LOSS OF NUTRIENT VALUE OF DUNG

India has 19 per cent (257 million) of the world's cattle (including buffaloes). Apart from providing milk for human consumption and draught power for agriculture, cattle also provide fuel in the form of dung (Table 2.2) and nutrients as manure. In some parts of India, such as the hilly regions, cattle are often maintained mainly for manure purposes. Cattle-dung manure provides organic matter and nutrients, such as nitrogen for crop production. Farmers have long recognized this and practise dung manure application in crop production. To analyse the loss of manurial value from the use of dung as fuel, the loss of only nitrogen (N) value is estimated here for convenience, and we fully recognize the contribution of dung manure to improving the organic matter content of soil. The total nitrogen content of dung production in India is estimated to be 1.35 Mt annually, out of this 0.4 Mt is lost annually through the burning of dung as fuel (Table 3.4). The loss of nitrogen on a

Table 3.4 The contribution of cattle dung as a source of nitrogen (N) to crop production

	Values for total dung production	Values for the dung quantity burnt as fuel	Values for the dung manure applied currently for crop production
National total dung production[a] (Mt/yr)	443.00	132.00	311.00
N content in dung in Mt[b]	1.35	0.40	0.95
N actually applied through dung per ha of cropland (kg/ha)[c]	9.60	2.80	6.80
Assuming 45 per cent of N in dung applied as available (kg/ha)[d]	4.30	1.30	3.00
Quantity of leaves required to substitute N currently provided by dung/ha (in fresh kg)[e]	717.00	217.00	500.00

[a] Dung burnt as fuel = 37 Mt (Table 2.2) with 18 per cent solids and 10 per cent moisture in dung cake at use stage = 132 Mt of fresh dung.
Dung production data from Table 2.19.

[b] N content in dung; 1.7 per cent on dry basis; N content in leaves 3 per cent on dry basis.
443 Mt × 0.18 solids × 0.017 kg N/dry kg of dung = 1.35 Mt N.

[c] Taking a net cropped area of 141 Mha.

[d] From Rajabapaiah *et al.* (1993).
9.6 kg N/ha × 0.45 of N available/kg of N = 4.3 kg/ha N.

[e] Green leaf biomass required = 4.3 kg of N/0.03 N/kg of dry leaves and 80 per cent of moisture in fresh leaves = 717 kg of fresh leaves.

per hectare of cropland basis is estimated to be 2.8 kg/ha. To supply the amount of N lost through combustion of dung, 217 kg of green leaves are required/ha. In one sense the nutrient aspect has been overstated in the literature, as the current consumption of N fertilizer (inorganic) is 68 kg/ha/yr at the national level (TSL 1992).

Furthermore, not all the nitrogen in the dung applied will be available, as only 40–45 per cent is retained and the rest is lost during aerobic decomposition (Rajabapaiah *et al.* 1993). If a farmer maintains a cow or buffalo with a mean dung yield of 5 kg (of fresh weight)/day, the annual N contribution would be 5.5 kg/animal/yr. Thus dung as a by-product of maintaining cattle for milk production or draught power seems rational. To obtain 5.5 kg N from tree leaves, 916 kg of fresh leaves have to be harvested. This partly explains the rationale for rural communities attaching importance to the manurial value of cattle dung. However, currently many farm households are forced to depend on dung as a fuel in the absence of fuelwood. The environmental impact of the use of dung as fuel, in terms of loss of nutrient value and organic matter to crop production, is important. Therefore the provision of quality fuels such as liquid or gaseous fuels is necessary to prevent dung from being used as fuel.

3.9 CROP RESIDUE USE AS FUEL AND LOSS OF NUTRIENT VALUE

In India the main uses of crop residues are as fodder and fuel, in that order. Crop residue use as a fuel in the domestic sector was estimated to be 40 Mt (Table 2.2, 'low estimate'). The environmental impact of burning of crop residues, particularly the impact on crop production due to loss of organic matter, has been recognized. In many farming systems the proportion of crop residues being ploughed back into crop fields is limited due to physical difficulties in incorporating them into soil (Barnard 1990). In India burning of the residues of cereals and pulses as fuel (except for pigeon pea) is small since cereal residues are used as fodder and pulse residue is left in the field or fed to cattle. The residues of crops which cannot be used as fodder and which are likely to have significant quantities of lignin (such as the woody residues of pigeon pea, coconut, cotton, and coffee) are burnt. Thus the loss of nutrient value in the burning of such residues as fuel is insignificant, although some loss takes place for reasons other than fuel purposes (such as burning sugarcane leaves and tops after harvesting).

3.10 CONCLUSIONS

The main focus of this chapter was to highlight the environmental implications of the use of biomass as fuel. It was shown that even though the contribution of forests via deforestation and clearfelling of trees to fuelwood supply is small (9 per cent of total fuelwood used), a large forest area has to be clearfelled to obtain it. Thus it may not be correct to say that fuelwood demand does not contribute to forest degradation or deforestation. But it is possible to say that the urban and industrial demand for fuelwood is likely to contribute more to forest degradation than rural fuelwood demand. Thus the promotion of efficient stove programmes in rural areas cannot be based on the premise that adoption of efficient stoves will lead to the large-scale conservation of forests. Such improved stove programmes are, however, required for other reasons, such as to remove smoke in the kitchen, to reduce drudgery in gathering fuelwood, and to reduce cooking duration.

The forestry sector in India has been shown to make no net contribution to the atmospheric build up of carbon dioxide. The emission of carbon from the combustion of wood used as fuel is more than offset by forest succession and the annual increment of woody biomass of trees in forests, plantations, and villages. In this context it is necessary to recognize the importance of trees outside the forests (in farms, homestead gardens, village commons, and plantations) in sequestering carbon and in reducing the pressure on forests by significantly contributing to fuelwood provision. The area under forests has been stabilized around 64 Mha since 1980. The recent achievement of India in conserving forests, in spite of having one of the lowest forest areas per capita in the tropical world, needs to be recognized. The contributory factors could usefully be studied by other tropical countries experiencing high deforestation rates.

The emission of pollutants and indoor air pollution is a cause for concern and needs to be addressed not just by improved stove programmes but by exploring the provision of gaseous and liquid fuels to improve the quality of life of women. The importance given by Indian farmers to cattle dung as a valuable manure is shown to be rational. This further reinforces the need to stop the burning of cattle dung as fuel and the need to switch to higher-quality fuels such as biogas. In the next chapter an attempt is made to understand the efforts aimed at conservation of biomass fuels, provision of biogas fuel for cooking, production of woody biomass to augment its supply, and conversion of biomass into modern fuels.

Photo 3.1 Regulated conversion of forest area to non-forest uses such as cultivation of crops and provision of other services.

Photo 3.2 Typical scene of degraded wasteland in South India.

4

Assessment of bioenergy programmes in India

The dominance of biomass energy in the energy consumption pattern of India has been highlighted in earlier chapters. The population in India is growing at a rate of about 2.12 per cent annually and it is observed that biomass use is keeping pace with population growth. The use of biomass as a traditional fuel is associated with a lower quality of life and environmental degradation. It is thus necessary to understand what India is doing to alleviate such problems associated with the dependence on and use of biomass and to draw lessons therefrom before exploring the potential role for bioenergy in promoting sustainable development.

India made a beginning with programmes related to bioenergy even before the first oil crisis of 1973. For example, during the 1950s it started biogas (for cooking) and improved stove programmes (mainly smokeless stoves), and various afforestation programmes to grow fuelwood. In the aftermath of the energy crisis of the seventies, India wanted to reduce its dependence on oil imports and was among the first countries to realize the need for alternative sources of energy, especially of renewables. This led to the realization of the role of renewable sources of energy (called non-conventional energy sources in India), and a Department of Non-Conventional Energy Sources (DNES) was set up in 1982. It was intended that these sources of energy would be given an identity and receive attention, and DNES would work independently in the large Ministry of Power and Energy, which concentrates on centralized power-generation programmes. In 1992, realizing the need to promote renewable energy sources further, an independent Ministry of Non-Conventional Energy Sources (MNES) was created. India is probably one of the few countries in the world which now has such an independent ministry for renewable energies. It has grown from a small budget of Rs140 million in 1982 to Rs1280 million in 1992–93; $US1 = Rs31 (1993/4), (MNES 1993*a*).

During the Seventh Plan period (1985–90) the main aims of the DNES were to support intensive research and development for developing indigenous technologies, setting up a large number of demonstration projects, testing devices under field conditions, and creating demand through government support (PC 1985). There was a change in focus in the post-1990 period and the role of MNES is now the 'development and utilization of renewable sources of energy along with conventional sources of energy to meet the growing needs of energy in the light of added significance; they are renewable and indigenous and these sources provide clean energy and help in preserving the environment and ecology' (MNES 1993*a*). It is important to note the emphasis on environmental benefits of renewable energy technologies and its recognition by government and energy planners in India.

The major programmes of the ministry can be grouped as: power generation using renewables, energy conservation, and biomass production. Here we concentrate on bioenergy:

1. Bioenergy programmes:
   (a) conservation
      (i) biogas
      (ii) improved stoves;
   (b) production
      (i) fuelwood production
      (ii) energy plantations;
   (c) conversion
      (i) biomass gasification
      (ii) liquid fuels
      (iii) cogeneration.
2. Other renewable energy programmes for power generation:
   solar energy, wind energy, micro-hydro energy.

The relative importance of bioenergy (biogas + biomass gasifiers + biomass energy production + improved stoves) programmes in the ministry can be seen in Fig. 4.1, which shows that they account for 62 per cent of the

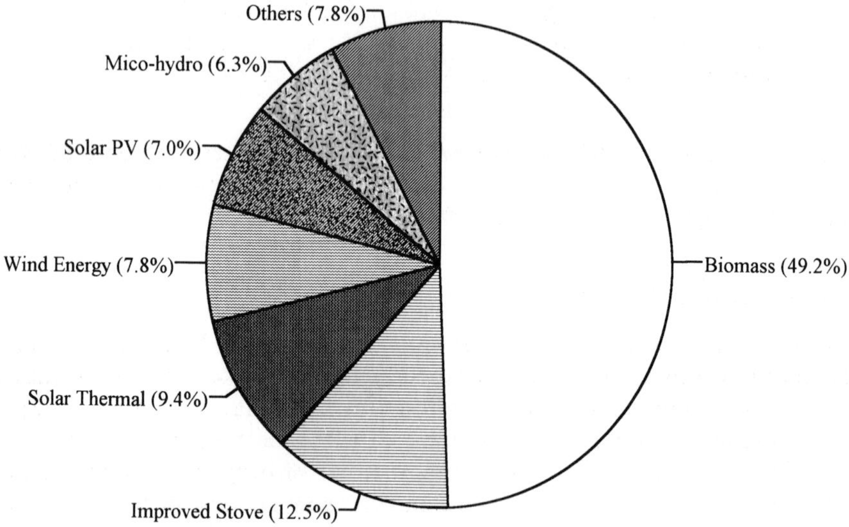

Fig. 4.1 Expenditure on different renewable energy programmes for 1992–97, indicating the share of bioenergy technologies in the total budget (Rs14.65 billion) for renewables in India (PC 1992). Biomass includes biogas + biomass gasifier + biomass production.

budget for the Eighth Five-Year Plan (1992–97). However, solar, wind, and micro-hydro energies are projected to play more important roles in the next decade (PC 1992). The major bioenergy programmes implemented during the years 1985–90 were biogas, improved stoves, biomass gasification for mechanical and electrical applications, and biomass energy plantations. India provides a unique opportunity to learn from all the bioenergy programmes, which include technology development, dissemination, and socio-economic aspects. The achievement and performance of the bioenergy programmes are discussed in detail in the following sections.

4.1 BIOGAS

India is a pioneer in biogas programmes, with the floating-drum biogas plant being developed in 1952. It was the first renewable energy technology to be launched in India, much before the global recognition of the need for alternate energy sources. By 1980 about 80 000 family biogas plants had been built. Biogas is presently the dominant bioenergy programme in India, accounting for 37 per cent of the funding allocation for MNES for the period 1992–97 (MNES 1993*a*). The biogas programme received a boost with the launching of the National Programme on Biogas Development (NPBD) in 1981–82 with funding and administrative support from the launching of DNES. The main goals of the programme are to:

(1) provide cooking energy in a clean and non-polluting form;
(2) produce enriched manure to supplement the use of chemical fertilizers;
(3) improve the quality of life of rural women; and
(4) improve sanitation and hygiene.

Thus the goals of NPBD were beyond the mere saving of fuelwood (or forests). Recently, biogas is also being promoted as a substitute for diesel in shaft-power applications (DNES 1992).

4.1.1 Technology and designs

Biogas (a mixture of about 60 per cent methane and 40 per cent carbon dioxide) is a combustible gas which is the product of anaerobic fermentation of cellulosic materials, such as animal dung and plant leaves, and also wastes from food processing and the household. Biogas can be combusted directly as a source of heat for cooking (calorific value 23 MJ/m^3) or used with internal combustion engines for mechanical or electrical applications. The slurry produced after digestion can be used directly as valuable manure for crop production.

The Indian biogas programme is wholly based on cattle dung and the proportion of rural households owning cattle is high, varying from 59 per cent

Box 4.1 Features of fixed- and floating-drum biogas designs in India (Ravindranath *et al.* 1992*a*)

Fixed dome	Floating drum
Digester and gas holder, masonry or concrete structure	Digester, masonry
	Gas holder, mild steel or fibre glass
Requires high masonry skills	Low masonry or fabricating skills
Low reliability due to high construction failure	High reliability, gas holder prefabricated
Variable gas pressure	Constant gas pressure
Digester could be inside the ground	Requires space above ground for three tanks; inlet, digester, outlet
Low cost (2 m^3 = Rs5000)	High cost of plant due to gas holder (cost for 2 m^3 = Rs8000)

in Bihar to 80 per cent in Uttar Pradesh state (ABE 1985). Cattle are reared for draught power, milk, and in some locations, such as in the Himalayas, for dung as a manure source alone. The Indian biogas programme includes small family-sized biogas plants (2–8 m^3) and community (village-level) or institutional biogas plant programmes.

Two major types of biogas designs are promoted in India—floating drum and fixed dome. The main features of these designs are given in Box 4.1 and are depicted in Fig. 4.2 (a and b). The floating drum (Fig. 4.2a) is an old Indian design with a mild-steel, ferro-cement or fibre-glass drum which floats along a central guide frame and acts as a storage reservoir for the biogas produced. The fixed dome design (Fig. 4.2b) is of Chinese origin and has a dome structure made of cement and bricks. It is a low-cost alternative to the floating drum, but requires high masonry skills and is prone to cracks and gas leakages. Currently, apart from the floating-drum design called the KVIC (Khadi Village Industries Corporation) model, five or six different indigenously developed fixed-dome designs are being disseminated. Greater emphasis is now being placed on fixed-dome designs due to their lower cost.

Family biogas plants come in different sizes depending on the availability of dung and the quantity of biogas required for cooking. The average size of the family is 5–6 persons, and thus a biogas plant of capacity 2–4 m^3 is adequate. The biogas requirement for cooking two meals is estimated to be 150 l/capita/day (about 1200 litres for a family (Dutt and Ravindranath 1993)). If the making of tea and snacks is included, about 200 litres of biogas is required /capita/day.

The cost of a biogas plant depends on its size and the design (Table 4.1). The subsidy component accounts for over 63 per cent in the north-east and

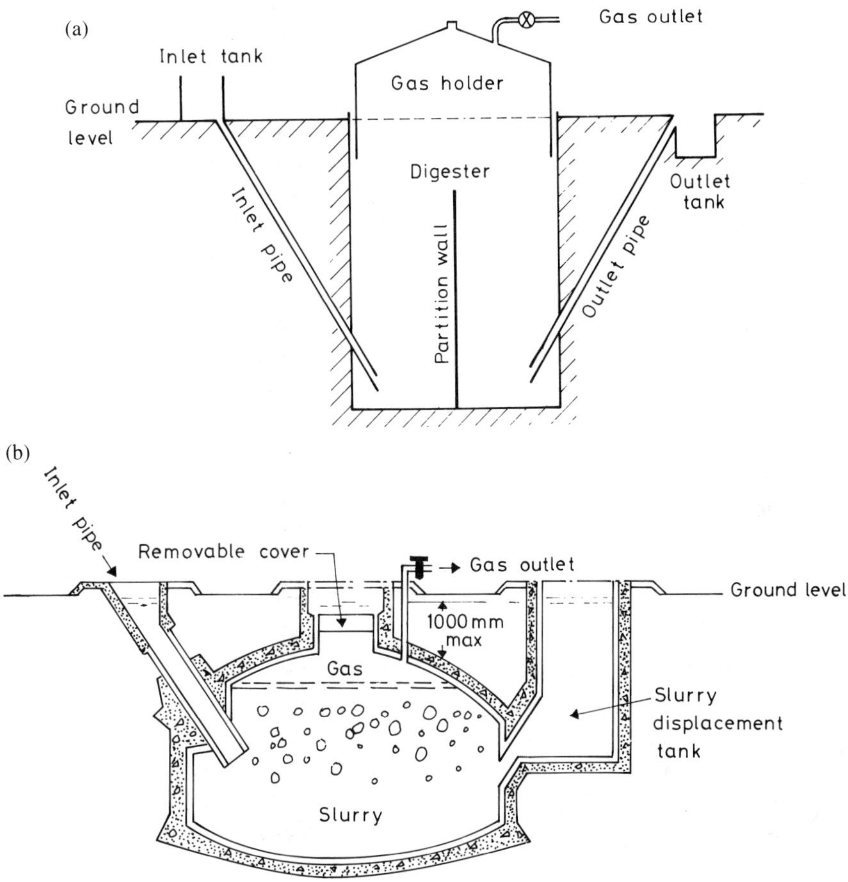

Fig. 4.2 Design of fixed-dome and floating-drum biogas plants. (a) Floating gas-holder drum design—the conventional Indian design; (b) fixed gas-storage dome design—low-cost modified Chinese design.

hilly forest districts of Uttar Pradesh state and to about 28 per cent of the total cost of a 2 m^3 KVIC design in the plains of India. The rationale for higher subsidies in forest districts is the assumption that their dissemination will reduce the pressure on forests and decrease deforestation. Households wishing to build a biogas plant can obtain a commercial bank loan at concessional rates of interest (at 11.5–13.5 per cent compared to 17.5 per cent interest on normal commercial loans) and also claim the subsidy component.

Community biogas plants involve the same technology as the family biogas plants (floating-drum design) at varying capacities up to several hundred cubic metres. So far, these are also based on cattle dung.

Table 4.1 Cost of family biogas plants (in Rs) for different capacities (in m^3) (MNES 1993*a*)

Design	2 m^3	3 m^3	4 m^3
Floating-drum KVIC type	8030	9500	10 700
Fixed dome (Deenabandhu type)	5325	6400	7800
Subsidy from government in plains	2200	2600	2600
Subsidy as % of KVIC floating-drum cost	28	29	33
Subsidy as % of fixed-dome cost	41	40	24
Subsidy in north-east and hilly districts of Uttar Pradesh	5100	6200	6200
Subsidy as % of floating-drum KVIC design in north-east and hilly districts	63	65	58

4.1.2 Achievements of the biogas programme

India has pursued the biogas programme in two modes—family and community levels—although the focus has been mostly on the family biogas programmes.

Community biogas (CBG) programme

This is at the village community level and was initiated in 1982. Up to the end of 1992, 875 community and institutional biogas plants were built. The goal of community biogas plants is to provide the benefits of biogas even to the poorer sections of the rural community who do not own sufficient cattle or have no funds for a family biogas plant. It is also aimed at using alternative feedstocks and to expand biogas use for power generation.

Family biogas (FBG) programme

The goal of the family biogas programme is to provide clean and convenient fuel for cooking for the family, apart from national goals such as conserving forests. It can be seen from Fig. 4.3 that the biogas programme had a modest beginning and that the number of biogas plants built annually has been increasing since the 1970s from a modest 14 320 plants built in 1975 to 188 000 in 1992–93. India now has the capability of building 220 000 plants/yr at current levels of financial allocation. The cumulative number of plants in December 1993 was 1.85 million (CMIE 1994)

4.1.3 Performance of the programme

The biogas programme can be assessed with respect to the acceptance or use levels, benefits derived, physical achievement as compared to its potential, and analysis of the barriers to expansion of the programme.

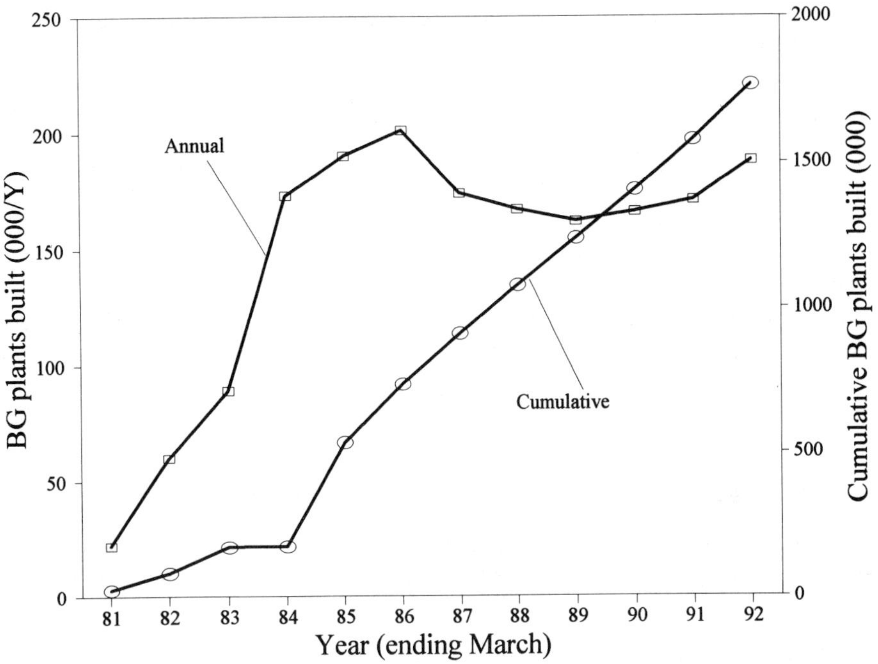

Fig. 4.3 Annual and cumulative number of family biogas plants built in India since the launching of the intensive programme in 1982; BG, biogas (MNES 1993*a*).

Performance of community biogas plants

There are very few studies describing the performance of the 875 community biogas plants. A study that compiled all the available evidence on community biogas plants reported that information is available for only 37 of them (Venkataramana 1991). Of these 37, 65 per cent are operational, 27 per cent had failed, and 8 per cent were not commissioned.

Even the plants that are reported as operational were functioning only partially due to technical and social problems. Details of performance such as dung collection, gas generated, and the extent of cooking energy provided are not available, and even the details of the reasons for failure are not clear. Some of the broad reasons for failure are given in Table 4.2. Inadequate dung is the main reason, followed by social or organizational problems. The issue of adequacy of cattle dung for community biogas plants is considered in Chapter 6. There are three successful community biogas plants in the Ungra area (South India) for power generation, and these are considered as case studies in Chapter 5.

Socio-economic aspects of the family biogas programmes

Given the positive relationship between the ownership of land and the owner-

Table 4.2 Achievement and performance of community biogas (CBG) plants in India

Total number of CBG plants built until the end of 1992	875
No. of CBG plants for which information is available	37
CBGs in operation[a]	24 (65%)
Failed CBGs	10 (27%)
Non-commissioned CBGs	3 (8%)
Reasons for failure	
Inadequate dung	46%
Structural failure	15%
Social or village organizational problems	39%

[a] Even CBG plants considered operational are operating at partial capacity and some face technical and/or institutional problems or shortage of dung.
Source: Venkataramana (1991) plus data from two other CBG plants in Karnataka.

ship of livestock (since cattle require fodder from cereal crop residues), one would expect that only larger farmers with sufficient land could own cattle and thus build biogas plants. However, a national study of over 5000 biogas plants showed that small farmers with less than 2 ha of land accounted for a third of the biogas plants and a small percentage of landless families also owned biogas plants (NCAER 1989). It is likely that these households rear dairy cattle.

Extent of use of family biogas plants

The rate of use of biogas plants is a simple and useful indicator of the performance of the programme. The assumption is that the family will operate the biogas plant only if it is providing significant benefits, since the operation of biogas plants involves human effort for obtaining water, mixing the dung and water, charging the plant, and removing the sludge nearly every day. A number of studies have been conducted in India to monitor the performance of the biogas programme; some instituted by the ministry and others by research groups.

A major study (NCAER 1992), sponsored by the ministry for the period 1985–86 to 1989–90, monitored 27 000 biogas plants in 3600 villages in 251 districts in India. The level of use of the biogas plants in major states is given in Fig. 4.4. It is estimated that at the national level 66 per cent of biogas plants are in use and there are variations at the state level; some states (such as Karnataka and Gujarat) with over 90 per cent of biogas plants in use, and some states (such as Bihar and Uttar Pradesh) with less than 60 per cent in use. The non-use rate of 34 per cent is not small, since it implies that of the 147 000 biogas plants built during 1991–92, 49 980 are not operational. In

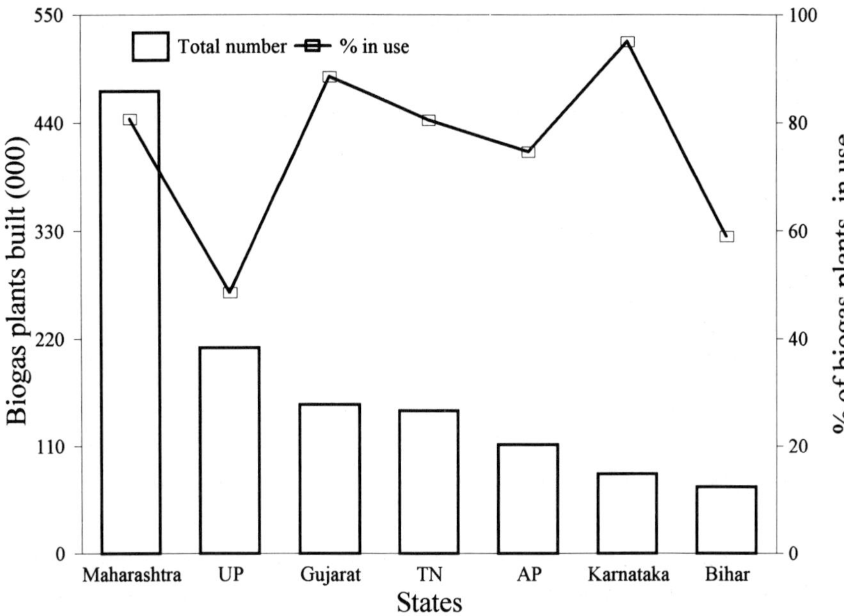

Fig. 4.4 Total number of family biogas plants built and percentage of them in use in seven major states; UP, Uttar Pradesh; TN, Tamil Nadu; AP, Andhra Pradesh (NCAER 1992; MNES 1993*a*).

terms of investment, about Rs535 million is therefore wasted (Rs10 700 per plant). What are the possible reasons for this significant rate of rejection? Two studies, one a success story from Sirsi, Western Ghats, Karnataka, and the other with poor performance from Maharashtra state are considered, to understand the reasons for success or failure.

Family biogas programme—study at Sirsi

Sirsi is an administration block in the western Ghat (hilly) region of Karnataka in South India (Fig. 2.4). In the Sirsi block, out of the 15 000 rural households, 11.5 per cent (1730) already have biogas plants, whereas at the national level only 1.4 per cent of rural households have biogas plants. In Sirsi, the biogas programme has been intensively implemented and a total of 497 biogas plants were built during 1992–93. A study was conducted during August 1993 with the aim of understanding the performance of the biogas programme (Bhat, P. R. and Ravindranath 1994). A sample of 10 villages was randomly selected and all the biogas plants in the selected villages were studied. There were 186 biogas plants in the 10 selected villages.

Biogas designs in use Four designs of biogas plants were in use:

(1) KVIC—floating drum with mild-steel drum = 32 per cent;
(2) KVIC—floating drum with fibre-glass drum = 50 per cent;
(3) KVIC—floating drum with cement drum = 15 per cent;
(4) Fixed dome (Deenabandhu design) = 3 per cent.

Since 1991 all the biogas plants built, in the sample studied, were of KVIC floating-drum design with fibre-glass drums. All four designs are available to users and the shift in preference for different designs over the years can be observed in Fig. 4.5, particularly the choice of the fibre-glass design in the recent past as it requires minimal maintenance. The KVIC model with a mild-steel drum is out of favour, as is the fixed-dome design, even though it is cheaper.

Biogas plants in use In August 1993, 99 per cent of the sample biogas plants were in use in the 10 villages.

Cooking energy provision Benefits derived in terms of the extent of cooking energy need met:

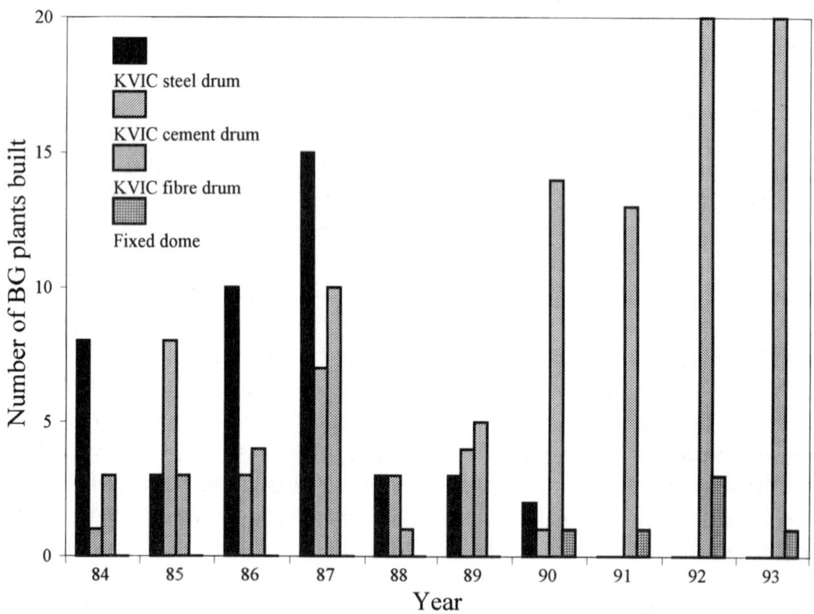

Fig. 4.5 Changes in the preference of biogas plant designs over the years in the Sirsi area of the western Ghat region of Kamataka; KVIC, Khadi Village Industries Corporation; BG, biogas (Bhat, P. R. and Ravindranath 1994).

(1) 100 per cent cooking energy needs met from biogas; 85 per cent of families with biogas plants;
(2) 75 per cent cooking energy needs met from biogas; 11 per cent of families with biogas plants.
(3) 50 per cent cooking energy needs met from biogas; 4 per cent of families with biogas plants.

Thus most families owning a biogas plant are meeting all their cooking energy needs from biogas.

Repair and maintenance When asked about maintenance problems, the families reported the following frequency of repairs:

1. Drum repairs: 9.5 per cent of biogas plant owners have had the drums repaired, only 4 out of 186 drums were replaced; the maintenance included mainly the welding of holes in steel drums.
2. Other maintenance problems included cracks in civil structures, leakage in gas transmission pipes, and collapse of central guiding rods. The occurrence of these was rare.

Impact on biomass conservation At 3.13 kg of fuelwood use/capita/day (Ranganathan *et al.* 1993) in the Sirsi area, assuming a family size of six, and 85 per cent of biogas plants providing the complete cooking energy requirement, the 1730 biogas plants in Sirsi block annually conserve 5465 t of wood. At a mean woody biomass productivity of about 6 t/ha/yr (in Sirsi area, Chapter 7) it is equivalent to a net annual increment of woody biomass from 910 ha of tree plantations. Thus the biogas programme in Sirsi has made a significant impact on local biomass resources (conserving wood and trees in forests or plantations) and improving the quality of life of rural women.

Reasons for high success rates It is always easier to list reasons for failure than to pinpoint the reasons for success. However, a survey of the households and discussions with biogas contractors revealed the following special features of the biogas programme in the Sirsi block. There are professional contractors who undertake biogas plant construction on their own and through institutions such as co-operative societies. All the families had their biogas plants built through contractors directly, or indirectly through co-operative societies, and they provide the following services:

(1) undertake turnkey projects, thereby avoiding the problems involved in getting a biogas plant built on one's own;
(2) provide a 5 year guarantee for the operation of the plant with free repair and service backup;

(3) accept payment only after successful completion and commissioning of the plant.

Institutions such as co-operatives do not require full pre-payment, so households are permitted to pay in instalments, thus reducing the up-front financial burden. There are several contractors in the block and there is competition among them. Thus there is practically no involvement of government agencies, except for certification at the end of the construction for release of the subsidy. The block-level officers recognize the contractors involved and even provide a fee to them for undertaking turnkey jobs.

Case study of Marathwada region in Maharashtra state

As a part of the development of management information systems to identify operational problems and constraints in the effective functioning of biogas plants, a large study of 1670 biogas plants in the Marathwada region of Maharashtra state was conducted in 1987 (Dyalchand and Murthy 1988). Table 4.3 gives data on the performance of these plants.

Table 4.3 Performance of biogas plants in the Marathwada region

	No. of biogas plants installed	Plants functioning	Effective functioning
Feasible plants (based on size of family and dung supply)	584	243 (41)	52 (9)
Non-feasible plants	1086	304 (28)	20 (2)
Total	1670	547 (33)	72 (4)

Figures in the parentheses indicate percentage of total number of installed biogas plants.

Performance of biogas plants

1. Sixty-five per cent of the biogas plants were found to be non-feasible, with inadequate cattle or dung to meet the biogas requirements of the family.
2. Irrespective of non-feasibility, biogas plants were built.
3. Only 33 per cent of the sanctioned biogas plants were functioning.
4. Thirteen per cent of the functioning plants (or 4 per cent of the total biogas plants approved and built) were functioning optimally.
5. Overall failure rate was 96 per cent of the total biogas plants built.

Reasons for poor performance

1. Inadequate dung (insufficient cattle owned). Non-adherence to the feasibility criterion, for example, households not owning cattle or owning only one or two were also included in the programme.
2. The administrative block and district-level government agencies were given a large target for physical achievement (number of biogas plants in a year) without considering the infrastructure.
3. Inadequate training of builders; in fact, biogas plants were built as part of the training programme by inadequately trained builders.
4. Absence of follow-up and user education was obvious from the fact that many families did not even know how to do the initial loading of the feedstock.
5. The major reason given for non-commissioning of the biogas plants, after they were constructed, was the non-availability of dung for initial loading. After commissioning of the plants, the main reasons for failure were inadequate feeding, cracked domes, absence of pipelines, and pipe leakage.
6. There was no feasibility survey, quality control, follow-up, user education, or shouldering of responsibility for failures. These problems are aggravated since government officers are frequently transferred to different regions or districts.

In the Marathwada region, the biogas programme was a wholly government-managed and target-oriented programme with no responsibility being taken by any functionary for ensuring the success of the biogas programme. Possibly the Marathwada case study could be an extreme case of failure. But there are lessons to be learnt.

Reasons for failure of the biogas plant

Both structural and operational reasons seem to contribute to the failure of biogas plants (Table 4.4 and case studies).

Feasibility of design Decisions about acquiring a biogas plant and its size depend mainly on the number of cattle owned and the availability of dung. As an extreme case, in Maharashtra 65 per cent of the biogas plants built were not feasible, indicating that families did not have adequate information regarding the dung resource requirement.

Structural defects (Table 4.4) The common defects for failure of fixed-dome biogas plants are: damage to the foundation and cracks in the dome and digester wall. The major reasons for failure of floating-drum biogas plants are: corrosion of the gas holder and collapse of the central guide frame.

Table 4.4 Structural and operational reasons for failure of family biogas plants (NCAER 1992)

Structural defects		Operational problems			
	% of plants	Technical	% of plants	Non-technical	% of plants
Broken defective foundation	29.1	Blocking of inlet/outlet	2.2	Shortage of dung	25.7
Cracks/defects in fixed dome	26.3	Water accumulation in gas pipe	13.9	Shortage of water	1.3
Cracked digester wall	20.2	Corrosion/leakage in pipeline	23.1	Shortage of manpower	1.3
Corroded gas holder	10.2	Defect in burner	12.9	Wrong selection of plant site	2.6
Broken central guide frame	4.5	Scum formation on digester slurry	2.8	Plant owner has lost interest in the plant	9.5
Broken central wall	3.5				
Other structural defects	6.2	Other operational problems	0.6	Other problems	4.1
Total % of FBGs with structural defects	100.0	Total of technical and non-technical 100.0			

Problems with the fixed-dome design arise out of the lack of skills to build leak-proof domes or strong walls and foundations. In the case of the floating-drum design, lack of funds to replace the steel drum or absence of infrastructure to repair the drum contribute to failure.

Operational problems (Table 4.4) Among the technical problems, corrosion, leakage and water accumulation in gas pipelines, and defective burners are the dominant reasons. Shortage of dung is the main non-technical reason for failure.

High cost of repair Damage to the civil structure is a common reason for failure in fixed-dome plants; the cost of repairing a cracked or damaged dome is as high as 30 per cent of the full cost of a biogas plant and repairing the inlet and outlet tank costs 5–10 per cent (Joshi *et al.* 1992*a*). Thus even if repair services are available the cost is so high that the owner of a biogas plant will hesitate to get it repaired.

Cheap and accessible repair and maintenance services and user education seem to be the main features lacking in the existing post-construction implementation phase. Several minor problems, which can be rectified in the village itself or even by the household, such as replacing damaged pipes, solving water accumulation problems, and getting the biogas stove repaired, are often not attended to because of lack of information. Even information on initial feeding of the biogas plant, and the correct ratio of dung to water, is often not given to the biogas owner. The social, economic, and institutional problems related to the dissemination of biogas plants are discussed later in this chapter. The Sirsi case study showed that given the right circumstances it is possible to ensure a high rate of success in biogas implementation, with little failure in construction and relatively few minor operational problems.

4.1.4 Impact of the biogas programme

This can be analysed considering the aspects of biomass conservation, improvement in quality of life, and economic benefits.

1. The impact on biomass conservation is estimated to be the reduction in fuelwood use by 2.1 Mt annually (1.85 million biogas plants × 0.66 in use × family size, 6 × 0.288 t of wood/capita/yr). This is equivalent to annual increment in woody biomass in 500 000 ha, at a productivity of 4.2 t/ha/year. Thus the fuelwood conserved is significant.
2. Reduction in drudgery for women. First, removal of smoke in the kitchen by a shift to biogas leads to the removal of problems associated with cooking in smoke-filled kitchens and thus to improved health. Secondly, savings in fuelwood or dung-cake use lead to savings in time involved in

cooking, gathering, and preparation of the biomass fuel. Thirdly, control of flame intensity, ease of lighting and switching-off, and reduction in cooking time also contribute to the reduction of drudgery.

3. Improved nutrient status of sludge. The total quantity of dung involved in feeding 66 per cent of 1.85 million biogas plants is 10 Mt annually (5 cattle/house × 4.5 kg of dung/animal/day × 365 days/year × 1.85 million biogas plants × 0.66 in use). The loss of nitrogen in aerobic decomposition from conventional composting is avoided for 10 Mt of dung.
4. Economic benefits. The saving in human labour for gathering and preparation of biomass and cooking time could be productively used elsewhere. Households purchasing fuelwood would save considerable sums of money.

4.1.5 Achievement compared to biogas potential in India

The success of the biogas dissemination programme could be gauged by comparing the biogas potential with actual achievement. In India the programme is over 40 years old, but only in 1981 was an intensive programme launched. According to estimates from the ministry, the potential for family biogas plants in India is 12 million. The total number of family biogas plants built by the end of 1993 was 1.85 million, implying that only 15 per cent of the potential has been achieved (MNES 1993*a*). At the current rate of funding and mode of dissemination, about 0.15 million biogas plants could be built per year. At that rate of implementation it would take at least 60 years to build 10 million family biogas plants. In the community biogas programme, the potential is theoretically 557 000 as each village has the potential for a community plant. At present the actual number of large biogas plants including community biogas is insignificant. Thus biogas potential will remain largely untapped in the near and medium term if no significant technological improvements and organizational changes in dissemination are brought about.

4.1.6 Barriers to implementing family and community biogas programmes

The biogas potential tapped so far is low. So what are the barriers to its expansion? Technical, financial, institutional, and social aspects of supply-side and demand-side barriers should be considered for family biogas and community biogas options. A detailed listing of barriers is given in Boxes 4.2 and 4.3. The crucial constraints are presented below.

Family biogas plants (Box 4.2)

The following barriers are considered for those households that have adequate dung resources for a biogas plant.

Box 4.2 Barriers to utilizing family biogas potential (Ravindranath *et al.* 1992*a*)

	Demand-side barriers	Supply-side barriers
Economic barriers	High capital cost/first cost compared to zero cost wood stoves Inadequate credit and high cost of credit, as own capital is scarce even for economic activities No income generation from investment in biogas or cash savings; competition from income-generating activities for investment High repair and maintenance cost and failures demotivate users Easy availability of wood or no cash cost involved in procurement of wood	Inadequate funding for biogas at national level for organizing dissemination programmes Inadequate subsidy arising out of low funding of the programme, especially for the plains Low market volumes for the entrepreneurs and government agencies, thus high cost of organizing biogas programmes; cost per unit of construction and maintenance high
Technical barriers	Inadequate dung resource/animal ownership Problem of initial feeding of plant, which requires large amounts of dung; lack of information on options for initial feeding Water scarcity for daily feeding of plant; especially in summer season and drought years in some regions High rates of failures and frequency of repairs in some regions No insurance against failures	Inadequate supply of raw materials and absence of quality control: cement, bricks, steel in rural areas Inadequate workshop facilities for fabricating components Inadequate or absence of trained manpower for construction and maintenance Inadequate funding for research and development (R&D)
Institutional barriers	Complex procedures for obtaining loan, subsidy, and repair services No institutions to provide back-up services for repair and maintenance problems No responsibility for failures leads to lack of faith by the users Cooking energy; a problem for women but men take investment decisions	Absence of NGOs in most areas and limited scope for dissemination through NGOs Absence of private entrepreneurs, no incentives and rigid policies for their participation Target-oriented nature of programme leads to lack of quality control Non-availability of women as builders and skilled persons for biogas dissemination programme

Box 4.3 Barriers to utilizing the full potential of community biogas plant for cooking (Ravindranath *et al.* 1992*a*)

Demand-side barriers	Supply-side barriers
Fuelwood available at zero cash cost in most locations	Limited funding at national level
Large funding required at village level does not enthuse local communities	Inadequate funding for creating support infrastructure for construction and maintenance of CBGs
No incentive for construction and maintenance by NGOs or private entrepreneurs	
Skewed ownership of cattle holdings, thus inequities lead to unwillingness to share benefits equitably	
Inadequate demonstration units to create interest in communities, NGOs, entrepreneurs	No organization at village level to take leadership and accept responsibility
Most rural leaders are men but gathering fuel and cooking is the women's problem	
Social inequities often make sharing of benefits difficult unless good leadership and organization exist	
Inadequate R&D in field; technology package not yet easily available to NGOs or entrepreneurs	Inadequate technical knowledge to handle slurry
Water shortage in some areas	Technical skill required for planning and implementation but such personnel not available
Fear of losing dung or nutritive value of dung; unacceptability of slurry due to inadequate information	

1. The high cost of biogas plants is the primary problem, coupled with availability of fuelwood at almost no cash cost to rural households. Since interest rates in rural areas are very high, it is difficult to expect all households with biogas potential to invest in a plant that does not provide any direct financial benefit.
2. In the government programme there is little or no responsibility for ensuring the success of the plants built. There is neither any guarantee on the successful functioning of the biogas plant nor any efficient, reliable repair and maintenance service.
3. There is a severe shortage of trained manpower for building and servicing the biogas plants and for educating households on the operation and maintenance of plants.

4. There is inadequate funding at the national level to promote the programme.

Community biogas plants (Box 4.3)

1. A large amount of funding is required for each village and there is lack of funding at the national level, even though the economies of scale operate at the community scale.
2. There is inadequate dung to meet the cooking energy requirements of the whole village (discussed in Chapter 6), and thus lack of interest in the programme at the village level since the ownership of cattle is unequal and skewed.
3. Government is not the right agency to implement the programme and the involvement of NGOs is marginal. No organizational arrangements exist to disseminate community biogas plants.
4. No institutional arrangement exists for village communities to take the initiative and responsibility for operating and managing the systems.

4.2 IMPROVED STOVES

The main driving force for launching improved stove programmes was to conserve fuelwood and to protect forests. The socio-economic analysis in Chapter 2, on the use of fuels for cooking in rural areas, indicated that biomass fuels dominated the cooking energy scene irrespective of the economic status of the household. Even the type of biomass fuel used (fuelwood, crop residues, and dung cake) did not vary much with increase in income or land-holding size in a given region (Fig. 2.8). Thus the governments of developing countries are justified in relying on efficient stove programmes to alleviate the cooking energy problem to some extent in the near term. India and several other developing countries, such as China, Kenya, Indonesia, Mexico, and Sri Lanka, have launched improved stove programmes. The Indian programme was initially launched as a demonstration project in 1983 and later converted into a fully fledged National Programme on Improved Chulha or stoves (NPIC) from April 1985. The main goals of NPIC (MNES 1993*a*) are:

(1) fuel conservation;
(2) removal/reduction of smoke in the kitchen and to improve health status;
(3) to check deforestation;
(4) reduction in drudgery of women and children;
(5) reduction in cooking time; and
(6) employment opportunities for the rural poor.

Thus unlike in the initial phase where the goal was mainly to conserve wood or forests, the current stove programmes have multiple goals aimed at improving the quality of life for women and the environment, and even as an opportunity for creating employment in rural areas.

4.2.1 Technologies, designs, and achievement

Traditionally, fixed mud stoves dominate the rural scene in India, and portable stoves are used only among a small segment. Both fixed (site-built) and portable efficient stove models are being disseminated. The number of improved stove designs currently disseminated are:

(1) Family: fixed designs = 35; portable designs = 16.
(2) Community/commercial application stoves: fixed design = 10; portable = 5.

A minimum efficiency threshold of 20 per cent for fixed and 25 per cent for portable stoves is fixed by the ministry for inclusion in the national programme (MNES 1993*a*). This is based on the assumption that traditional stoves have a low efficiency of 5–10 per cent. However, it was noted in Fig. 2.6 that traditional stoves have efficiencies in the range of 10–15 per cent. There are some highly efficient stove designs, such as the ASTRA stove. It has a thermal efficiency of 45 per cent in the laboratory, and even in the field a 35 per cent efficiency has been achieved (Ravindranath *et al.* 1989*a*). In the southern part of Karnataka state, the traditional stove is a three-pan mud stove with a thermal efficiency of 14.7 per cent (± 3.7 per cent); simultaneous cooking in all three stove openings is possible since three dishes are traditionally cooked in the region (Ravindranath *et al.* 1989*b*). The ASTRA Centre in Bangalore developed a high-efficiency improved stove with the following specifications (Fig. 4.6):

(1) mud stove built *in situ*;
(2) three pan openings;
(3) thermal efficiency of 45 per cent in the laboratory;
(4) smoke removal through chimney;
(5) an enclosed fuelwood feeding port;
(6) snugly fitting pan seats for reducing heat loss;
(7) vessels immersed in the pan seats for maximum vessel contact with flame and hot gases;
(8) a grate on which fuelwood burns;
(9) can use all fuel types traditionally used in villages;
(10) cost around Rs150 with components accounting for Rs100 and labour for Rs50.

The stove programmes are implemented by various agencies: state government departments, autonomous bodies (like agro-industries corporations, Dairy Development Board), voluntary agencies, and self-employed workers. The state government is the dominant agency disseminating the stoves. State

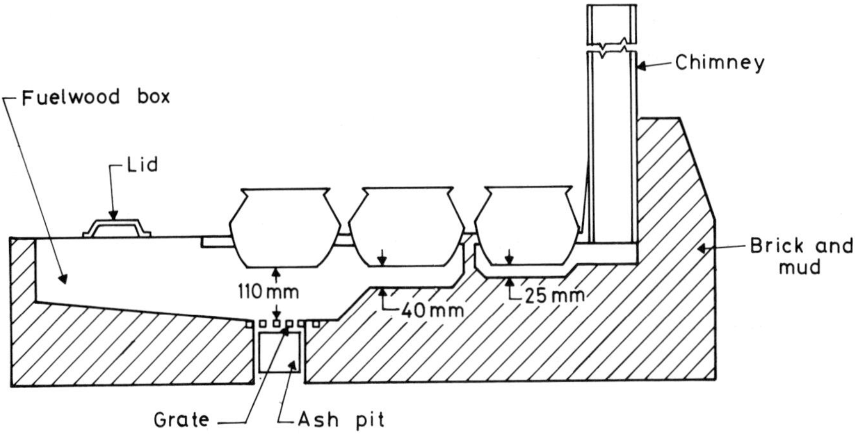

Fig. 4.6 ASTRA stove design and dimensions (Lokras *et al.* 1983).

governments provide subsidies ranging from 25 to 75 per cent of the cost of the stove depending on the location and socio-economic status of the household.

Targets and achievement

The targets set for the stove programmes in India and the physical achievements are shown in Fig. 4.7. Targets have been surpassed for most years. Annually about 2.5 million stoves are now being built and about 14 million improved stoves have been built in the past 10 years, an achievement which seems to be significant. The rate of stove construction in Karnataka state over the years is also given in Fig. 4.7. Starting in 1984–85, it increased up to 1986–87 and again sharply increased from 1989–90 to 1991–92. It is interesting to note that as the targets were surpassed, they were enhanced and more money flowed in. In India the success of a programme is measured in terms of achievement with respect to targets and even the ability to spend money budgeted for a given programme.

4.2.2 Evaluation of the improved stove programme

The performance of the stove programme needs to be evaluated in terms of the parameters set out in the main goals of the programme, namely: fuelwood conservation, removal or reduction of smoke in the kitchen, reduction in cooking time, and employment generation.

Stove use level

The assumption in using this measure is that if the stove is beneficial to the housewife, she will use it; otherwise if it consumes more fuel or emits more

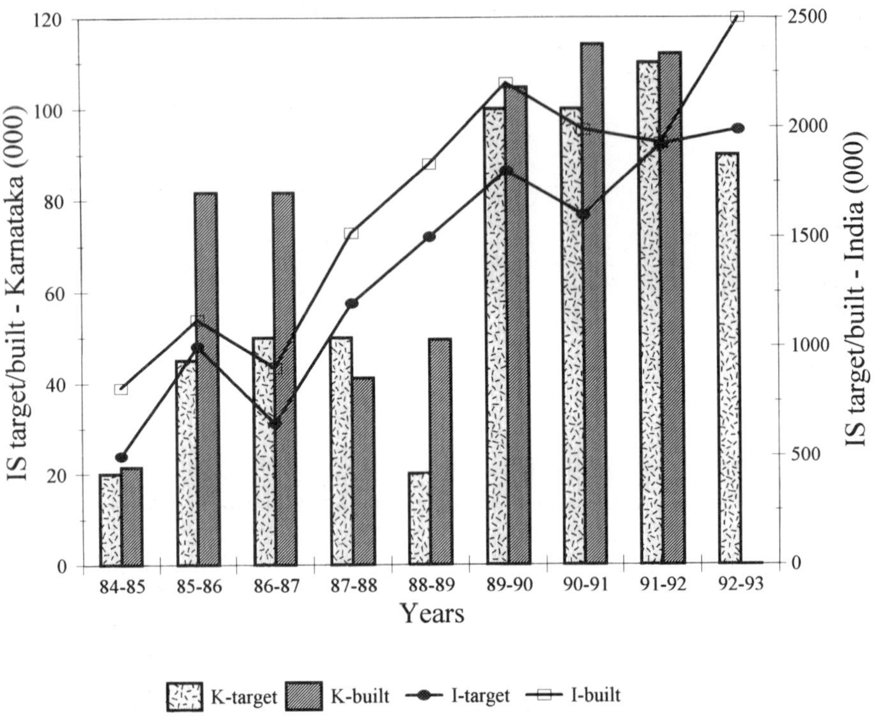

Fig. 4.7 Dissemination of improved cooking stoves; targets and achievements in Karnataka state and at national (India) level; IS, improved stove; K, Karnataka; I, India.

smoke into the kitchen or is inconvenient, she is likely to dismantle the stove and/or discontinue its use. The findings of a number of macro-level and detailed micro-level studies are presented in Table 4.5. MNES (1993*a*) data show that nationally 70–75 per cent of improved stoves are in use. Other studies show use levels varying from 33 per cent in West Bengal to 79 per cent in Rajastan. A study sponsored by MNES (1993*a*) shows that 60 per cent of stoves built during 1988–91 are in use, but the use level was 86 per cent for the stoves built during 1991–92, suggesting that when the stoves are new the rate of use is higher than when they age.

Karnataka state is an interesting case since the use of ASTRA stoves has been monitored over many years through different studies at the micro- and macro-level. The level of stove use was as high as 79 per cent at the macro-level in 1986 when the stove programme was new, but in later years large monitoring studies have shown a decline, with the level of stove use stabilizing around 60 per cent. A micro-level study in Ungra area shows a declining trend with age from 100 per cent of ASTRA stoves in use in 1984 to 56 per cent in 1994, when the stoves were about 7–9 years old, since all the stoves

Table 4.5 Performance of improved stoves; level of acceptance or use

State	Year[a]	No. of improved stoves monitored	% Improved stoves in use	Reference and monitoring agency
Uttar Pradesh	1987	816	50	[b]
Tamil Nadu	1986–87	311	65	[b]
Rajastan	1986–87	786	79	[b]
West Bengal	1986–87	418	33	[b]
Gujarat	1987	1347	51	[b]
Maharashtra	1987	1321	39	[b]
Karnataka[c]	1986	2201	79	ASTRA
	1987	351	65	ASTRA
	1988	3294	66	KSCST
	1989–90	3876	62	KSCST
	1992–93	1645	67	KSCST
	1992–93	4075	66	KSCST
Ungra area[c] (Karnataka)	1984	150	100	ASTRA
	1985	450	97	ASTRA
	1986	450	91	ASTRA
	1988	450	84	ASTRA
	1994	250	56	ASTRA
National[d]	1988–91	Large	60	NCAER
	1991–92	Large	86	NCAER
	1992	Large	70–75	MNES

[a] Percentage of stoves in use refers to the year of monitoring.
[b] Compilation by Ramakrishna (1991).
[c] KSCST (Karnataka State Council for Science and Technology), Bangalore; ASTRA data from Ravindranath *et al.* (1989*a*, 1989*c*); KSCST and ASTRA survey reports, unpublished; 1994 monitoring of stoves carried out by the authors for the present analysis.
[d] NCAER and MNES data from DNES (1992) and MNES (1993*b*).

were built during 1984–85. The stoves are made of mud and thus have limited life. There is a range for level of stove use; nationally we assume a use level of about 60 per cent.

Efficient ASTRA stove: case study of Sirsi

Sirsi was one of the first administrative blocks in Karnataka where the ASTRA stove programme was launched intensively in 1984 and where over 2000 such stoves have been built. This means that 13 per cent of all the rural households have improved stoves, compared to the 12.7 per cent at the national level. It must also be noted that an additional 11.5 per cent of households are already covered under the biogas programme.

Table 4.6 Performance of efficient ASTRA stoves as a function of their age, in the Sirsi area of South India monitored during August 1993 (Shastry and Ravindranath 1994)

Year when stoves built	Age of stove (years)	No. of stoves in sample	% of stoves in use in 1993
1993	<1	4	100
1992	1–2	10	60
1991	2	15	80
1990	3	39	38
1989	4	38	47
1988	5	33	66
1987	6	11	54
Before 1986	>7	62	58

Level of ASTRA stove use A study of the stove programme was conducted in 10 randomly selected villages of Sirsi during August 1993, and all the 212 ASTRA stoves present were investigated (Shastry and Ravindranath 1994). The sample consisted of stoves with ages ranging from 1 year to over 7 years (Table 4.6). The level of use ranged from 100 per cent for stoves built during 1993 to 58 per cent for stoves over 7 years of age. The most interesting feature of the study is that in Sirsi block over 54 per cent of stoves built 5–7 years ago were still in use in 1993. The study further analysed the reasons for the non-use of stoves. Of the 93 stoves not in use, 31 per cent were not in use due to reasons unrelated to stove performance, e.g. family switching over to biogas, changing houses, or rebuilding the kitchen.

Performance of the ASTRA stove The opinion survey on the performance of ASTRA stoves in use showed the following:

(1) 94 per cent reported a saving in fuel when compared with the traditional stove;
(2) 91 per cent reported no smoke in the kitchen;
(3) 81 per cent reported a reduction in cooking time.

Measurement of fuel saving

Fuel use was measured by taking a sample of 25 traditional stoves and 25 ASTRA stoves, which were selected at random from the same village. A kitchen performance test was conducted for 2 days continuously to obtain the mean fuelwood use. The mean fuelwood use/capita/day was:

(1) 1.67 kg/capita/day for the traditional stove;

(2) 1.43 kg/capita/day for the ASTRA stove;
(3) 14 per cent reduction in fuel use.

The annual fuelwood saving from using an ASTRA stove is 88 kg/capita.

Impact on biomass conservation In Sirsi block 56 per cent of the 2000 ASTRA stoves in use are estimated to save 588 t of fuelwood annually (0.24 kg/capita/day saving × 365 days × 6 (family size) × 2000 stoves × 0.56 in use). Thus even at low levels of performance (with respect to the percentage of stoves in use and lower saving rates) the quantity of fuelwood saved is significant.

Biomass conservation

This was, and still is, the main motivating force in promoting improved stove programmes, although claims to the contrary are made. There is a large literature base on the need for and the methods of monitoring the fuelwood conservation achieved from improved stove programmes. It is desirable to use the term biomass rather than fuelwood since a significant percentage of households use crop residues and dung cakes as fuel (Table 2.2). Different methods for monitoring the biomass conservation are considered here:

1. Thermal efficiency test. One of the standard methods to evaluate the performance of a stove is to measure its thermal efficiency as reflected by the water boiling test or PHU test (Lokras *et al.* 1983). This method is good for comparing the performance of stoves with respect to heat transfer efficiency, but may not fully reflect actual cooking conditions which include boiling, simmering, baking, and roasting.
2. Opinion survey. Asking the housewife about the saving in biomass use in an improved stove compared to a traditional stove; or asking her how many days a headload of fuelwood lasts with the improved stove as compared to a traditional stove; or the number of sticks or dung cakes used now and before.
3. Visual observation of biomass used for cooking per day or per meal. Asking the housewife about the number of sticks or the size of the bundle of fuelwood or number of dung cakes used previously with the traditional stove as compared to that with the improved stove.
4. Kitchen performance test. Here the housewife is given a weighed quantity of fuelwood and is asked to use it for the next 24 hours. The remaining wood is weighed and the number of persons for whom food was prepared is noted.

Kitchen performance tests should give a good estimate of biomass use in traditional and improved stoves for cooking on a per capita and household

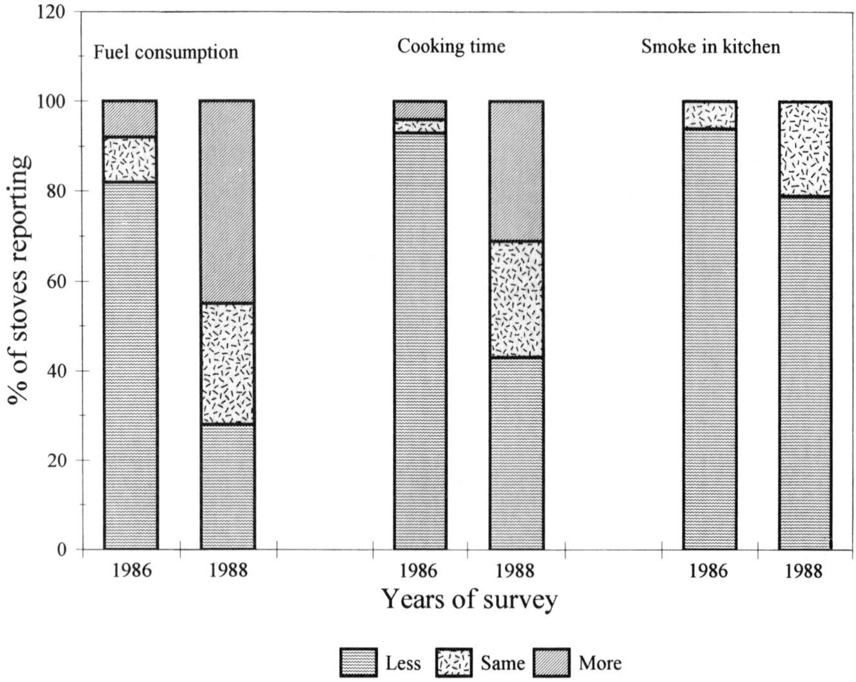

Fig. 4.8 Performance of improved cooking stoves in Karnataka, based on opinion surveys in 1986 and 1988. Traditional stoves were compared with efficient ASTRA stoves on the basis of relative fuel consumption, cooking duration, and smoke levels in the kitchen (Ravindranath *et al.* 1989*c*).

basis. However, in the field there are several constraints in following the above procedure for kitchen performance tests. What may be feasible in many situations is a combination of opinion surveys backed up by kitchen performance tests on a smaller sample size. Unfortunately, there have been very few studies using kitchen performance tests with significant sample sizes.

Opinion survey on biomass conservation The findings of two studies in Karnataka using opinion surveys carried out at different periods are given in Fig. 4.8. It can be seen that the households reporting fuel savings declined from 82 per cent in 1986 to 28 per cent in 1988. It is also important to note that 45 per cent of households reported higher fuel consumption in the ASTRA stove compared to traditional stoves. However, in a survey of improved stoves in Sirsi, where they were 5–7 years old, 94 per cent of housewives reported a reduction in cooking fuel use.

Biomass conservation based on measurement Conservation estimates based on actual measurements are available only for a few locations and some

Table 4.7 Fuel savings in improved stoves based on measurements

State	Village	Fuel savings in %	Fuel savings kg/capita/yr
Karnataka[a]	Ungra		
	1985	48	153
	1986	33	98
	Sirsi[b]	14	88
Karnataka	State level, 1986[c]	19	73
	Intensive study, 1987[d]	19	69
Rajastan[e]	Mithiberi	6	25
Haryana[e]	K Balwas	7	29
Tamil Nadu[e]	Pottal	24	73
	Panagudi	−5	−14
MNES[f]	National		116

[a] Ravindranath *et al.* (1989*a*).
[b] Shastry and Ravindranath (1994).
[c] Jagadish and Ravindranath (1990).
[d] Ravindranath *et al.* (1989*b*).
[e] Joshi *et al.* (1989).
[f] 700 kg wood saved/family = 116 kg/capita/yr. Wood savings from MNES (1993*a*).

of the findings are given in Table 4.7. A micro-level study in Ungra showed that initially the reduction in fuel use was 48 per cent when the improved stoves were 6 months old and it declined to 33 per cent when the improved stoves were 30 months old (Ravindranath *et al.* 1989*a*).

Even multi-locational measurements in Karnataka state showed a 19 per cent reduction in fuel use. Micro-level studies in three other locations showed fuel conservation in the range 6–24 per cent. However, in one location in Tamil Nadu state, the fuel use was marginally higher in improved stoves compared to traditional stoves (Joshi *et al.* 1989). Based on studies in various locations at the national level, MNES (1993*a*) concluded that the fuelwood saved is about 700 kg for a family of six per year. The ministry estimate may be too optimistic when compared to field study results quoted in Table 4.7.

Thus opinion surveys as well as biomass use measurements in India show mixed results. It seems reasonable to assume that at the national level 60 per cent of stoves are in use and that biomass conservation could be in the range 19–33 per cent, the lower value being more likely. A potential for conservation could be safely taken as 33 per cent at the national level.

Reduction in cooking time

A study in Ungra has shown that women spend 8.7 hours/day on domestic chores such as cooking, gathering fuel, fetching water, and livestock grazing (Shailaja and Ravindranath 1990). Thus any saving in cooking time is crucial for improving the quality of life of women. According to an opinion survey in Karnataka, 93 per cent of housewives reported a reduction in cooking time using an improved stove in 1986, but this declined by about a half according to a 1988 study (Fig. 4.8). A measurement of cooking duration in Ungra village in South India showed that there was a 36 per cent reduction in cooking time (Ravindranath *et al.* 1989*a*). The opinion survey in Sirsi showed that 81 per cent of improved-stove users experienced a reduction in cooking time. Therefore improved stoves seem to have reduced the cooking duration significantly. Very rarely has the reduction in cooking time been considered important by the researchers since hardly any studies have looked at this aspect.

Removal or reduction of smoke in the kitchen

The two Karnataka studies (Fig. 4.8) showed that over 79 per cent of improved stove households reported no smoke in the kitchen, which implies a significant reduction in the level of smoke. Even the periodic micro-level study in Ungra village (Ravindranath *et al.* 1989*a*) as well as Sirsi revealed that over 80 per cent of households reported no smoke in the kitchen.

4.2.3 Understanding the acceptance and performance

From the Karnataka studies it is possible to conclude that elimination of smoke is an equal or more important consideration than conserving biomass in using improved stoves. A housewife is likely to use the improved stove as long as it provides one or more benefits such as savings in fuel, cooking time, and removal of smoke, and does not cause problems. We conclude that, nationally, 60 per cent of improved stoves in use are providing some benefit to housewives, which need not necessarily be fuel saving. It is incorrect to assume that discontinued use of all stoves is due to stove failure or to stove-related problems. Monitoring in Sirsi (Shastry and Ravindranath 1994) and Ungra (Ravindranath *et al.* 1989*a*) over a long period revealed that some of the reasons for dismantling of the improved stoves were: moving the kitchen, shifting to a new house, repairs to the house, splitting of joint families, the stove having been built in the wrong location, switching to biogas plants, etc.

Stove performance depends on several structural and operational parameters. Not many studies provide detailed explanations for the poor performance of improved stoves. Field studies in Karnataka have shown the following structural and operational reasons for poor performance of ASTRA stoves in use (Ravindranath *et al.* 1989*a*,*c*):

(1) larger clearance below vessels than optimum;
(2) use of incompatible vessels;
(3) non use of a pan opening or seat during cooking;
(4) blockage of the chimney with soot;
(5) damage to the pan seat.

The structural and operational problems arise due to inadequate training of builders or lack of communication with the housewife before and after construction of the stove. She often does not know what to expect from a stove or how to use and maintain it. Based on a number of studies, Ravindranath *et al.* (1989*c*) list the following reasons for poor performance of improved stove programmes:

(1) inappropriate stove design;
(2) moderate efficiency of some of the improved stoves;
(3) target chasing by the administration;
(4) absence of quality control and user education;
(5) inadequate number of trained builders and inadequately trained builders;
(6) the programme is government sponsored with no person being responsible for the performance of the programme.

In the government-managed dissemination programme, officers are busy just achieving targets as their performance is measured by the physical targets achieved and their ability to spend the budget allocated in the shortest possible period within the financial year. The stove dissemination programme is largely dominated by men.

4.2.4 Potential compared to achievement and barriers

According to projections made by the ministry, 120 million improved stoves are required in India (MNES 1993*a*). However, assuming a stove dissemination rate of 2 million/year (Fig. 4.7), a further 53 years will be required to provide stoves for all the remaining households. The majority of stoves built are mud stoves which have a life of around 5–8 years, even if properly maintained. Thus the stoves have to be replaced and, at the current rate of provision, all rural households may never obtain an improved stove. The barriers to expansion can be considered from both the demand and the supply point of view. Some of the financial, technical, and institutional barriers to expansion of the stove programme are presented here.

Cost Improved stove costs vary depending on the design. For example, the cost of the ASTRA stove is Rs150 and households are expected to finance the installation and claim a part of the cost later as subsidy. A poor household would have to spend the wage earnings of one week (at Rs20/day). Thus the first cost of a stove is still high and acts as a deterrent, especially for the poor. The possibility of providing loans is yet to be explored. The cost seems high in the face of uncertain direct benefits to the household.

Fuel gathered at no cash cost As seen in Chapter 2, the majority of households gather fuelwood, while dung cakes and crop residues are available as by-products with no cash expenditure, although their preparation usually involves significant human effort. However, there is little financial incentive to invest in an improved stove. In urban areas, households purchasing fuelwood are more likely to buy an improved stove. But presently the improved stove programme is concentrated mainly in rural areas.

Inaccessibility of the technology In Karnataka state there are about 29 000 villages. No data are available on the number of stove builders but the field experience of the authors shows that the number of trained builders is very small as there is only one master trainer per district (of about 1500 villages). The master trainer is expected to perform several functions, such as certifying the stoves, keeping records of the programme, attending meetings, apart from training builders. To contact a builder one has to go to the town and meet several officials. Furthermore, the components of the stove, such as the chimney, grate, and primary vent door, are not available in the village. Thus, even if a family is prepared to pay the full cost of the improved stove, it is not readily accessible.

Lack of information on stoves In many villages the households are unaware of the improved stove design, and even if they know about its existence they may not know about its potential benefits, the cost, whom to contact, or the availability of a subsidy.

Cooking is a woman's problem The saying goes 'If men were to cook, the fuelwood or smoke problem would have been solved long ago'. Obtaining an improved stove involves expenditure and many procedures and men may not take much interest in a problem that concerns mainly women.

Poor performance The performance of the improved stoves is poor in many locations and thus may not motivate householders. Quality control is absent in the programme.

Absence of service facilities Currently at the village level, there are no trained persons available for assistance or for repairs. This makes households even more hesitant to spend money on a facility for which maintenance is a problem.

Inadequate funding Currently about 2 million stoves are built per year. If a subsidy and expenditure of Rs50/stove is assumed, then currently the ministry is spending Rs160 million/year. This accounts for about 7 per cent of the total budget for all renewable energy sources. If the programme was to be stepped up to an annual target of, say, 20 million stoves, the allocation would have to increase commensurately. Therefore unless funding for the MNES is greatly increased, there is no possibility of significantly enhancing the targets. The mode of dissemination has to be changed to encourage self-employed workers to promote high-performance stoves.

4.3 BIOMASS GASIFIER PROGRAMME (KILOWATT-SCALE)

Biomass, or more particularly wood, can be converted to a high-energy combustible gas for use in internal combustion engines for mechanical or electrical applications. The process is generally known as gasification and the technology has been known for decades, but its application to power generation is of recent origin. A biomass gasifier (Fig. 4.9) consists of a reactor where, under controlled temperature and air supply, solid biomass is combusted to obtain a combustible gas (consisting of H_2 and CH_4). This gas passes through a cooling and cleaning system before it is fed into a compression-ignition engine to run in dual fuel mode for generation of mechanical power or electricity (by coupling to a generator). An assessment of its potential concluded that India presents a unique opportunity for large-scale commercial exploitation of biomass gasification technology to meet a variety of energy needs, particularly in the agricultural and rural sectors (Jain 1989). The large potential of biomass gasification for water pumping and power generation for rural electrification was recognized and a research, development, and dissemination programme was launched during the period 1980–85.

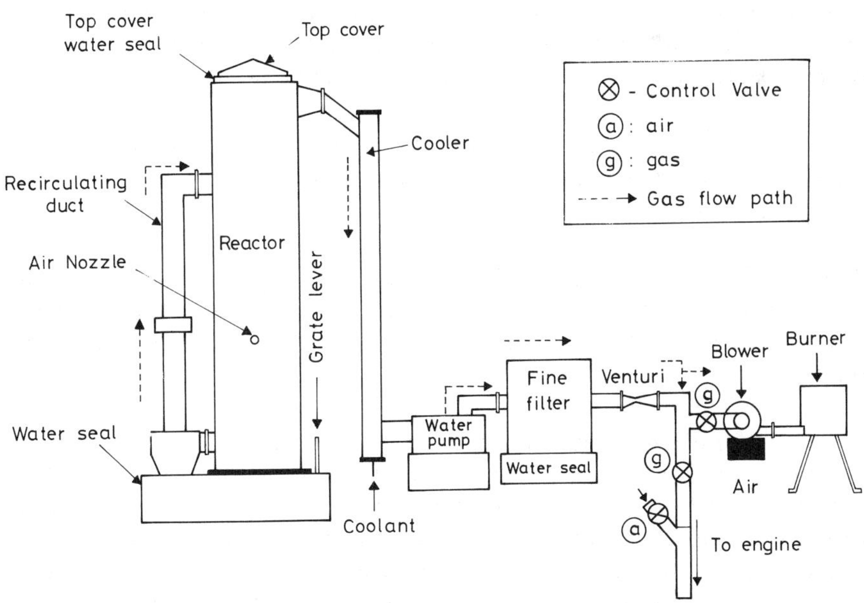

Fig. 4.9 A 20 kW gasifier with cooling and cleaning system.

The recent ministry report estimates the potential for biomass-based power generation at 17 000 MW (MNES 1993a).

Although initially a few programmes were launched based on imported designs, the Indian programme is now largely locally driven. A number of research and development (R&D) groups are working on the programme, and a number of designs of varying capacities have been developed (Paul and Mukunda 1993). The biomass gasifier programme in India is along the following lines,

(1) R & D in biomass gasification technology;
(2) pilot testing and evaluation, including prototype production;
(3) demonstration and evaluation of biomass gasifiers;
(4) training and manpower development;
(5) commercialization and manufacturing.

Research and development

The ministry reports a number of achievements:

1. Wood gasifier designs of 5–100 kW capacity are developed and available for mechanical and electrical applications.
2. Gasifiers based on crop and plantation residues, such as rice husks, groundnut shells, cotton stalks, and coconut shells, are at various stages of development and field testing.

Demonstration programme

A demonstration programme was launched in 1986 and about 800 gasifier systems of various ratings were disseminated, with a total capacity of 6.5 MW, until 1992 (Table 4.8). The majority of the systems are small wood gasifiers for irrigation-water pumping, and some are used for electricity generation.

Table 4.8 Dissemination of biomass gasifier systems in India (Subramanya *et al.* 1993)

	Number	Total capacity (kW)
Total number of systems installed (up to 1992)	800	6500
Total number of systems with details	668	4025
Capacity of system		
3.7 kW	538	2018
7.5 kW	89	667
20.0 kW	24	480
40.0 kW	14	560
100.0 kW	3	300

Table 4.9 Status and use of biomass gasifiers disseminated in the major states (Subramanya *et al.* 1993)

States	Total systems installed	Systems inspected	Running on dual-fuel mode	Running on diesel mode	Running on both modes	Not used or not commissioned
Gujarat	125	23	5	7	0	11
Karnataka	188	25	5	6	3	11
Madhya Pradesh	90	18	1	10	0	7
Maharashtra	136	13	4	2	2	5
Total	539	79	15	25	5	34
		(100)	(19)	(32)	(6)	(43)

Monitoring gasifier performance

Unfortunately, no detailed investigations are available at present on the performance of the gasifier programme. The findings of a short study (Subramanya *et al.* 1993) in four major states are presented in Table 4.9, which gives mainly the extent of gasifiers in use. Out of the 539 gasifiers only 79 were inspected: of this only 19 per cent of the systems were being used in the dual-fuel mode and 6 per cent of the systems were used in both dual-fuel and diesel-alone mode. The rest were used in a diesel-alone mode, not used at all, or not yet commissioned. The use level was poor with respect to the other capacity systems.

Case study of wood gasifiers supplied in Sirsi

In the Sirsi administrative block of Karnataka in South India, seven wood gasifiers of 3.7 kW capacity were installed out of the national programme of about 800. All seven were installed to pump irrigation water during the summer months in the Areca garden locations away from the grid. Out of the seven we were able to contact six farmers to establish the level of use, the benefits derived, and problems faced.

Out of the six gasifiers, four had been used for 4 years, and two were used for 2 years. Three of the systems were used for nearly 2000 hours. Sirsi is a high-rainfall area and receives 300 cm annually. Thus irrigation is required mainly during the summer months. All the farmers reported substantial diesel replacement.

Use of all the gasifiers was discontinued after they developed problems, such as cracks in the nozzle and for other, unknown, reasons. All the farmers complained of the cumbersome procedure of operation, especially the wood preparation. Interestingly, however, all wanted to use the system if their gasifier had performed well and if local servicing and repair facilities were available. This gives hope that if a user-friendly design, coupled with adequate servicing and repair infrastructure, is provided, farmers will be keen to use them as they see clear advantages for locations away from grid transmission lines.

Potential and achievement

When compared to the potential for power generation of 17 000 MW (MNES 1993*a*), the present installed capacity (6.5 MW) is insignificant and the performance is even poorer. In India there are over 9 million electric pumpsets and 5.5 million diesel engines used for pumping irrigation water. If a user-friendly and reliable gasifier system was available with appropriate incentives, many of the 14.5 million pumping systems could be candidates for potential replacement by wood gasifiers. The reasons for low achievement and poor performance are presented next.

Subsidized electricity and diesel Small gasifiers (5 kW) were supplied to individual farmers for pumping water for irrigation. But farmers have a choice of options for irrigation-water pumping—electricity-powered pumps or diesel engines. Diesel fuel and electricity are subsidized in India; in fact, electricity is virtually free for agricultural pumping. Both diesel and electric systems are widely used and an extensive infrastructure is available for supply and servicing in all parts of the country. Thus any new option would have to be better than the existing systems in terms of cost, convenience of operation, and reliability. It is difficult for any alternate system to compete with the highly subsidized grid system. These issues are discussed in Chapters 6 and 12.

Inappropriate locations for gasifier demonstration units Although it is not documented in any report, it is common knowledge among researchers in India that many of the gasifiers were supplied to farmers who already had electric pumpsets or diesel engines and who did not really need a gasifier. The farmers accepted them since the gasifier-diesel engine unit was highly subsidized. If the demonstration gasifier units had been supplied to farmers in remote areas, who had no access to grid power, the rate of acceptance or use level would certainly have been different.

Pumping—a wrong end use? Possibly pumping irrigation water was an inappropriate end use to start dissemination as the competing option of electricity is highly subsidized. The gasifier system would have possibly worked in situations where there was no significant subsidy available and grid electricity was not present. For example, small-scale industries in remote areas and community lighting, as in case of Hosahalli village where there was a need for a reliable electricity supply (Srinivas *et al.* 1992; Chapter 5).

Botheration factor In the case of electric pumpsets, all that the farmer has to do is press a button to lift water. For diesel engines the farmer has to buy diesel, fill the tank, and start the engine. Gasifier operation involves a range of activities, the most cumbersome processes being: obtaining wood, cutting it to the appropriate size, drying it in the sun, and loading it into the gasifier before starting the system. This botheration factor is a deterrent, especially in locations where easier options are available.

The technology is still in the development stage It has yet to be made user-friendly and there are potentials for cost reduction and improving the life of the systems.

4.4 LARGE BIOMASS-BASED POWER-GENERATION SYSTEMS

The systems discussed above were mostly small systems for use in the mechanical mode, except for a few 100 kW systems which were for power generation. The potential for power generation using biomass gasification is estimated by the ministry to be 17 000 MW (MNES 1993*a*), and the potential for co-generation using sugarcane residues alone is estimated to be 3500 MW (MNES 1993*c*).

In India two co-generation units in sugar mills have been installed recently in Tamil Nadu state, with a capacity of 3 × 2.5 MW for efficient utilization of bagasse to produce about 4 MW of surplus power for feeding to the grid. The system at MRK Co-operative Mills was commissioned in mid-1992 and the second plant at Cheyyar Co-operative Sugar Mills was in the process of being commissioned in 1994. No information is yet available on the performance of these systems. Thus the Indian experience is very limited in large-scale power generation systems using biomass feedstocks. The potential for power generation using the biomass gasification route, the advantages, and an approach to bioenergy are discussed in Chapter 6.

4.5 BIOMASS PRODUCTION FROM AFFORESTATION PROGRAMMES AND IMPACT ON FUELWOOD SUPPLY

The main goal of afforestation programmes or social forestry in India is to meet the fuelwood needs of local communities. The area brought under afforestation between 1980 and 1992 was about 17 Mha. About half of it is considered as community forestry and Forest Department plantations (Ministry of the Environment 1990) and the rest as farm forestry. By assuming an annual woody biomass productivity of 4.2 t/ha/yr for farm forestry and 2.6 t/ha/yr for the other two categories (from Seebauer 1992, discussed in Chapter 7), the annual woody biomass production from such afforestation programmes for 1993 is estimated to be 58 Mt.

Next it is necessary to estimate the contribution of this annual production to meeting the fuelwood requirements. The annual fuelwood use is estimated to be about 227 Mt (Table 2.2). Wood or timber production from afforested plantations is used as fuel, as poles in building construction, as structural timber (in rural housing), and as softwood for industry. The softwood and structural timber requirement for 1986 was estimated to be about 18 Mt (FSI 1988). Even if all the requirement is assumed to be met from social forestry plantations, only 18 Mt out of 58 Mt is adequate, leaving 40 Mt as available for use as fuel. Thus the contribution of social forestry plantations to fuelwood supply is significant.

The contribution of Forest Department plantations raised under the social forestry programme analysed for 45 randomly selected villages in Sirsi block (Bhat, D. M. and Ravindranath 1994) showed:

(1) the annual productivity from the tree plantations in village common lands is higher than the requirement for 31 per cent of the villages;
(2) the fuelwood available annually from tree plantations is in the range of 50–100 per cent of the fuelwood requirement for 33 per cent of villages; and
(3) the fuelwood available is less than 50 per cent of the annual requirement for 36 per cent of the villages.

Thus woody biomass production from forestry plantations in some locations, such as Sirsi, is adequate to meet the fuelwood needs of most of the population.

4.6 CONCLUSIONS

This chapter aimed to analyse the current programmes related to bioenergy, since India was one of the first and few countries to recognize the role of bioenergy and to initiate programmes in the aftermath of the 1973 oil crisis. India has formulated and implemented bioenergy programmes on a significant scale, which the developing world may find of practical interest. The lessons learnt in implementing biomass conservation, production, and conversion programmes are too valuable to ignore. For example, the high success rates of biogas programmes under a private initiative in Sirsi, the poor performance of government-managed stove programmes, the target-chasing feature of government-managed programmes irrespective of the performance, the lack of responsibility or accountability for success or failure, and the higher cost due to involvement of government machinery are all worth considering in programme formulation. One of the problems of all government-managed programmes is the chasing of physical or financial targets. Often the success of a government functionary is measured by his or her ability to spend the money in the given (often short) period. In spite of many drawbacks, however, India's achievement in biogas, improved stove, and afforestation programmes should be recognized.

There seem to be separate departments or sections to promote stoves, biogas, producer gas, and afforestation programmes, and each section has its own agenda, priorities, and funding, irrespective of the requirements on the ground. For example, there is no village-level attempt to meet the energy services of all activities in an integrated and sustainable way, although bioenergy provides such an opportunity (as will be seen later). The findings of this chapter will be further augmented in the next chapter with specific case studies from both developing and industrialized countries.

Photo 4.1 Family-sized biogas plant using cattle dung in a village in Sirsi (W. Ghats).

Photo 4.2 Cooking with biogas generated by a family biogas plant.

Photo 4.3 Fuel-efficient ASTRA stove (compare with traditional stove in Photo 2.1).

5
Bioenergy projects

The importance of bioenergy as a modern fuel has been recognized over the past decade in industrial as well as developing countries. Attempts made in India on biomass conservation, production, and conversion to modern energy forms were broadly analysed in the previous chapter. However, there are many specific examples in India, as well as in other developing and industrial countries, of converting biomass into modern energy carriers, and it is useful here to consider some of these case studies to demonstrate the feasibility of bioenergy programmes and to learn from these experiences. The selection is basically determined by the availability of information to the authors. It is not possible to go into details of individual projects, but instead the approach adopted and performance of the projects are presented briefly. A sample of small village-level systems and large systems are considered to cover a broad spectrum.

1. Case studies from India:
   (a) producer-gas electricity
      (i) energy forest for electrification of Hosahalli in South India
      (ii) for electrification of villages in the hilly areas of North India
      (iii) for community irrigation in Ungra
      (iv) from rice-husk-based co-generation in Punjab;
   (b) biogas for cooking as well as electricity through
      (i) community biogas for village electrification—Pura village in South India
      (ii) community biogas for cooking—Methan village in western India.
2. Case studies from outside India:
   (a) producer-gas in South-East Asia;
   (b) electricity from biomass in the USA;
   (c) co-generation in sugar mills in Mauritius;
   (d) ethanol production in Brazil;
   (e) Large biogas programme for electricity in Denmark.
3. Proposed projects:
   (a) urban waste biogas programme—Tanzania;
   (b) Brazilian biomass power project;
   (c) energy plantations in Brazil.

5.1 CASE STUDIES OF PRODUCER-GAS ELECTRICITY IN INDIA

5.1.1 Integrated producer-gas and energy forest system for decentralized power for Hosahalli village

Background

India has about 70 000 villages yet to be connected to the electricity grid. The supply of grid power to rural areas is characterized by:

(1) low loads;
(2) power shortages;
(3) low reliability;
(4) low and fluctuating voltage; and
(5) high transmission and distribution costs, and power losses.

Decentralized power generation based on renewables is an attractive option to meet village energy needs. ASTRA, at the Indian Institute of Science, has developed a small-capacity open-top wood gasifier (Mukunda *et al.* 1993) and implemented a field demonstration programme in the non-electrified South Indian village of Hosahalli (Ravindranath *et al.* 1990; Srinivas *et al.* 1992).

Approach

The steps adopted in implementation of the project are presented here.

Step 1	Selection of a non-electrified village—Hosahalli, 100 km from Bangalore.	1987
Step 2	Discussion with village community—explain the project; benefits, responsibilities, roles, costs, etc. Request for land for energy forest and for a small generator building; procured land from the village community.	1987
Step 3	Planting a mixed-species energy forest—six tree species, density 6600 plants/ha; 2 ha.	1987
Step 4	Generator building construction, installation of gasifier (5 kW), electrical wiring of all houses, training village youth.	
Step 5	Discussion with community and commissioning of lighting system; dual-fuel operation.	1988
Step 6	Village community requests a pump for drinking-water. Installation of water pumping unit connected to 5 kVA three-phase alternator. Commissioning of water pumping operation.	Sept. 1990
Step 7	Formation of village wood gasifier and energy forest management committee. Collection of charges for lighting.	Oct. 1990

Step 8	Request from village community for a flour mill connected to the gasifier electricity generator. Installation of flour mill and commissioning.	March 1992

The programme was implemented by scientists of ASTRA, Indian Institute of Science, Bangalore, with funding from the government of India.

Services provided with producer-gas electricity

The electricity produced was used for three services: lighting, pumping domestic water, and flour milling.

Lighting All 42 houses were provided with a 40 W fluorescent tube and a 15 W incandescent bulb, along with eight street lights. Connected load = 2.68 kW. Hours of operation: 6 p.m to 10 p.m.

Water supply A submersible pump of 3 hp capacity was connected to a deep tube well. Water is pumped to storage tanks to provide 2–3 hours of water supply per day.

Flour milling A 7.5 hp flour mill was connected to the producer gas diesel-engine generator. The flour mill, which operated for 2 hours/day, was operated for a few months in 1992, its operation was suspended as the rate of milling was lower than the desired rate by the village community due to limitations of the 5 kVA system.

Operation of the system and its performance

The lighting and water-pumping systems have been in operation since the date of commissioning; lighting from December 1988 and water pumping from September 1990. Lighting was provided for 83 per cent of the days and water pumping for 78 per cent (data from 43 continuous months). The reasons for non-operating days ranged from engine, generator, or gasifier problems to non-availability of diesel, absence of operator, etc.

Wood productivity The mean annual rainfall of Hosahalli is 70 cm. The wood productivity of the mixed-species energy forest was monitored and a productivity of 6 t/ha/yr was recorded. Annually about 5–7 t of wood is used for electricity generation. Thus the wood supply from the 2 ha of forest is sustainable. A productivity of 6 t/ha/yr was obtained without any fertilizer application or irrigation. Excess wood was sold in the form of poles for housing for the village community.

Wood consumption The gasifier used 1.2–1.6 kg of air dry wood/kWh, which is slightly higher than the laboratory determined level of 1.2–1.4 kg/

kWh. The higher moisture content of the wood could be one of the reasons for the slightly higher wood use.

Diesel substitution According to the design specification of the dual-fuel system, 85 per cent diesel replacement is feasible. In the village an overall diesel replacement of 67 per cent was achieved (based on 32 months' data). The relatively higher moisture content of the wood, particularly during the rainy season, and operation on diesel mode initially for 10–15 min are reasons for higher diesel consumption.

Gasifier performance A major problem with the gasifier was an air-nozzle crack in the stainless steel reactor which occurred periodically. A new ceramic shell gasifier was installed (in 1993) which also developed some problems. An alternative design is being tested. It should be noted that small gasifier designs are still in the development phase even though they are now available commercially.

Operation of system Village youth were trained in the operation and minor maintenance of the system.

Management of the decentralized system

The village community is consulted through a village council meeting at every stage before taking decisions. A small management committee supervises the day to day operation. The responsibilities of the village council and management committee are:

(1) ensuring monthly payments for the lighting service;
(2) supervision of operators;
(3) protection of the energy forest and the generation system;
(4) taking decisions on the operation and management of the whole system.

Conclusions

(1) Technical and operational feasibility of the decentralized power-generation system based on an energy forest–biomass gasifier system has been demonstrated.
(2) Electricity from a decentralized source was accepted by the village community even though the neighbouring villages are connected to the grid.
(3) The community was involved at all stages through village council meetings, ensuring their commitment and participation. For example, protection of the energy forest by the entire community with no fencing or paid guard. Community commitment was particularly forthcoming from the day domestic water was pumped, since this was their primary priority need; lighting was their next priority.

(4) This indigenous technology had full technical back-up from a local institution.
(5) The reliability, local management, and self-reliance features of the Hosahalli project had the 'demonstration effect' on neighbouring villages. They are now requesting similar producer-gas systems for their villages, even though they are already connected to the electricity grid.

5.1.2 Producer-gas electricity for six villages in North India

A village electrification programme using biomass gasification has been implemented in a cluster of six villages near Kotdwara (Garhwal region, North India), by the Himalayan Environmental Studies and Conservation Organization and the Indian Institute of Technology (IIT), Delhi. A brief summary of the biomass energy project in a cluster of villages around Kumbichar village is given here, based on Joshi *et al.* (1993). A 3.5 kW gasifier was installed in Kumbichar village, which uses mainly *Lantana camara*, a wild shrub growing in the village, as feedstock. It is a small village and all 18 houses are provided with electricity for lighting from a 3.5 kW system. During the day, the system is used to operate a flour mill, spice grinding, and a leaf plate making machine. The system is operated by trained village youth. Each household pays Rs20/month for lighting. The operational cost of producer-gas electricity generated was Rs0.93/kWh compared to the grid electricity rate of Rs1.38/kWh (1992 price). A diesel substitution of 70 per cent has been achieved. The system has been expanded to five more villages.

An interesting aspect of the project is the use of the relatively low-density shrub *Lantana camara*. *Lantana* grows wild in many parts of India and thus using it for power generation in this demonstration project provides an excellent example of using such a hardy, quick-growing, and coppicing weed for economic gain. It is also relevant that the households collect and cut *Lantana* sticks for gasification each day, thus avoiding the growing and collection of wood. This contribution of the village community demonstrates their commitment to the programme and reduces the investment and operational costs.

5.1.3 Producer-gas electricity for a community irrigation system

Ungra case study

Irrigation is a critical input to agriculture for increasing and stabilizing crop yields and is very important to prosperity in Indian agriculture. Currently only 35 per cent of the cropped area is under irrigation, with 9.1 million electric pumps and 5.5 million diesel pumps for lifting irrigation water (TSL 1992). Producer-gas systems can be used to pump water at the individual level as well as the community level. Here we briefly present the case study of a producer-gas-based community irrigation system which is still in the initial phase of implementation.

Approach Ungra is a semi-arid South Indian village which receives a mean annual rainfall of 67 cm. A community irrigation system powered by a 20 kW producer-gas system was installed in July 1993 to pump water from a borewell for nearly 12 hours/day. Twenty-one farmers receive water and the area irrigated is about 5.25 ha. The main crop irrigated is mulberry, which produces about 10 dry t woody biomass residue/ha/yr (Ravindranath and Somashekar 1991). The farmers have agreed to provide the mulberry stalks as feedstock to the gasifier. Thus the electricity generation system is going to be self-sufficient for biomass feedstock and may not require any dedication of land. The farmers are paying the full cost of operation for the energy used for water pumping. The system is still under standardization.

The main lesson from this experiment is that it is possible to install village community irrigation systems powered by higher-capacity systems, and farmers will pay for electricity at commercial rates. Crop residue can also be used as feedstock for gasification to reduce the dedicated land as well as the investment costs.

5.1.4 Rice-husk-based co-generation power plant in Punjab

Rice husk is one of the important crop residues with a large energy potential for power and heat generation. The total rice husk production in India is 16 Mt (for 1991). Rice husk is produced in rice mills of different sizes in decentralized locations in a large proportion of villages in India. Research is in progress in India on small kilowatt-scale rice-husk-based gasifiers (MNES 1993*a*). A rice-husk-based 10.5 MW co-generation plant was installed in 1990 in Punjab by Oswal Agro Finance Ltd (OAFL) for power generation. This is the first such large biomass-based power generation system in India's private sector. Raju Associates (1992) of New Delhi prepared a techno-economic study of the OAFL co-generation unit.

Background

OAFL is an integrated rice-processing plant at Dhuri in Punjab. The company is an export-oriented firm. OAFL processes rice, retains the rice husk as a part of the processing fee and uses it to produce furfural for export.

Power generation

Out of the total installed capacity of 10.5 MW, the processing units use 5.5 MW and the surplus is sold to the state utility grid. The generation unit has two step-grate boilers, each having the capacity to produce 35 t/hour of steam at 505°C, 63 bar. About half the steam is fed to the extraction/condensing turbine and the remaining half is for process use; the balance of the steam is condensed. The unit operates on a topping cycle of co-generation.

Power generation and sales

The plant was commissioned in March 1991. The plant had initial problems, and after it started producing electricity, it had difficulty in selling power to the state utility. The state utility is unwilling to pay an economic price, even though the state is suffering from power shortages. The utility sells electricity at Rs1.10/kWh, which is a subsidized price, and is unwilling to buy from the plant at an economic price. Also, the state utility does not permit the sale of electricity to other consumers directly.

The estimated cost of producing the electricity at the time of project formulation was Rs0.35/kWh and the actual cost of electricity during 1990–91 was Rs1.10/kWh. The two crucial factors that determine the cost of electricity are the price of rice husks and the plant capacity utilization. Rice-husk feedstock constitutes 40–45 per cent of the total cost of production of co-generated electricity. The cost escalation of rice husk has been crucial as it increased from a market price of Rs250/t in 1988 to Rs800/t in 1991–92 and then to Rs1100/t in 1992–93. The capacity utilization of the integrated plant, and thus the co-generation unit, was also low (<25 per cent). A sensitivity analysis by Raju Associates (1992) showed that the cost of electricity generation increased from Rs1.05/kWh at 100 per cent capacity utilization to Rs2.25/kWh at 25 per cent capacity utilization.

Lessons learnt from the project

1. The import of turbines and non-availability of spare parts delayed commissioning and led to cost overruns.
2. The feasibility report failed to make a realistic forecast of prices of inputs and outputs, particularly the price of rice husk, which escalated over threefold.
3. The monopoly of the state utility and its policy of not fixing electricity prices based on the long-run marginal cost method. The unwillingness of the state utility to buy electricity at an economical price is a crucial factor for the viability of such power-generating units.

Most of the problems and performance failures faced by the Punjab unit arise from the fact that it is the first such unit in the country. The study concluded that commercial viability of rice-husk-based co-generation units would require:

(1) better planning (forecast of prices of inputs, outputs, and operational costs), implementation, technology choice, and organization of spares and services; and
(2) economical pricing of electricity and freedom for generation units to sell electricity in the market at commercial rates.

Overall it can be concluded that the poor performance of the rice-husk-based co-generation system was mainly due to factors external to the power generation system. Thus, when new units are planned the Punjab experience indicates that appropriate institutional policies are essential to make such units financially successful.

5.2 CASE STUDIES OF COMMUNITY BIOGAS PROGRAMMES IN INDIA

5.2.1 Biogas electricity in Pura village

In India generally, the biogas option is considered largely as a cooking fuel (Chapter 4). The need for considering decentralized electricity options and the potential of biogas is analysed in Chapter 6. A field-demonstration programme was implemented in Pura village in South India to use cattle dung in a community biogas plant to generate electricity for services such as pumping drinking-water and home lighting. In phase I, an attempt was made to provide biogas for cooking through a community biogas plant. The programme objectives were changed after phase I, duc to two major factors. First, the available dung resource was inadequate to meet the cooking energy requirements of all the households. Secondly, the priority need of the community was not cooking, but the provision of domestic water, followed by home lighting. A brief summary of phase II, which has been in operation since 1987, is discussed here, based on Rajabapaiah *et al.* (1993).

Approach

1. A village survey was conducted to study the energy sources and end uses, and to identify problems related to energy supply and utilization.
2. The energy needs of the village were identified.
3. Discussions were held with the village community, which agreed to become a partner in the project.
4. Land was obtained for the construction of the biogas plant and generation unit.
5. Biogas plants were constructed and a diesel engine generator set installed.
6. A submersible pump, water storage, and water distribution systems were installed. All the households requiring lighting were connected to the generator.
7. The water pumping system was commissioned in September 1987 and the lighting system in October 1988.
8. Village youth were trained in the operation of the system.
9. The village community contributed cattle dung and collected sludge in return; a nominal service charge was paid to the persons bringing dung.

10. Technologists from ASTRA, Indian Institute of Science, Bangalore executed the project and provided all technical inputs, with funding from the Karnataka state government.

Technology

The Indian floating-drum design with modified dimensions for cost reductions was used. The Pura biogas plants have a capacity to digest up to 1.2 t cattle dung/day and produce 42.5 m^3 biogas/day. Sand bed filters were installed to remove excess water and convert the sludge to a dung-like consistency for subsequent use as a fertilizer. The filtrate, which contains the required anaerobic micro-organisms, is mixed with the input dung. A 5 kW diesel engine is connected to a 5 kVA, 440 V three-phase generator for electricity generation.

Lighting

Out of 87 households in the village, 39 already had grid electricity. There are 103 fluorescent tube lights of 20 W capacity (97 in homes, two street lights, four in the biogas complex) connected to biogas-generated electricity. Forty-seven houses opted for one tube light and 25 houses have two tube lights. Lighting is provided in the evening for 2.5 hours/day. Even homes connected to the grid had lighting connections from the biogas system.

Water supply

A submersible pump is connected to a tube well and water is pumped to storage tanks for 1 hour and 40 minutes/day. The majority of the households have opted for private taps at their doorsteps.

Performance of the system

The power generation system is operated for 4 hours/day with the diesel substitution averaging 77 per cent. Maintenance of the biogas system includes periodic gas-holder painting to prevent corrosion.

Payment for services

The households pay for the lighting service as well as private taps; currently Rs5/lighting point and Rs5/tap per month.

Management of the system

The management of the system and the decision-making power rests with the community. A village development society has been formed. All major decisions are taken at the village council meeting.

The community biogas electricity system has been in operation in the field for nearly 8 years. The critical factors for the success of the system are:

(1) involvement of the village community and meeting their priority needs;
(2) appropriateness of the technology;
(3) high reliability;
(4) self-reliance and local control;
(5) full technical support by technologists from ASTRA; and
(6) adequate funding.

Future

The success of Pura biogas electricity has led to similar demands from neighbouring villages, so the project is now being expanded to four more villages. An identical system has already been commissioned in two neighbouring villages (during 1993) and two more villages were to be brought on stream during 1995. There will be a cluster of four community biogas electricity systems within a radius of less than 5 km in South India. Based on the success of these decentralized biogas electricity systems, about 100 more such systems are being planned.

5.2.2 Community biogas plant for cooking in Methan village

The Pura community biogas system is an example of the use of biogas for electricity generation. Biogas is an ideal fuel for cooking, and Methan is an example of the use of the community biogas potential for cooking. Methan village is in Mehasana district, Gujarat State, western India, and houses 600 families. The case study is based on Venkataramana (1992) and MNES (1993*b*).

Approach

With the confidence developed by the successful and ongoing milk co-operative movement, a community biogas plant system was planned for the supply of cooking gas. The entire capital cost was borne by a project funded by the government of India. GEDA (Gujarat Energy Development Agency) organized the entire installation through a private firm of consultants on a turnkey basis. The dung required for the initial charging of the plants (about 600 t) was contributed by the villagers, but they were paid for it by the contracting firm.

All the members who contribute dung daily to the system receive the digested slurry back on a pro rata basis. Biogas is provided for cooking in 326 households. The gas users are charged at a flat rate of Rs40/month. The daily operation of the plant has been entrusted to a contractor who supplies manual labour equivalent to eight workers to charge the plant, remove the slurry, and clean the plant.

Technologies

Biogas plant There are eight plants with a total gas production capacity of 630 m^3/day. Six have a 85 m^3/day capacity, and the other two 60 m^3/day. They are of the conventional Indian floating-drum design, fitted with a mechanical stirrer at each site for mixing the inlet dung slurry. The biogas plants are spread over three sites in the village to ensure uniform pressure in the gas distribution.

Piping Underground piping exists for gas distribution from the plants to individual households. Fourteen water traps are built at various points of the piping in the village to remove any condensed water.

Performance

Since its commissioning in 1987, the system has run more-or-less continuously without any major problems. A supervisor and a driver collect the dung daily from the households, using a tractor. All cooking fuel requirements of 326 out of 600 families (54 per cent) have been replaced with biogas. Nearly 13 t of fresh cattle dung is transformed into fertilizer every day. For the women in these 326 families, the drudgery of collecting fuelwood and exposure to smoke have been avoided. In addition, about ten persons, one supervisor, one driver, and eight plant operators, are being employed in their own village.

Management

A registered society, the Silver Jubilee Biogas Producers and Distributors Cooperative Society (SJCS) was established with all the biogas beneficiaries as members. It has been demonstrated that by forming a co-operative society the village community can successfully manage the decentralized energy systems in the village environment.

Methan is the example of the potential of a community biogas option for cooking, even in a village where only a part of the cooking energy requirement can be met. Some of the factors responsible for the success of the project are:

(1) a successful co-operative effort, an extension of the milk co-operative movement, and management by the local community;
(2) installation of the system through a commercial enterprise on a turnkey basis;
(3) operation and maintenance through a commercial enterprise;
(4) meeting all the energy needs of a basic activity such as cooking for 54 per cent of households;
(5) adequate funding for installation of the system.

5.3 BIOENERGY PROJECTS OUTSIDE INDIA

5.3.1 Producer-gas programme in the Philippines

In 1981, the government of the Philippines launched an ambitious programme to retrofit 1150 diesel pump-powered irrigation systems with producer-gas systems. The gasifiers were supplied to agricultural co-operatives and their purchase was supported by soft loans. A survey of the operational status of gasifiers by BTG (1993) in 1989–90 showed that only 5 per cent of over 1000 gasifiers installed were still in use. The reasons for the poor performance included technical, financial, and institutional factors (Bernardo and Kilayko 1990; BTG 1993).

Technical factors The gasifier design, which was not fully developed, was imported.

Financial factors The price of the feedstock, charcoal, increased six fold from 1977 to 1985. In some situations charcoal was more scarce and expensive than diesel.

Institutional factors The operators were ill-trained and poorly paid, and thus had no incentive to maximize performance. In order to meet the target fixed by the government all the criteria and procedures for identifying suitable locations and infrastructure were flouted. The implementing agencies were also inadequately funded.

This example of the Philippines gasifiers is frequently quoted to point out the low feasibility of biomass gasifier options. However, one should keep in mind the fact that in spite of all the problems some gasifiers are still in operation where conditions are suitable, i.e. those which were suitably located, where the personnel were adequately trained, and adequate feedstock was available. There is a need to learn from the experience of such a programme where failure was inevitable from the outset.

5.3.2 Electricity from biomass: USA

Background

The Public Utility Regulatory Policy Act (PURPA) was introduced in 1978, under which the US electricity utilities were required to purchase electricity from other suppliers at the cost they 'avoided'. The 'avoided cost' was calculated as the marginal cost of electricity generation from a new conventional power station (coal, gas, or oil). The utilities were obliged to buy alternative sources of electricity regardless of internal economic considerations even if the most economical way of providing base-load and peak demand was through the use of electricity supplies from other sources, including their own stations (Turnbull 1993). Electricity generation systems needed to have a capacity of less than 50 MW in order to qualify under PURPA. The Act led to the establishment of a number of co-generators based on waste materials and byproducts from saw mills, abattoirs, food processing, and paper mills. There were 149 stand-alone and 367 co-generation facilities. The installed capacity of stand-alone plants was 2459 MW, while co-generation utilities dominated by accounting for 5962 MW out of the total 8421 MW of biomass-powered installed capacity in 1989. California alone has 94 plants with an installed capacity of 991 MW.

Future

The USA currently obtains about 4 per cent of its energy from biomass. According to the Renewable Intensive Energy Scenario (Johansson *et al.* 1993) the contribution of biofuels in USA could rise from 3 EJ in 1985 to 22 EJ by 2025. The US government has also implemented a renewable electricity supporting subsidy of 1.5 US c./kWh, which will go a long way in promoting renewables, particularly the energy options.

5.3.3 Electricity from co-generation in sugar mills: Mauritius

Sugarcane is the dominant crop in Mauritius, accounting for 88 per cent of the cultivated area, 20 per cent of the country's GDP, and over 40 per cent of its export earnings. Sugarcane production during 1990 was 5.5 Mt and out of this only one-third (29 per cent) of the potential excess energy from bagasse was used (Comarmond 1992). Electricity production from bagasse increased from 27 GWh in 1980 to 71 GWh in 1991. In the past, the majority of sugar mills used inefficient low pressure and temperature back-pressure technology. This technology has the capacity to produce 300 kg of steam/t cane. In 1985 Flag United Estate Limited (FUEL), the largest sugar mill in Mauritius, commissioned a modern steam boiler capable of burning both bagasse and coal, sufficient to deliver high temperature and pressure steam to power both the factory and a 24 MW CEST alternator. The dual-fuel boiler system also enables burning of coal during the off-season. Year-round electricity supply is more valuable to the Central Electricity Board (CEB) which pays a premium for such supplies, for example 100 c./kWh for permanent generation, 45 c./kWh for seasonal, and only 16 c./kWh for intermittent (e.g. wind) supply. The electricity cost is determined by the 'avoided cost' that has to be incurred if the demand were to be met from the CEB's own generation systems, particularly its diesel-powered generating plant.

Two 24 MW bagasse/coal plants have been planned. The goals of the bagasse energy development programme in Mauritius are to increase cane-residue-derived electricity production. According to one estimate, if Mauritius decides to exploit the full potential of all sugarcane residues (bagasse + tops + leaves) along with other crop residues, the potential is 3500 GWh/yr (at 40 per cent conversion efficiency of biomass to electricity) or 2500 GWh at 30 per cent conversion efficiency, compared to the projected electricity demand of 1678 GWh for year 2000 (Comarmond 1992). According to another estimate, the sugarcane residue potential for the year 2027 is 14 300 GWh compared to the 1991 total electricity use of 737 GWh (Williams and Larson 1993). This latter estimate assumes that sugarcane production increases by 3.1 per cent annually and advanced biomass-integrated gasification–intercooled steam-injected gas turbine (BIG/ISTIG) technology is used with a conversion efficiency of 38 per cent. Thus Mauritius could become self-reliant by using only a part of the bagasse potential.

5.3.4 Large biogas plants in Denmark

Background

Denmark is a leading dairy and pig-farming country in Europe. It also leads the world with respect to deployment of renewables for electricity generation and central heating. In addition to wind power, Denmark has been working on biogas technology since 1970, both for environmental and energy reasons. The Danish experience is briefly presented here from DEA (1991). In the 1970s most biogas plants built were small single-farm designs. Out of the 40 such biogas plants built, about 30 were abandoned. This led to the evolution of the existing designs and development of large-scale biogas plants.

Large biogas plants

The first large-scale biogas plant was built in 1984, and by 1992 a total of 10 such large plants and 10 smaller single-farm biogas plants were in operation. One of the first large plants, at Vester Hjermitslev, had a capacity of 1500 m^3/day; the latest plant, at Lemvig, has a capacity of 7600 m^3/day and was commissioned in May 1992. The Lemvig plant has a capacity to generate 9500 MWh of electricity and 15 000 MWh of district heating per year; it is owned by a co-operative society consisting of 80 farmers who provide dung and receive digested fertilizer in return.

Technology

These plants mainly use animal dung (75–90 per cent of the feedstock) in addition to organic wastes from other sources, such as households. Four plants operate at thermophilic process temperatures (50–55 °C) using a highly automated woodchip heating process to maintain the temperature of the digester. The rest of the plants use mesophyllic digestion processes at 35–40 °C. Most biogas plants are equipped with heat exchangers in which the heat from the digested manure is used to heat the raw manure, thus reducing the process heat requirements.

Performance

The plants built initially encountered some problems. From the experience gained, recent plants, including the large Lemvig plant, are operating without any major problems. The total biogas produced annually is 15 Mm^3 (in 1991), equivalent to 0.5 PJ.

Future of the biogas programme in Denmark

Encouraged by the success of the biogas programme and its relevance to environmental benefits, the Danish Energy Agency (DEA) has committed further research and follow-up actions:

(1) 6–7 new large biogas projects are planned for commissioning; and
(2) currently only 2 per cent (0.5 PJ) out of the total potential of 25–30 PJ is used, this could be increased 5–10 times by 2005.

According to DEA the centralized biogas plants could be profitable without public grants. However, an indirect state subsidy on current energy taxation is necessary. Denmark has established its leadership in biogas technology and, based on its experience, it is involved in the UN global environment facility (GEF) funded large biogas plant for Dar es Salaam, targeted to produce 11 GWh of electricity annually using urban wastes.

5.3.5 Biogas programme in China

Rural communities in China depend largely on biomass, particularly on straw and fuelwood for cooking and heating. In the 1960s China launched a large-scale family biogas programme which, at present, is estimated to have about 5 million small, family biogas plants in working condition, using animal and human wastes. This is nearly three times the number in India and is the largest biogas programme in the world. There is, however, a lack of recent information on the performance of such family biogas plants (Shangbin 1993). In addition to family biogas plants, according to incomplete statistics, 1700 large- and medium-sized biogas digesters have been installed. Out of this, 1300 have a capacity of less than 100 m^3 and 300 digesters have a capacity of over 100 m^3. These use predominantly poultry droppings and distillers' grain, and the biogas is usually supplied to households as cooking fuel. Each house is assured of 565 m^3 of biogas or 1.51 m^3/day (Shangbin 1993). In addition to using biogas as cooking fuel, there are 186 biogas-based power generation units with a total installed generation capacity of 3459 kW and an annual generation of 3 GWh of electricity.

In China there has been a move towards financial self-reliance, which has led to reduction in installation subsidies for family biogas plants from two-thirds to one-third. This has resulted in a decline in the construction of family biogas plants. There is also a shift towards centralized medium and large biogas plants which are operated on commercial terms using advanced technology. Significant research efforts have been invested in biogas and there are 116 biogas research centres (Daxiong *et al.* 1990). China still has a serious commitment to biogas and continues to have the world's largest biogas programme.

5.3.6 Liquid fuels from biomass

Bioethanol is produced from crops with high sugar or starch contents. Sugarcane is the world's largest source of fermentation ethanol. Bagasse, the stem fibre by-product of sugarcane crushing, can be used as a convenient on-site raw material for electricity generation, along with other unused parts of the plant. An efficient ethanol distillery using sugarcane by-products can therefore be energy self-sufficient and may also generate a surplus of electricity, in addition to animal feeds and a range of chemical-based products. Alcohols have favourable combustion characteristics, namely clean burning and high octane-rated performance. Internal combustion engines optimized for operation on alcohol fuels are 20 per cent more energy efficient than those operated on gasoline (Johansson *et al.* 1993), and an engine designed specifically to run on ethanol can be 30 per cent more efficient (EPA 1990). Furthermore, there are many environmental advantages of using ethanol, particularly with regard to lead, CO_2, SO_2, particulate, hydrocarbon, and CO emissions.

Several countries have pioneered both large- and small-scale ethanol fuel programmes. Latin America, dominated by Brazil, is the world's largest producer of bioethanol (Rosillo-Calle 1990). Alcohol fuels have also been aggressively pursued in a number of African countries, such as Zimbabwe, Kenya, and Malawi. Some countries have modernized their sugar industries and have low production costs. Many of these countries are landlocked so it is often not feasible to sell molasses as a by-product in the world market, while oil imports are expensive and subject to disruption. In the USA the current fuel ethanol production capacity is over 4.6 billion litres annually and there are plans to increase this capacity by more than 2.3 billion litres. Worldwide fermentation capacity for fuel ethanol has increased eightfold since 1977 to about 20 billion litres/yr in 1990 (Rosillo-Calle 1990). Bioethanol trials have been carried out in Germany, Italy, and Sweden, and a small amount is already being blended with gasoline in France. In Europe, promising feedstocks include cereals, sweet sorghum, maize, and surplus grapes or wine.

Brazil

Brazil created the National Alcohol Programme (ProAlcool) in 1975. Currently, under this programme about 13 billion litres of ethanol are produced from sugarcane. In 1989, over 12 billion litres of ethanol replaced about 260 000 barrels of imported oil/day in Brazil. Nearly 50 per cent of the gasoline used by the 10 million cars in Brazil has been replaced by ethanol from sugarcane (Goldemberg and Macedo 1994). From 1976 to 1987 the total investment in ProAlcool reached $US6.97 billion and the total savings equivalent in imported gasoline was $US12.5 billion. The government established several support mechanisms by ensuring a market, guaranteeing a price, providing financial incentives to ethanol producers and car owners, and investments in research and development. Over 700 000 direct jobs, with perhaps 3–4 times this number of indirect jobs have been created. Introduction of new equipment, increases in productivity, and the number of new plantations has given a continued boost to alcohol production, while the costs of ethanol declined at an annual rate of 4 per cent between 1979 and 1988 (Goldemberg *et al.* 1993). The costs of ethanol production could be further reduced if sugarcane residues were to be fully utilized. For example, in Sao Paulo the use of stillage (the fermentation effluent) as a fertilizer in sugarcane fields has increased productivity by 20–30 per cent while reducing costs. Furthermore, an overall cost reduction of about 23 per cent could be achieved in the next few years simply by adopting the available technologies (Goldemberg and Macedo 1994). However, ProAlcool has certain economic difficulties (mainly due to changes in the oil and sugarcane markets) which will have to be overcome; without a stabilized supply of and demand for alcohol, the long-term future of the industry may not be secure.

5.4 EXAMPLES OF PROPOSED PROJECTS

A large number of bioenergy projects have been planned, and some are being implemented in industrialized and others in developing countries (Patterson 1994). Here only two projects are presented briefly; first, a large biogas project in Tanzania, followed by a power generation project through advanced biomass gasification in Brazil.

5.4.1 Takagas (waste gas) project in Tanzania for electricity generation

Tanzania is one of the few countries in Africa with an established biogas programme, with over a 100 biogas units, and has now embarked on a large Takagas project ('Taka' means waste). A pilot plant is being implemented in Dar es Salaam to process nearly 160 t of organic wastes/day (Mrindoko 1993). The inputs comprise wastes from markets, households, breweries, hotels, etc. The first three sources account for 85 per cent of the feedstock. An initial goal of the project is to demonstrate the environmental and energy benefits of such a biogas programme, which is expected to be operational in 1996.

The Takagas project is expected to produce 6400 m^3 of biogas/day. The biogas will be used to generate 11.8 MWh/day of electricity, about 1.9 MWh for internal use and 9.9 MWh to be sold to the grid. The project is being jointly funded by the Tanzanian government and UNDP (GEF programme) with a total investment of $US3.9 million. The surplus income from the sale of electricity is proposed to be used for subsequent replication of the project. Biogas projects using wastes from the sisal and coffee industries are also being planned, making Tanzania a major biogas promoter in Africa.

5.4.2 Brazilian biomass power project

Brazil, apart from being a world leader in bioethanol production and its use as liquid fuel, is also taking leadership in promoting another bioenergy technology, namely biomass power based on advanced gasification technology: biomass integrated gasification–gas turbine (BIG/GT). A 25–30 MW BIG/GT system is being implemented in Brazil (Elliot and Booth 1993). A group of companies in Brazil, Sweden, Finland, USA, and UK are collaborating in this project to build the world's first BIG/GT power plant (Elliot 1994). It is projected to cost about $US70 million, with a part grant from GEF (World Bank). It is the world's first commercial-scale BIG/GT demonstration project. The project involves participation by Brazilian as well as multinational companies and is expected to be operational by 1997. In the long term, BIG/GT systems are projected to require an investment of $US1300–1500/kW and the final cost of electricity is estimated to be 5.1–5.6 c./kWh.

5.5 WOODY BIOMASS PRODUCTION POTENTIAL AND COSTS: AN EXAMPLE FROM BRAZIL

Brazil is one of the few countries with large-scale experience in managing tropical tree plantations. Unlike in India, the Brazilian tree plantation programme is largely industry-driven. The Brazilian experience is summarized using information from Betters *et al.* (1991) and Carpentieri *et al.* (1993). Brazil conducted extensive research and adopted high-yielding industrial tree plantations for paper, fibre, charcoal, and electricity. In 1965, Brazil passed the Forest Code which provided reforestation incentives, tax incentives, and easy credit for industry to promote reforestation programmes. Following the passage of the Forestry Act, nearly 1.5 Mha of *Eucalyptus* and pines were planted in just 2 years. A total of about 7 Mha were planted up to 1990. The Belgo Mineria Steel Company alone, for example, planted 0.5 Mha for charcoal production, thereby indicating the industrial interest. During the 1970s the Brazilian government leased large blocks of land to companies to encourage investment.

Eucalyptus has been grown extensively in Brazil since the nineteenth century, but only in the past decade or two was there a concerted effort to select species and clones from around the world which are particularly suited to the climatic and soil conditions of Brazil. Now plantations often have 20 or more different *Eucalyptus* clones and species to match the site quality. This has resulted in large increases in yields along with maintenance of biodiversity, since different species and clones are intermixed with blocks of indigenous vegetation—up to a third of the plantation area being left for forest succession or rehabilitated with indigenous species. Reforestation programmes involve land preparation using machines, but the bulk of planting is done manually. The planting density is generally in the range of 1300–2200 trees/ha. Most short-rotation commercial plantations are fertilized with fertilizer applied to individual trees at the time of planting and later it is broadcast over the entire plot, depending on soil characteristics and rotation policy. After the logged slash is burned and the area between the rows cut with a disc harrow, fertilizer is broadcast for the coppice shoots. Due to intensive research and plantation management by private companies, high yields are obtained routinely; in the early 1970s *Eucalyptus* yields averaged around 29 t/ha/yr, and now yields up to 56 t/ha/yr are obtained (Betters *et al.* 1991).

Potential of tree plantations for energy in north-eastern Brazil

Having realized that the economical hydroelectric resources in semi-arid northeastern Brazil could be exhausted by about the year 2000, the hydroelectric company CHESF, a federally owned company responsible for generation and distribution of electricity in the north-east, initiated studies to estimate the potential for biomass-based electricity generation systems as an alternative to hydro expansion. As a part of the plan, CHESF conducted an extensive study to estimate the woody biomass potential and costs of production by adopting a biogeoclimatic regional assessment basis. Here some of the findings of that study (Carpentieri *et al.* 1993) are summarized, showing how it is possible to analyse comprehensively the potential and cost of biomass production for energy.

The study mapped the soil and climate in north-eastern Brazil and collected yield and cost data from various commercial plantations and research plantations focusing on the north-east. Five 'bioclimatic regions' (BCRs) were defined, based on average temperature, elevation, precipitation, and water deficit (Table 5.1). Region I provides the best conditions for tree growth, and region V the poorest. Region III with a rainfall of 70–130 cm/yr accounted for nearly half the total potential land. Woody biomass productivity was estimated by combining the BCR/soil map with data on wood yields from existing plantations in identical bioclimatic soil regimes. *Eucalyptus* productivity data were obtained from large commercial plantations as well as small-scale experimental plantations. The average yields were estimated for different BCRs, and ranged from 44 m^3/ha/yr in region I to 6 m^3/ha/yr in region

Table 5.1 Analysis of the potential of *Eucalyptus* biomass for energy and its economics in north-eastern Brazil by bioclimatic regions I to V

Costs	I	II	III	IV	V	Weighted average
Plantation establishment ($/GJ)	0.33	0.33	0.52	0.98	2.44	0.55
Plantation maintenance ($/GJ)	0.76	0.79	0.81	0.92	1.27	0.81
Total cost ($/GJ)	1.09	1.12	1.33	1.90	3.71	1.36
Projected average yield (m^3/ha/yr)	44.00	33.00	28.00	15.00	6.00	26.60
Potential plantation area (Mha)	4.05	7.74	25.94	11.56	1.17	50.46
Precipitation (m/yr)	1.5–2.3	1.0–1.7	0.7–1.3	0.5–1.0	0.25–0.6	

Discount rate 10%.
Potential yields (stem wood only) estimated at the hydroelectric company CHESF, Recife, Brazil by correlating yields from existing *Eucalyptus* plantations (commercial and experimental) with climate, soil, and other characteristics of the plantations' bioclimtic regions. The potential yield for a bioclimtic region is taken to be the simple average of (1) the average yield of commercial plantations in that bioclimtic region, and (2) 0.75 times the average yield of experimental plots for that region. For bioclimtic region I, the average yield for commercial plantations was assumed in this calculation to be the best yield of commercial plantations in region II, as there are no commercial plantations in region I.
The potential plantation area is approximately the total area of the nine states in north-eastern Brazil, land that is generally unsuited for agriculture, primarily deforested land.

V. The 'mean estimated' yield for the whole of north-eastern Brazil was 27 m^3/ha/yr. The CHESF analysis shows that precipitation is the dominant factor affecting the yield. The projected yields for north-eastern Brazil are lower than those routinely obtained in other regions of Brazil that have higher rainfall.

Total biomass production potential

Assuming the area available in the north-east (not including any Amazon region) and the mean *Eucalyptus* yields, the total woody biomass production potential is estimated to be 1340 Mm^3/yr, giving 12.6 EJ/yr. This is high when compared with the total primary energy use in the north-east, which was 1.1 EJ (in 1990).

Cost of biomass energy

The cost of different operations is expressed in terms of levelized cost in $/GJ of woody biomass energy and is given in Table 5.1. The cost of production increases as productivity declines ranging from $1.09 per GJ in region I to \$3.71 per GJ in region V. Such cost and supply analysis would be very useful in all countries where modern bioenergy programmes are being contemplated.

5.6 CONCLUSIONS

The main goal of this chapter was to show that there are successful bioenergy field projects on small (village level) and large scales in developing as well as industrialized countries. Significant experience is already accruing on various aspects of bioenergy. For example, we can conclude that the success of decentralized bioenergy systems will depend on the involvement of local communities and meeting their priority needs. For successful adaptation of bioenergy systems it is best if the technologies are indigenously developed, but if they have to be imported, they must be accompanied by the involvement of local research and possibly commercial institutions, adequate training of local technicians in operation, repair, and maintenance, and by a supply of spares or, preferably, the means of making them.

As long as there are restrictions on the sale of energy at economic prices by bioenergy enterprises, and if they have to further compete with subsidized energy supplies, bioenergy programmes are unlikely to succeed. Under market situations, as in many industrialized countries, a financial subsidy or incentive could promote bioenergy systems in a significant way, unlike in many developing countries where governments largely control energy supplies and pricing. A number of countries are launching new bioenergy projects with confidence. Next it is necessary to explore the potential of bioenergy for India with some confidence based on field experience, however limited it is at present.

Photo 5.1 A new community biogas plant for electricity generation in Suggenahalli, similar to the system in Pura village.

Photo 5.2 Producer gas system for electricity generation in Hosahalli village (South India) for lighting and shaft power application.

Photo 5.3 A very large alcohol-from-sugarcane plant which cogenerates heat and electricity from bagasse (Sao Paulo, Brazil).

Photo 5.4 Large-scale biogas digester using animal and other agricultural waste (Lemvig, Denmark).

6
Bioenergy—a modernized fuel option for India

The analysis of energy use patterns for India has shown the continued dominance of biomass, particularly in rural areas. Biomass use is usually characterized by low efficiency of use and lower quality of life. In order to promote development it is necessary to provide quality fuels (liquid and gaseous fuels) and electricity in a sustainable, equitable, and affordable way. The main focus of this book is on exploring the potential of bioenergy in meeting the energy needs of developing countries such as India. This is attempted here by defining the goals of development, the services requiring energy, the forms of energy required, and analysing the potential of bioenergy in comparison with other options.

6.1 ENERGY AND DEVELOPMENTAL GOALS

6.1.1 Goals of development and energy planning

India is a large country (328 Mha) with a large population of 844 million (1991) which is projected to reach 1.4 billion by 2025 (UNPF 1992). The Indian economy is growing at moderate rates and is expected to grow at 5.6 per cent annually during 1992–97 (PC 1992). The goals of development during the Seventh Five-Year plan (1985–90) centred around eradication of poverty and employment generation. The Eighth Five-Year development plan for the period 1992–97 has the two following major goals for development (PC 1992):

(1) generation of adequate employment to achieve near full employment by the turn of the century;
(2) strengthening infrastructure (energy, transport, communication, irrigation) in order to support the growth process on a sustainable basis.

In many developing countries, conventional wisdom equated development with growth and further equated economic growth with energy consumption, especially of electricity and oil. By pursuing such a strategy in India power generation has increased fivefold from 56 TWh in 1970–71 to 300 TWh in 1993–94 (CMIE 1994). The growth rate in electricity generation during the 1980s was 9 per cent annually. However, this electricity growth has neither led to equitable development nor has all the electricity demand been met.

For example, the inequitable distribution of power is reflected in the extent of electrification of rural and urban households; 75 per cent of the population lives in rural areas. In urban areas over 92 per cent of households use electricity for lighting, whereas in rural areas only 27 per cent of households have an electricity supply (NSS 1992; TERI 1992*b*). The number of rural houses electrified is as low as 4 per cent in Bihar and 9 per cent in Uttar Pradesh (NSS 1992). Electricity use varies from 631 kWh/capita/yr in Punjab to 102 kWh/capita/yr in Bihar state (CMIE 1992). A further example of inequity is that nearly all the LPG used for cooking in India is in urban areas since no supply network exists in rural areas, even though richer rural households are prepared to buy LPG. The poorer sections of the population, particularly in rural areas, continue to use low-quality fuels and low-efficiency devices. The general conclusion of studies on rural electrification is that it has accentuated inequity (Ranganathan 1993).

Conventional wisdom in energy planning has led to the neglect of the crucial aspects of the lives of the rural and urban poor, their basic human needs (drinking-water and cooking-energy), their settlements (villages and slums), their fuels (fuelwood, crop residues, dung), the end uses (cooking and lighting), and their energy-using devices (cooking stoves, kilns). Instead, the energy supplies of the élite—oil and electricity—were overemphasized (Goldemberg *et al.* 1988). With this in mind, the focus of this book is on rural energy needs.

The conventional approach in energy planning has also resulted in environmental degradation, such as indoor air pollution (due to fuelwood use in cooking stoves), local and regional air pollution (from coal-based thermal power-plant emissions and automobile exhausts), waste disposal (e.g. coal ash), deforestation or forest degradation (dependence on fuelwood and inefficient stoves), and global problems (such as greenhouse gas build-up leading to global warming). Furthermore, excessive dependence on conventional energy sources (particularly oil and coal-based electricity) is considered non-sustainable. Thus there is a growing realization that the conventional supply-focused approach to energy planning is failing on economic and environmental grounds (Reddy *et al.* 1991).

Change is, however, occurring slowly since, starting with the Eighth Development Plan (1992–97), there seems to be a new perception among planners in India. Energy planning is considered at three levels: short, medium, and long term. The goals are given in Box 6.1 (PC 1992). Some of the features of the goals of planning of interest to the focus of this book are:

1. Energy efficiency; the intention of the planners is to concentrate on improving energy efficiency in the short term.
2. Basic energy needs of rural and urban poor; it is proposed to promote equitable distribution of benefits especially favouring the rural poor.
3. Renewable sources of energy; to accelerate development of renewable

Box 6.1 Goals of energy planning of the government of India as given in the Eighth Five-Year Development Plan (PC 1992)

Short-term goals

1. Maximize returns from the assets already created in the energy sector.
2. Initiate measures for reducing technical losses in production, transportation, and end use of all forms of energy.
3. Initiate action to reduce the energy intensity of the different energy-consuming sectors of the economy and promote conservation; demand management through appropriate organizational and fiscal policies.
4. Initiate steps for meeting fully the basic energy needs of the rural and the urban households, so as to reduce the existing inequities in energy use.
5. Maximize satisfaction of demand for energy from indigenous resources.

Medium-term goals

1. Initiate steps towards progressive substitution of petroleum products by coal, lignite, natural gas, and electricity, so as to restrict the quantum of oil imports to the current level.
2. Initiate action for accelerated development of all renewable energy resources, especially the available hydroelectric potential.
3. Promote programmes to achieve self-reliance in the energy sector.
4. Promote R & D effort on decentralized energy technologies based on renewable resources.
5. Initiate appropriate organizational changes in the case of different energy sub-sectors consistent with the overall energy strategy.

Long-term goals

1. Promote an energy supply system based largely on renewable sources of energy.
2. Promote technologies of production, transportation, and end use of energy that are environmentally benign and cost efficient.

sources of energy in the medium term and to promote an energy system based on renewable sources in the long term.

4. Environmentally sound technologies; in the long term, at least, it is proposed to promote environmentally benign technologies. Sustainability of energy sources is the main focus of all future energy programmes.
5. Participation of local communities and self-reliance; an important feature of all future development, starting from the Eighth Plan, is the proposal on decentralization of decision-making to ensure participation of local communities.

To summarize, the long-term goal would be to meet the energy needs of development (especially those of the rural and urban poor), to promote energy efficiency, the deployment of environmentally sound energy technologies (particularly renewable energy sources), and the promotion of community participation in the management of energy systems.

6.1.2 Intention versus action of the planners

Very often the intentions of planners are genuine but their actions are different due to various factors. It is therefore necessary to understand the attempts of conventional planners in translating intentions into action. Consideration of a few parameters is necessary to test their resolve.

Share of renewables New power generation capacity, to be added from conventional sources during the Eighth plan (1992–97), is projected to be 30 538 MW. Coal-based thermal power is expected to account for 66 per cent. In comparison, the target for renewable energy is only 750–1000 MW (PC 1992). However, with the upgrading of the Department of Non-Conventional Energy Sources to a Ministry, the targets have been increased to 2000 MW. One can view this positively or negatively. Positively, because if the 2000 MW level is achieved by 1997, it would give a large impetus to the medium- and long-term shifts to renewable energy technologies. Negatively, because even though most of the renewable energy technology programmes were initiated during period of the Seventh plan period 1985–90, the contribution of renewable energy technologies was still marginal during that period. One could also pessimistically suggest that even during the post-1997 period, the role of renewables will be marginal.

Funding allocation Out of the total projected allocation of Rs1155 billion for the power sector, the allocation for non-conventional energy sources (mainly renewable energy sources) is only 1.2 per cent (PC 1992). Programmes such as biogas and improved stoves, which are targeted towards rural areas and the rural poor, receive only a part of this 1.2 per cent. The broad goals of energy planning, which aim at shifting to renewable energies and meeting the energy needs of the poor, remain only as good intentions. However, this could alter with the changed perceptions of energy planners at national and global levels regarding issues related to local air pollution, deforestation, land submergence, displacement of people, and climate change. The situation may also change due to reallocation of international assistance favouring energy-efficient and renewable energy technologies, not only because of environmental concerns but also because renewables are likely to become financially competitive with conventional energy technologies (Jackson 1992; WB 1992).

6.2 RURAL ENERGY

The rural population accounts for 75 per cent of India's population and is projected to be 64 per cent by 2011 (TSL 1992). India will continue to be dominated by the rural population. The energy needs of rural areas are different from those of urban or industrial centres, and some of the features of rural energy are listed below.

6.2.1 Features of rural energy

Small human settlements

Rural communities basically comprise small human settlements (nearly 557 000 villages) with about 73 per cent having fewer than 1000 people (or 200 households; NIRD 1991). The energy requirement, or the load factors, of villages are small compared to urban and industrial centres, leading to a high cost of transmission and high transmission and distribution losses.

Geographically dispersed villages

The half million villages are geographically well spread out in India. The houses in each village are in clusters in the plains, and are dispersed in hilly or forested areas (such as the Western Ghat mountain ranges). In order to provide electricity, transmission lines have to criss-cross the landscape, entailing high costs. It is quite common to find power transmission lines that are several hundred or thousand metres long providing electricity for domestic lighting to just a single or a few houses with a load of a fraction of a kW. In comparison, urban centres and industries have large, concentrated power demands.

Dominance of cooking and agricultural pumping

As seen in Chapter 2, cooking is the dominant heating energy requirement and pumping irrigation water is the dominant electricity or shaft power requirement.

Geographically dispersed locations of irrigation pumps Since the pumps are connected to open wells and tube wells, it is common in rural areas to see transmission lines of 1 to several kilometres for energizing one 5 kW irrigation pumpset.

Seasonality of loads During the peak monsoon months, the power requirement for agriculture drops to low levels and then peaks in summer and premonsoon months.

Low purchasing power

Landless and marginal farmers (owning <1 ha of land) account for over 50 per cent of the rural population (NIRD 1991) and their purchasing power is low. Thus the incentive for electricity utilities to connect individual houses is low. The poorer households cannot afford upfront costs for obtaining electricity or LPG supplies.

Absence of infrastructure

In rural areas there is a limited infrastructure for supplying modern energy carriers and devices and for their maintenance. For example, LPG use in cooking is wholly restricted to large urban centres. However, maintenance facilities exist for diesel engines and irrigation pumps at the small town level since they are dispersed all over the country.

Low efficiency of use of fuels

The efficiency of biomass use for cooking is low (Fig. 2.6). Kerosene use for lighting is associated with a poor-quality light output. The luminous efficiency of a kerosene wick lamp is 0.1 lumens/W, compared to 12 lumens/W for a 60 W incandescent bulb (ADB 1993). The efficiency of agricultural pumps has also been shown to be low. A complete retrofitting of irrigation pumps is estimated to be able to save 47 per cent of electricity (ADB 1993).

These features mean that rural energy needs have thus to be considered differently from the needs of urban and industrial centres, where there is a large, concentrated demand for electricity and petroleum fuels, backed up by purchasing power.

6.2.2 Current and projected energy use

The major sectors using energy in rural areas are domestic, agriculture, lighting, and rural industries. The major activities requiring energy in any typical village can be categorized as:

(1) heating at low (heating bath-water), medium (cooking), and high temperatures (kilns);
(2) lighting (domestic), streets and community facilities;
(3) stationary power (water pumping, flour milling, agro-processing);
(4) mobile power (agricultural operations);
(5) transport (agricultural and housing materials).

A list of end uses of energy is given in Box 6.2. The extent of energy use in selected villages and at national level was discussed in Chapter 2. It showed that in the six village examples, biomass sources accounted for 79–99 per cent of the village energy supply (Table 2.6). After biomass, kerosene is the fuel used in all villages for lighting, whether electrified or not. Electricity is used to varying extents for lighting and pumping irrigation water in all electrified villages (Table 2.21). Diesel is used in significant quantities in non-electrified villages (Maulaganj) as well as electrified villages (Dhanwas) for pumping

Box 6.2 Existing sources, end uses, and devices of energy in rural areas (Reddy and Ravindranath 1987)

End use	Source of energy	Devices	Remarks
I Heating			
Low temperature			
Water heating	Fuelwood, crop residues	Traditional stove	Low thermal efficiency, fuelwood becoming scarce
Medium temperature			
Cooking	Fuelwood, crop residues, dung	Traditional wood stove	Low thermal efficiency, long cooking time, smoke-related health hazard, cooking fuels (wood) becoming scarce
High temperature			
Brick industry	Fuelwood, rice husk	Kilns	All these traditional systems have a low thermal efficiency; fuelwood is becoming scarce; residues may have other valuable uses
Lime burning	Fuelwood	Kilns	
Pottery	Fuelwood	Kilns	
Tiles	Fuelwood	Kilns	
Tobacco curing	Fuelwood	Barns	
Jaggery processing	Surgarcane bagasse, fuelwood	Traditional stove	
Bread baking	Fuelwood	Ovens	
II Stationary power			
Pumping	Electricity/diesel	Engines	Electricity shortage; large T & D losses; high transmission line cost; low load factors
Flour mill	Electricity	Engines	
Crushing oil seeds and sugarcane	Electricity/animal	Engines/ crushers	Diesel, a non-renewable imported source
III Lighting			
Lighting	Kerosene	Open-wick lamps	Low luminous efficiency, supply unreliable and expensive
	Electricity	Incandescent lamps	Low luminous efficiency; high transmission costs and T & D losses

IV Mobile power			
Agricultural operations	Animal	Animal-drawn implements	Adequate for single cropping but inadequate for multiple cropping; low efficiency of energy conversion
Transport	Animal	Animal cart	Suitable for short-distance transport

T & D, transmission and distribution

water and post-harvest operations (threshing) (Tables 2.6 and 2.21). Total energy use at the national level in rural areas is:

Wood	= 187 Mt (Table 2.2)	Electricity	= 67.0 TWh (Table 2.22)
Crop residues	= 40 Mt (Table 2.2)	Diesel	= 4.7 Mt (Table 2.22)
Dung cakes	= 32 Mt (Table 2.2)	Kerosene	= 3.4 Mt (Table 2.22)

Several studies have made future energy projections based on various assumptions and methodologies. Some projections are given in Table 6.1 to indicate the ranges involved. By the turn of the century electricity and oil requirements are projected to be 517 TWh and 102 Mt, respectively (CMIE 1994).

According to the projections by ABE (1985), for 2004–5 the fuelwood requirement is estimated to be nearly 330 Mt, resulting in severe scarcity so that much more dung may have to be burnt. Under the 'business as usual' scenario, assuming a 5 per cent GDP growth rate, the demands for all fuels are projected to increase at least threefold by 2009–10 (TERI 1992*a*).

Given the current financial problems, it is not clear how India will be able to finance power projects and large imports. Electricity scarcities and foreign exchange problems are projected to continue (PC 1992). Mere enhancement in the supply of fossil fuels and electricity is unlikely to change the current inequitable energy distribution. The dominance of biomass energy in rural India is projected to continue, along with its low efficiency of use and the resulting poor quality of life to the rural population. There appears to be a need for an alternative approach, which will result in the energy needs of rural areas being met in a sustainable and equitable way.

6.2.3 Energy needs and bioenergy options

Wood (for cooking), kerosene and electricity for lighting, and diesel and electricity for shaft-power needs together account for 85 per cent of village energy use in the case of Ungra and 84 per cent in BNPura, when all energy sources are included (including animal and human) (Table 2.6). Our subsequent discussions on bioenergy options will only include heating, stationary shaft power, and lighting energy requirements of rural areas.

Table 6.1 Some projections of energy demands for India

	Current consumption level (1991–2)	1996–7	2000	2004–5[a]	2009–10	2009–10[b]
Fuelwood (Mt)	227	–	–	330	–	–
Coal (Mt)	217	311	–	472	648	626
Electricity (TWh)	287 [78][c]	433	517 (1999–2000)	501–592	190[c]	217[c]
Oil (Mt)	54	81	102 (2001–2)	90–110	175	186
Source	Table 2.1	PC (1992)	CMIE (1994)	ABE (1985)	ADB (1991)	TERI (1992*a*)

[a] 5% GDP annual growth rate assumed.
[b] Assuming business as usual.
[c] Only hydro- and nuclear electricity included.

Animate energy and mobile power.

The motive power requirements of agricultural operations and rural transportation are met largely by draught animal power. Animal energy systems are complex and there is inadequate understanding of the desirability of substitution of animal power by mechanical power. There are nearly 77 million draught animals in India (CMIE 1992). Draught animal energy availability and use is linked to fodder production and availability, milk and meat production, seasonality of power requirement, and labour employment. These factors are part of a complex village ecosystem which is usually self-reliant (Reddy 1981).

1. Bullocks or cows used as draught animals are bred and reared in the village ecosystem itself. Any other options would involve large investments and import of machinery. Often milch animals (cows) are used for draught power to save investment on bullocks.
2. Fodder energy inputs for draught animals come from within the village ecosystem as a by-product of crop production.
3. The dung output is valuable manure input for crop production.
4. Animal energy is a renewable and sustainable local source of energy.
5. An attempt to displace animal energy with mechanical power may lead to unemployment, especially in the dominant rain-fed agriculture areas where single cropping is practised. Thus the employment aspect is critical to animal energy systems.

The analysis of alternatives to animal energy is a complex ecological and socio-economic issue due to the above factors. Thus, animal energy issues are left aside from the scope of the main focus of this book.

Base year for the analysis

The latest year for which most data are available is 1991, which is the year used for estimating the energy, investment requirements, and the potential for bioenergy. The population census data refer to the same year. The Eighth Five-Year development plan in India began in 1992 and a review of all programmes is available up to the end of the financial year 1991–92.

6.2.4 Energy end use approach

Approach

In considering the potential of bioenergy, six village case studies (Chapter 2) are used, since these represent some geographical and socio-economic situations. These cases are supplemented by other village case studies in different sections, depending on the availability of information. Finally, national-level projections are made using the size of the village grouping criterion, and the national population and land-use statistics.

Energy end use approach

The conventional approach of energy planners is to estimate the current consumption levels and growth rates, make population projections, assume a different (usually higher) GDP growth rate and GDP elasticity of energy demand, and then plan for enhancing the energy supply. This approach has indeed led to a growth in the rates of energy supply but not in the delivery of energy services to the dominant end uses (such as cooking, lighting, and pumping drinking-water) and sections of society (rural communities and even the urban poor). No significant improvement in the quality of life of the rural population has resulted from such an approach. Alternative approaches are now being debated, such as the end-use-oriented approach (Goldemberg *et al.* 1988; ADB 1993) and the DEFENDUS (development-focused end-use-oriented) paradigm (Reddy *et al.* 1991). The DEFENDUS approach to energy demand and supply focuses on people-based development by identifying technological opportunities through an analysis of the end uses of energy and adhering to a least-cost approach to the planning of energy supplies (Reddy *et al.* 1991). The DEFENDUS, as well as other end use approaches, aims at meeting a defined set of energy services, which promote development through a least-cost set of energy technology packages. The DEFENDUS approach to the economy as a whole would involve issues such as load management, structural shifts in industry and transport, changes in transmission and distribution systems, linking with national grids, and so on (Reddy *et al.* 1991).

In this book, a modest attempt is made to adopt the energy end use approach basically for providing energy services to rural areas which have distinct features (section 6.2.1). The focus of this chapter is limited to exploring the potential and feasibility of bioenergy options in meeting rural energy needs. Urban, industrial, and transportation energy requirements are not considered. Meeting certain rural energy needs through alternative energies such as bioenergy would release some of the commercial energy presently used in rural areas. Rural energy needs should not be considered a burden by energy planners or electricity utilities. Provision of quality energy services to the rural population must receive as much, or more, attention from energy planners than urban and industrial needs, since 75 per cent of the population lives in rural areas and the majority of them are poor.

In rural areas there is significant potential for improvements in efficiency of energy use (biomass, kerosene, and electricity). Increases in energy efficiency alone will not lead to improvement in the quality of life of the rural population as their current energy use levels are very low and they depend on low-quality fuels. Instead, the energy services or end use requirements must be met by a shift to higher-quality fuels, such as liquid and gaseous fuels, and electricity, which should be used efficiently. Thus we adopt an end use analysis approach, where the potential of bioenergy options in meeting the energy needs of development is analysed. It is proposed to consider the bioenergy option in the context of the medium- and long-term goals of energy planning adopted by the Planning Commission (Box 6.1).

All the activities requiring energy in rural areas, the various energy options available, and their environmental and economic impacts are considered. We explore the possibility of making a single village or a cluster of villages self-reliant with respect to most of their energy requirements in a sustainable way. A previous study of a South Indian village has shown the potential of options based on biomass for meeting all the heating, lighting, and stationary shaft-power needs in order to promote the quality of life (Ravindranath and Chanakya 1986). The Rural Energy Centre concept currently gaining some acceptance in South India also has biomass energy as the focus for meeting the main energy requirements of villages (Rajabapaiah *et al.* 1993; Ravindranath 1993).

Rural energy needs or services of the household sector (cooking, lighting, drinking-water supply), shaft-power uses (pumping irrigation water and agro-processing industry), and process heat for agro-processing dominate rural energy requirements and pertain to the survival of the rural communities and their quality of life. These energy needs, especially household energy problems, are reflections of poverty and lack of investment in the poor. They persist not because they are insoluble but because they have not been addressed with sufficient commitment and vigour (Leach 1987).

6.3 COOKING

As seen in Chapter 2, cooking is the dominant energy-using activity in rural areas, since every rural household requires energy to cook each day, with direct implications for quality of life and environment. The goal should be to provide quality fuels to all people, especially the poor in rural areas, in a reliable, affordable, and sustainable way. Before considering various options it is useful to understand the impact of income on fuel choice in urban areas, where a set of options is already available. This may be an useful indicator of the choice of options, when they become available to rural households as the quality of life improves with increase in income. An urban survey of fuel choice is given in Fig. 2.7. From this it is possible to construct an energy ladder (Fig. 6.1), from least to most preferred fuel. An energy ladder constructed by Smith *et al.* (1994) also exhibits a similar trend. Thus households would usually prefer to switch fuels as follows:

Dung cake → fuelwood → kerosene → LPG → electricity

Solid fuels → liquid fuels → gaseous fuels → electricity

Even among solid fuels there is a preference, as observed in many locations (Chapter 2). If fuelwood is available, then crop residues or dung cakes are not preferred as fuel.

Dung cake → crop residues → twigs of trees → cut wood → charcoal

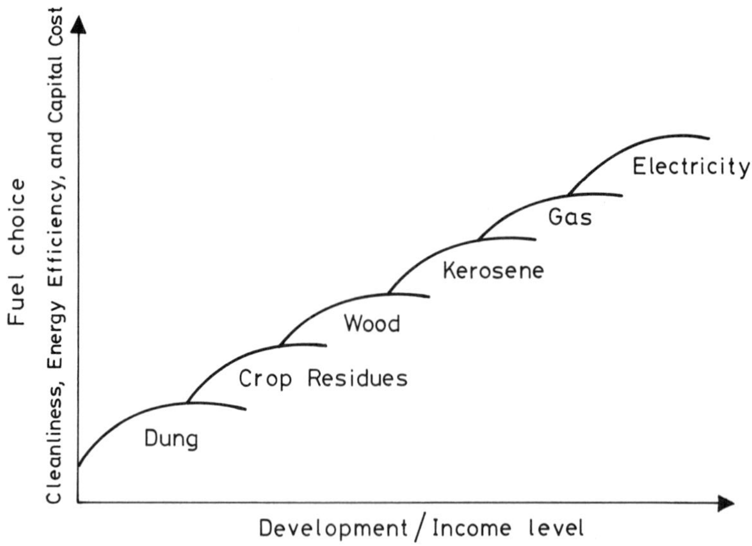

Fig. 6.1 Energy ladder typical in southern Asia indicating the shifts in choice of fuel with increase in income/development level (Smith *et al.* 1994).

Due to the high cost of charcoal in India, its market has disappeared even in urban areas, and there are also restrictions on charcoal manufacture. Among the other fuels, cut wood is the most preferred solid fuel; due to its high density it requires less tending by the housewife during combustion compared to twigs and crop residues. In rural areas in the absence of an LPG network and with electric power shortages, currently the following choices are available.

Dung cake → crop residues → fuelwood → kerosene → biogas

The choice of a fuel and conversion device depends on their suitability, the access to the fuel, and the cost of the fuel and device (Dutt and Ravindranath 1993). In this chapter we will consider the potential for shifting to bioenergy options and commercial fuels, and the first step is to estimate the requirement of different fuels.

6.4 FUEL REQUIREMENT FOR DIFFERENT OPTIONS

To obtain a complete picture of the cooking energy needs in different parts of India, it is necessary to conduct controlled cooking tests in different food zones. Cooking practices are not uniform and, for example, in Karnataka state alone three food zones have been defined, namely rice, ragi (finger millet), and jowar (sorghum) zones (Fig. 6.2). The details of cooking practices are given in Table 2.9.

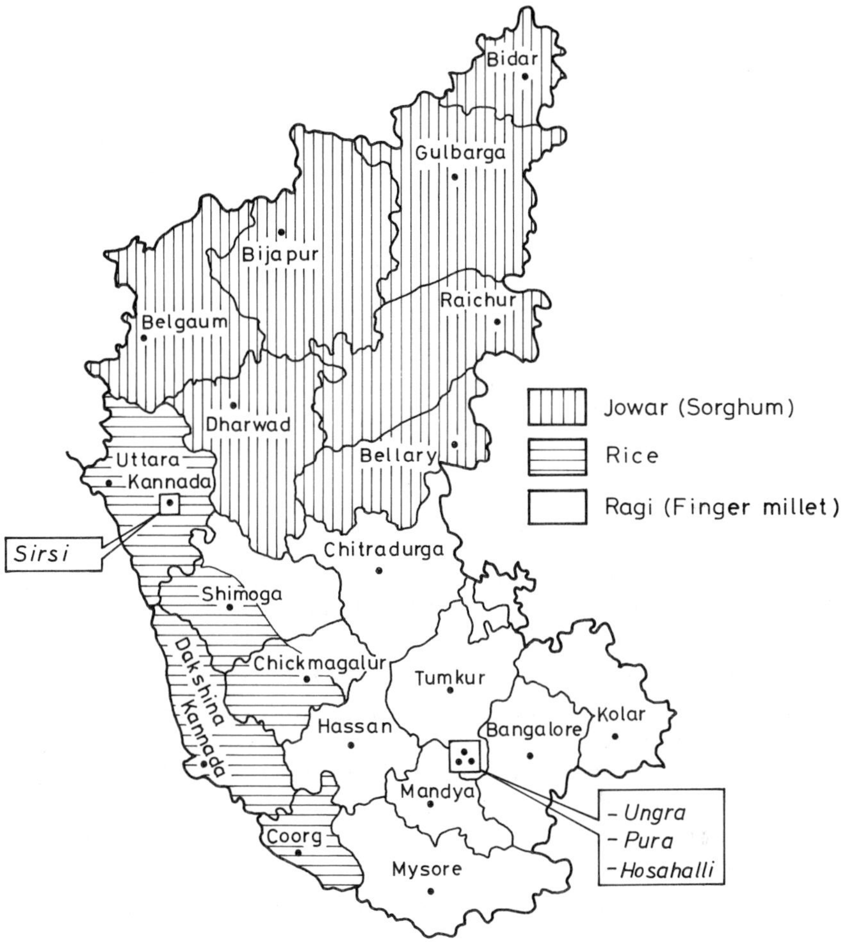

Fig. 6.2 Three food zones of Karnataka state in South India. Food zone indicates the geographical area with homogeneous cooking practices. Location of Ungra, Pura (community biogas system), Hosahalli (energy forest producer gas system), and Sirsi.

In the sorghum zone, cooking mainly involves baking the *roti* (flat bread) on a low heat for long periods. In contrast, in the ragi zone cooking involves bringing the contents of the vessel to a boil and letting it simmer for some time. Given such clear differences in cooking practices, the following method is suggested for estimating the cooking fuel requirements (Ravindranath and Ramakrishna 1994):

1. Stratify rural and urban India into major food zones based on staple cereals and main dishes cooked—at most 8–10 zones.

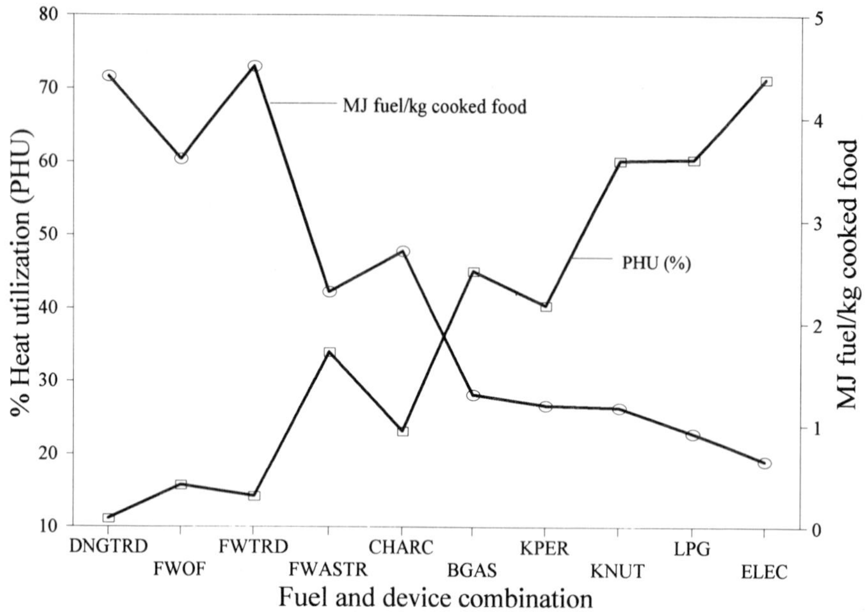

Fig. 6.3 Thermal efficiency (PHU; percentage heat utilization) and fuel used per kilogram of cooked food for different fuel–device combinations (Dutt and Ravindranath 1993). Fuel and stove combinations are: DNGTRD, dung traditional; FWOF, fuelwood open fire; FWTRD, fuelwood traditional; FWASTR, fuelwood ASTRA; CHARC, charcoal; BGAS, biogas; KPER, kerosene perfect design; KNUT, kerosene nutan design; LPG, liquid petroleum gas; ELEC, electric hot plate.

2. For each zone define the standard meal: typical family size, diet, cooking-vessel shape and size, sequence of cooking dishes, quantities cooked.
3. Select all feasible fuel–device combinations.
4. Conduct thermal efficiency and controlled cooking tests.
5. If possible, examine a full day's requirements including main meals, snacks, and tea preparation and repeat at least three times to obtain a good mean.

The above procedure was adopted for a typical meal in Ungra village in the semi-arid Tumkur district (ragi zone) of South India. The results of controlled cooking tests and thermal efficiency tests using different fuels and devices are given in Fig. 6.3

Efficiency tests and specific fuel consumption (SFC)

Thermal efficiency (percentage heat utilization, PHU) tests were conducted and the results were discussed in Chapter 2. Assuming that the PHU test may not reflect actual cooking conditions, controlled cooking tests were conducted. The energy use in terms of SFC in MJ/kg of food cooked was compared with the PHU values (Fig. 6.3). The results of PHU and controlled cooking tests show that there is a strong relationship between the two. The results show a high correlation coefficient of −0.926. However, it is difficult to make any suggestions as to fuel choice or quantities required using the findings of PHU tests alone.

The comparative controlled cooking tests help in estimating the per capita requirement of different fuels for a given population in the food zone. Using the findings of the Ungra study, the requirement for different fuels for cooking can be estimated on a per capita basis in physical and energy units (Fig. 2.1). The efficient ASTRA stove requires only 54 per cent of the fuel used in a traditional wood stove. Among all the solid fuels, the ASTRA stove requires the least energy source or wood (6.4 MJ/capita/day), while if the fuel source is kerosene or LPG, the fuel requirement is 3.2 or 2.4 MJ of fuel/capita/day, respectively. Thus it is possible to estimate the quantity of different fuels required for the village or for the population in the cooking zone.

Estimation of fuel required for selected villages and at the national level

The fuel options considered for rural areas are efficient stoves using wood, kerosene, biogas, and LPG. Electricity is not included as an option for cooking in rural areas because:

1. The usage of electricity for cooking should be discouraged on 'thermodynamic principles' according to ABE (1985).
2. Although the efficiency for electricity at end use level is high, the actual efficiency would be low if the efficiency of conversion of coal to electricity is included. Efficiencies would be even lower if the transmission and distribution losses are included.
3. Changing to electricity would require large investments at the household level in order to shift from lighting to heating cables.
4. Provision of electricity to all houses in rural areas would require large investments in the transmission and distribution systems, including changing transformers, capacitors, and cables.
5. There is the problem of reliability. If there is damage to a transformer or to a transmission line, all the connected villages would lose power. Such instant disruption is a problem when cooking with electricity and does not happen with any other fuel.
6. At the power requirement of 225 kWh/capita/year (180 kWh + 25 per cent added for cooking snacks, coffee, etc., from Fig. 2.1), the total power requirement for cooking in rural areas alone would be 141 TWh (at the 1991 population level) compared to the total power generation of 287 TWh in 1991–92, and thus may not be a feasible option.

The use of dung cakes is also not considered as an option due to:

1. loss of the nutrient value of dung;

2. the human effort required for preparing dung cakes, and higher emissions of pollutants;
3. the need for continuous attention on the part of the housewife when using dung cake in cooking;
4. the potential for conversion to biogas.

Thus kerosene, biogas, and fuelwood are the fuels we use for estimating cooking energy requirements. Biomass-based liquid fuel (methanol) is also included to get a comparative estimate of requirement.

6.5 BIOGAS OPTION FOR COOKING

Biogas can be used as a fuel for cooking and to power internal combustion engines for shaft-power applications or for electricity generation (Rajabapaiah *et al.* 1993). Thus we need to understand the advantages of using biogas for cooking rather than for other applications.

6.5.1 Rational for using biogas for cooking

Biogas is the only locally available, convenient, safe, environmentally sound, and sustainable gaseous fuel option for cooking. Although producer-gas could also be used, there are problems, such as the danger of exposure to carbon monoxide. Biogas is currently promoted as a fuel for cooking in India (MNES 1993*a*) with cattle dung as the feedstock, since the existing designs permit only the use of dung. Biogas would be the most preferred fuel for cooking in rural areas for the following reasons:

1. It is a gaseous fuel with the same convenience as LPG and thus leads to improved quality of life. The drudgery associated with solid fuels and smoke emissions is avoided.
2. It is a renewable resource, locally available, and the village household or community can become self-reliant in cooking fuel.
3. It reduces pressure on village trees and forests.
4. It leads to global benefits, such as reduction in CO_2 emissions, by avoiding fuelwood combustion, and it has a high efficiency of conversion.
5. Other bioenergy and renewable energy options are available for electricity generation at the village level (e.g. producer gas, solar photovoltaic (PV)).
6. The adequacy of biogas for electricity generation (which is limited by the availability of the dung resource) is discussed later in this chapter.

6.5.2 Options for using cattle-dung biogas for cooking

There are basically two routes for using biogas for cooking, namely individual family and community biogas plants, both of which are considered. It is necessary to estimate their potential in meeting cooking energy needs.

Family biogas potential

Data to estimate the potential are available for four villages in the semi-arid Tumkur district and three villages in the Western Ghats region of Karnataka state. It can be seen from Table 6.2 that in the semi-arid villages about 8–17 per cent of households have enough dung for a family-sized unit. In the high-rainfall regions the percentage of households with adequate dung for family biogas is in the range of 49–62 per cent. Thus there is a wide range of opportunity depending on the cattle population and its distribution.

Community biogas potential

If community biogas plants are installed in the same villages, to make full use of all the dung potential, what is the size of the population whose cooking energy needs can be met (Table 6.2)? It is interesting to note that in the semi-arid villages, the fraction of the population whose cooking requirements could be met with a community plant increases by 2–3 times. This indicates that many households own cattle but not enough for a family-sized biogas plant. In the heavy-rainfall region, the additional population whose cooking requirements could be met by installing community plants is only marginal.

The community biogas potential can be estimated for the six villages selected in this study (in Chapter 2) and, unfortunately, the distribution of the cattle population among the households is not available for an estimate of the family biogas potential. The percentage of the population whose cooking needs could be met through community plants ranges from 19 per cent in Maulaganj to 61 per cent in BNPura (Table 6.3). Thus village-level data from 12 villages (Tables 6.2, 6.3) in different regions show that a quarter to a half of the cooking energy needs of villages could be met through the community biogas option in the villages.

Biogas potential at the national level

Based on data on the distribution of rural households with five and six cattle, the potential for family biogas is estimated nationally to range from 12 to 17 million households (Table 6.2) and, correspondingly, the cooking energy needs of 12–17 per cent of the rural population (72–104 million) could be met. When the community biogas option is considered, assuming that one community plant is installed in every village, the cooking energy needs of 32 per cent of the rural population (or 200 million) could be met. Thus, with the adoption of a community biogas option, the cooking energy needs of an additional 96–128 million people could be met. An upper limit of one-third of the rural population's cooking energy need seems to be the potential of the biogas option (based on dung feedstock). Thus biogas based on animal dung alone will not meet all the rural cooking energy needs.

Table 6.2 Family and community biogas potential of cattle dung in villages of semi-arid (about 70 cm annual rainfall) and high-rainfall (about 300 cm annually) regions of Karnataka state

Village	No. of HH	Human population	Cattle population	Dung production (kg/day)[a]	Biogas production (m^3/day)[b]	Family biogas potential (no. of HH)[c]	Family biogas option[c] (% population covered)	Community biogas option (% population covered)
Semi-arid area								
Ungra	177	1047	478	2151	75	24 (14%)[d]	13	36
Suggenhalli	155	726	325	1463	51	13 (8%)	9	35
Pura	87	463	248	1116	39	15 (17%)	16	42
Hosahalli	43	191	127	572	20	8 (9%)	20	52
High rainfall								
Malenhalli	86	547	500	2250	78	44 (51%)	51	71
Tattikai	85	622	380	1710	60	42 (49%)	40	48
Sirsimakki	81	611	448	2016	70	50 (62%)	49	58
India[e](6)	104 M	627 M	257 M	1156 Mkg	40 Mm^3	12 M (12%)	12	32
India[f](5)	104 M	627 M	257 M	1156 Mkg	40 Mm^3	17 M (17%)	17	32

HH, households; M, million.
Source: villages of semi-arid and high-rainfall regions, Ravindranath *et al.* (1992*a*).
[a] Dung production: (cattle population × 4.5 kg dung/day).
[b] Dung yield × biogas production rate (35 l/kg/day of dung) = biogas production.
[c] Households who own adequate cattle to provide 200 litres biogas/capita/day are taken for estimating family biogas potential. Biogas requirement taken at 200 l/capita/day for cooking (from Dutt and Ravindranath 1993).
[d] Figures in parentheses: % of village HH whose cooking energy needs can be met from biogas.
[e] India(6) = considering HH with six head of cattle will have adequate biogas for a family size of 6.
[f] India(5) = considering HH with five head of cattle will have adequate biogas for a family size of 6.
Number and (%) of rural HH owning five and six head of cattle taken from ABE report (1985) and number of rural HH taken from NIRD (1991).

Table 6.3 Community biogas (CBG) potential in the selected villages

Villages	No. of cattle[a]	Dung production[b] (t/day)	Biogas production[c] (m^3/day)	% Population whose cooking needs can be met through CBG
Ungra	478	2.15	75	36
BNPura	463	2.08	73	61
Dhanwas	406	1.82	63	38
Nanuda	285	1.28	45	27
Maulaganj	274	1.23	43	19
Bishnapur	615	2.76	96	51

[a] Cattle population from respective studies for the villages (from Chapter 2).
[b] Dung production = cattle population × 4.5 kg fresh dung/animal/day.
[c] Dung production × 35 litres of biogas/kg of dung.

Community or family biogas option?

If we conclude that all the cooking energy needs cannot be met, what should be the strategy for using the biogas option for cooking. Given the suitability of biogas fuel for cooking, it is desirable to use all the potential for cooking. In India both family and community biogas programmes have been implemented and their performance was discussed in Chapter 4. The experience in India shows that the community biogas option for cooking has generally failed, due to shortage of dung and organizational problems. The performance of family biogas cooking is better, with an estimated 60–77 per cent of the plants being in use at the national level (MNES 1993*a*).

The merits of family and community biogas options are listed in Box 6.3. The major factors in favour of the community biogas option are the economy of scale and the possibility of doubling the percentage of population provided with cooking fuel. The poorer households with fewer cattle would not benefit even if the community biogas option is adopted, as has been seen in several village case studies. It is widely recognized that there are social and organizational problems in introducing community biogas plants (Table 4.2). One of the major factors that probably contributes to the social problem is the inability of the community biogas option to meet the cooking energy requirements of the whole village. If adequate dung was available, the social problems would certainly have been easier to manage. Otherwise the question is why should all cattle-owning households in the village contribute dung if only a few (the affluent one-third) in the village benefit? In contrast, in Pura village all the households willingly contribute cattle dung to community biogas, since the power generated is used to pump drinking-water to benefit all the households in the village (Chapter 5).

Box 6.3 Merits and demerits of cattle-dung family and community biogas options

	Merits	Demerits
Family biogas option (FBG)	No social or organizational problems Private initiative and benefit could ensure better performance of plant If cost reduction is achieved through R & D and technology transfer network, all FBG potential could be tapped	High first cost, thus limited potential for large-scale use Only rich farming households benefit from the programme, thus promoting inequity All the dung potential for biogas cannot be utilized. Households with inadequate dung will be left out
Community biogas option (CBG)	Full potential of dung feedstock available in the village could be used for biogas production Economy of scale operates and thus it is economically more feasible than the family system It is relatively more equitable since even poor households with one or two cattle can contribute and benefit by obtaining some form of compensation for contributing dung Biogas can be used for cooking or electricity generation, though only part of the requirement of either could be met	The needs and priorities of different houses may differ Inadequate dung in most villages, thus only a part of the village cooking energy needs could be met, leaving the rest to depend on other options Involves social and organizational problems due to inequities in ownership of cattle and inequitable flow of benefits Requires strong local institutions to manage the system

6.5.3 Community biogas potential—leaf litter plus cattle dung

Since the potential based on cattle dung is limited (Tables 6.2, 6.3), there is a need to consider alternative feedstocks to augment biogas production. Plant leaf biomass contains cellulose (and is low in lignin) and if subjected to anaerobic fermentation, produces biogas. The biogas production from typical leaf biomass is in the range of 60 l/kg of fresh leaves (Chanakya *et al.* 1993). The technology is available for fermentation of solid leaf litter. Large biogas plants in Denmark and China (Chapter 5) also use organic feedstocks other than cattle dung, particularly domestic and agro-industrial wastes.

Leaf biomass is widely available from forests, tree plantations, trees on farms, crop residue, and as weeds (shrubs and herbs) from grasslands or degraded lands in villages. One may argue that leaf litter from crop fields and forests should not be removed as it leads to depletion of soil nutrients by disrupting nutrient cycling. This does not, however, occur if the enriched slurry after the anaerobic digestion process is returned to the area from which it was collected. The question also arises as to the quantity of leaf litter required and its availability in typical villages. Among the types of leaf biomass available in any typical village, the crop residues of cereals are used as cattle fodder and pulse residues are partly used as fodder. The remaining litter available is not a large amount, given that all the leaves are shed in the field before harvest; also ligneous litter, such as that of pigeon pea, cannot be used as feedstock. Oilseed crop residues (cotton, castor, niger) are also ligneous. Thus the availability of cereal, pulse, and oilseed residues is limited and seasonal. The grass produced in village commons is usually fully grazed by livestock since the livestock density is high. Thus grass biomass would be unavailable as feedstock. The human effort required to collect shrub, weed, or herb litter would also be very high since its availability is low due to the grazing pressure and seasonality.

The availability of leafy feedstock should be year round, it should be available from a captive area to reduce the human effort required in collection, and it would be best if it is a by-product of some other activity so that the cost of gathering would be reduced. The approach adopted here is based on using the by-product of wood feedstock grown in dedicated energy plantations for producer-gas electricity generation (described in sections 6.11 and 6.12), where it is shown to be a feasible option for meeting the electricity requirement of villages. The unused leaf material could be used for biogas production. The digested leaf biomass slurry would be returned to the plantations. Here four village case studies are considered to estimate the biogas potential using leaf litter from such local bioenergy plantations along with cattle dung (Table 6.4). Through this concept of integrated energy forests for producer gas and biogas, the human effort required for collecting leaf biomass could be reduced, since the leafy material is removed when preparing wood for combustion in gasifier reactors.

The mean leaf litter production from plantations is estimated to be 6 dry t/ha/yr (Garg 1992; Singh *et al.* 1992). It is suggested that one-third of the litter be left in the plantation for *in situ* decomposition and only two-thirds, or about 4 dry t/ha/yr, be removed as feedstock for biogas production. Also, it may not be feasible practically to collect and transport 100 per cent of the leaf component of twigs and branches harvested for gasification.

The potential of leaf litter biogas could be considered by using an energy plantation area at a woody biomass productivity of 4 and 8 t/ha/yr for biomass electricity generation. If the higher woody biomass productivity of 8 t/ha/yr and the area required to be under energy plantation for electricity is assumed, it is possible to meet the biogas cooking energy requirement of 68 per cent

Table 6.4 Biogas potential of cattle dung and leaf litter (from producer-gas energy plantation) for meeting the cooking energy requirements of four sample villages and the national situation (land area under energy plantation estimated assuming a wood productivity of 8 t/ha/yr, from Table 6.8)

	Ungra	Nanuda	BNPura	Maulaganj	National level
Population	1047	838	593	1143	627 million
Biogas potential of dung (m^3/day) (Tables 6.2 and 6.3)	75	45	73	43	40 Mm^3
Land proposed for energy forest for producer-gas electricity @ 8 t/ha/yr (ha) (Table 6.8)	33	24	19	34	15 Mha
Leaf harvest potential of land under energy forest[a] in t/yr (fresh)	660	480	380	680	300 Mt
Biogas from leaf litter[b] (m^3/day)	108	79	62	112	49 Mm^3
Total biogas from leaf biomass from energy forest + dung (m^3/day)	183	124	135	155	89 Mm^3
% of population whose cooking energy needs can be met from biogas from dung + leaf biomass[c]	87	74	100	68	71%

[a] Leaf litter production dry t/ha/yr, Singh *et al.* (1992): *Shorea robusta*, 6.9 t/ha/yr; *Tectona grandis*, 7.7 t/ha/yr; *Eucalyptus*, 6.5 t/ha/yr; *Populus deltoides*, 5.3 t/ha/yr. Garg (1992): *Acacia nilotica*, 5.6 t/ha/yr; *Prosopis juliflora*, 6.1 t/ha/yr; *Dalbergia sisso*, 4.9 t/ha/yr; *Terminalia arjuna*, 5.3 t/ha/yr.
Leaf biomass harvest potential = mean yield of 6 dry t/ha/yr or 30 fresh t/ha/yr. Only two-thirds of 6 dry t/ha/yr (= 4 t/ha/yr) is taken as available for biogasification.
[b] Biogas production from fresh leaf biomass = 60 l/kg fresh leaves (Chanakya *et al.* 1993).
[c] Biogas requirement 200 l/capita/day for cooking.

(Maulaganj) to 100 per cent (BNPura) of the village population (Table 6.4). At the national level about 71 per cent of the population could be covered under the community biogas option. However, if the land to be dedicated for energy plantation at only 4 t/ha/yr is assumed for biomass electricity generation, the cooking energy requirements of all the population in all the villages could be met from leaf biomass (from energy plantations) plus cattle-dung biogas production. In fact in this case there would be surplus biogas for other end uses, after meeting the cooking energy requirement. Other herbaceous biomass (such as weeds and crop residue) could also be used as feedstock for biogas production.

A leaf-biomass-based plant by itself may not be feasible as villages may not have any forest land and the human effort required to collect and transport several hundred kilograms of low-density leaf litter every day from any dispersed source would be high and expensive. By using the energy plantation for biogas as well as producer-gas, the feasibility of both could be greatly improved. Thus technically it is possible to provide gaseous fuel for cooking in rural India along with the attendant social and environmental benefits. This approach of combining producer-gas and biogas systems illustrates the need for considering all the village energy needs together in an integrated way, rather than viewing the energy needs of each activity separately.

6.5.4 Producer-gas for cooking

Producer-gas (a mixture of combustible gases such as carbon monoxide and hydrogen) obtained through gasification of biomass under controlled conditions in a gasifier has potential as a fuel for cooking in rural areas. The producer-gas option is discussed in detail, as a fuel for generating electricity, later in the chapter. One of the major dangers of using producer-gas for cooking in rural areas is the presence of carbon monoxide, a poisonous gas. Often kitchens in rural areas are not separate rooms with doors but have only a separating wall, and the houses are also poorly ventilated. Due to these reasons, and to the fact that often children and old women cook, it is unsafe to suggest the use of producer-gas for cooking in rural areas. Any safety measures suggested could fail in some rural situations, leading to disasters. Because of these factors, producer-gas is not considered by researchers (Dutt and Ravindranath 1993) or planners (PC 1992). However, producer-gas could possibly be used as a cooking fuel in large establishments such as hostels, hotels, and institutions where safety measures could be enforced.

6.6 LIQUID FUELS FROM BIOMASS FOR COOKING

After gaseous fuels, liquid fuels are preferred for cooking. Ethanol and methanol, which can be produced from biomass, could be used as fuel for cooking in place of kerosene. In populous developing countries such as India, dedicating good agricultural land to crops such as sugarcane or sweet sorghum for ethanol fuel production is not considered feasible. However, lignocellulose from woody plants (trees, shrubs, etc.), which can be grown on degraded land, could be considered as a potential feedstock for liquid fuel production. Alcohols are not commonly used as fuel for cooking, but a study in Mauritius on the use of ethanol in cooking stoves showed an efficiency of 60 per cent (Baguant *et al.* 1991). Methanol is preferred as a fuel for cooking since ethanol has a higher value as motor fuel or chemical feedstock, and because of possible social problems arising from human use of the intoxicating liquid. Methanol is a highly toxic product and can be produced relatively cheaply through a thermochemical process using lignocellulose and involving three basic processes:

(1) synthesis gas (Syngas) generation via thermochemical gasification processes;
(2) upgrading of synthesis gas to produce a gas consisting purely of hydrogen and carbon monoxide;
(3) methanol synthesis from CO and H_2.

Biomass to methanol conversion plants are not yet commercially available and the costs are still uncertain (OTA 1992). However, the technical potential and the land requirements are considered here (Table 6.5). Low-pressure oxygen gasifier (LPO), high-pressure gasifier (HPO) and low-pressure indirect gasifier (IND) processes are analysed and the land requirements are estimated using data from Wyman *et al.* (1993). The land required to produce 37 Mt/yr of methanol for all the 627 million rural population is estimated with the objective of meeting the liquid fuel requirement for cooking in rural areas. The land required to produce methanol for meeting all the cooking energy requirements of the rural population is in the range 7.5–10.6 Mha at a productivity of 8 t/ha/yr. A high productivity of 8 t/ha/yr is assumed since biomass production would be on commercial terms, with technology coming from industry for obtaining high yields (discussed in Chapter 7). We conclude that the land requirement is unlikely to be a constraint for producing methanol. Methanol plant production capacities are around 1000 t/day and the land required per methanol production unit at 8 t/ha/yr is around 82 000 ha. The next question is how many such large, vacant, degraded tracts are available. Probably not many in a country such as India with its high population density. Thus it may not be possible to dedicate 100 per cent of the land in a contiguous area. What may be feasible is to expect that 5–10 per cent of the geographical area could be dedicated to energy plantation in any given area. The catchment area for the required feedstock, will come from a radius of 72 km when only 5 per cent of the land area is dedicated to plantations. If the community biogas option is adopted to provide cooking energy for about 32 per cent of the population (200 million people, Table 6.2), the methanol and consequent land requirements would be reduced by a third. With such land requirements the methanol systems may be feasible in many districts. However, moderate-capacity methanol-production systems have to be developed keeping the land availability and requirements in mind.

It appears feasible to consider the methanol option, provided that the production costs are favourable and the production capacities are correct. With technological improvements the costs may decline in the next 5–10 years so that oil-importing countries such as India could seriously consider the methanol option to reduce their dependence on imported oil. One may argue that lignocellulose would be better used for producing high-value liquid fuels for transportation rather than for production of methanol or ethanol as a

Table 6.5 Methanol production for meeting rural cooking energy needs (equivalent to 19 Mt of kerosene per year) and land required for growing woody biomass for methanol[a] production (Wyman *et al.* 1993)

	Biomass LPO[b]	Biomass HPO[b]	Biomass IND[b]
Methanol plant capacity (t/day)[c]	790.0	920.00	1110.00
Wood required (dry t/t methanol)[c]	2.3	1.97	1.63
Wood biomass required for producing 37 Mt/yr methanol (Mt/yr)	85.0	73.00	60.00
Land required at a productivity of 8 t/ha/yr (Mha)	10.6	9.10	7.50
Number of units required at the above capacity to produce (37 Mt of methanol/yr)	128.0	110.00	92.00
Total wood required for methanol production per unit (t/yr)	664 060.0	663 636.00	652 170.00
Land required per unit at a productivity of 8 t/ha/yr (ha)	83 000.0	82 954.00	81 521.00
Land area required in terms of radius for catchment area for wood[d] (in km)			
at 5% land area	72.0	72.00	72.00
at 10% land area	51.0	51.00	51.00

[a] Kerosene required = 24.5 kg/capita/yr for meal cooking (Dutt and Ravindranath 1993) + 25% of it for snacks, tea, etc. = 30 kg/capita/yr = 60 kg of methanol/capita/yr (at energy value of 20.9 MJ/kg of methanol and 42.1 MJ/kg of kerosene).
Methanol required by the rural population for cooking = 627 million × 60 kg/capita/yr = 37 Mt/yr.
[b] LPO, low-pressure oxygen gasifier; HPO, = high-pressure oxygen gasifier; IND, low-pressure indirect gasifier.
[c] From Wyman *et al.* (1993).
[d] Geographical area to be dedicated for wood production for methanol assumed to be 5% and 10%

source of heat energy for cooking. The government could decide in favour of cooking due to the large social and environmental impacts. However, biomass production and methanol synthesis must be driven by profit motives.

6.7 KEROSENE OPTION FOR COOKING IN RURAL INDIA

It was shown above that it is possible to meet rural cooking energy needs through the use of community biogas based on cattle dung and leaf litter or methanol from lignocellulose. It is useful to consider non-bioenergy options, such as kerosene, in order to obtain an idea of the quantities as well as the problems involved in large-scale use.

It was seen from the earlier discussion on the energy ladder that, after gaseous fuels, liquid fuels are preferred. In urban areas kerosene is the most popular fuel after LPG for cooking (Fig. 2.7). Currently in conventional energy analysis, kerosene is not considered as an option for cooking on the assumptions that:

(1) it involves large imports and outflow of foreign exchange; and
(2) rural households cannot afford its purchase.

According to an estimate in a South Indian village, 30 kg of kerosene capita/yr is required (footnote [a] of Table 6.5). A rough extrapolation to the national level for the total rural population implies that 19 Mt of kerosene should be adequate to meet the annual rural cooking energy requirement. The kerosene option has several advantages:

1. Familiarity with the fuel and conversion device, unlike LPG in rural areas.
2. A shift to kerosene does not involve any large expenditure on cooking devices, since kerosene stoves are available in small-town shops and only cost around Rs100. LPG requires large investments of nearly 10–15 times that of kerosene stoves on both cylinders and stoves.
3. Kerosene is supplied to nearly all villages in India and thus a distribution network already exists. For LPG, the whole network has to be established.
4. There is risk of accidents in the use of LPG cylinders unless accompanied by rigorous education and safety measures.

The quantity of kerosene required at the national level is not too large and unmanageable. In fact, options are available to reduce kerosene and diesel consumption in other sectors and also to reduce the kerosene requirement of 19 Mt/yr for cooking in rural areas:

1. If the family biogas option is adopted, the kerosene requirement will decline to 16 Mt.
2. As will be shown later in this chapter, the adoption of a bioenergy option such as biomass gasification for decentralized power generation and by providing electricity for lighting, a saving of 3.4 Mt of kerosene currently used for lighting in rural areas could be obtained (Table 2.22). Additionally 5.5 million diesel pumps consume 4.7 Mt of diesel fuel/yr (Table 2.22). It is possible to save the diesel used in pumping irrigation water by shifting to biomass electricity from decentralized power generation (as shown later in the chapter). Thus it may be possible to obtain an annual saving of about 8 Mt of oil.

The argument against the use of kerosene based on fears of large imports and increases in consumption of fossil fuels appears to be exaggerated. In a similar analysis, Foley (1985) concluded that if 2 billion users of wood for cooking shifted to petroleum fuels in developing countries, the additional demand would be about 100 Mt annually, which was just 3.5 per cent of the world's oil output in 1983. Thus even a major shift to petroleum fuels in the domestic sector could be accommodated without encountering any significant production or resource constraints. A similar analysis could be carried out using the LPG option. The country's policies on oil imports are crucial in considering the petroleum fuel option.

6.8 EFFICIENT WOOD STOVES

Over 100 million rural households use traditional wood stoves for cooking in India, with an efficiency in the range of 10–15 per cent (Fig. 2.6). Improved stoves have been shown to have a thermal efficiency of up to 40 per cent. As shown in Chapter 4, it is possible to conserve about one third of the biomass currently used as fuel and the potential for conservation can be up to 50 per cent. According to the fuel preference of households, solid fuels are least preferred and are used only if gaseous or liquid fuels are not available or not feasible (Fig. 6.1). However, it is pragmatic to consider efficient stoves as an intermediate option until liquid and gaseous fuels are made available (WB 1992). The advantages of efficient stoves are:

(1) familiarity of cooking with wood stoves and the availability of local skills enables the quick dissemination of efficient stoves;
(2) low-cost option; at Rs150 per stove, Rs15 billion would be adequate to provide 100 million stoves; against this, Rs10 000–15 000 are required per house for a family biogas plant;
(3) efficient stoves also reduce or remove smoke emission in the kitchen and reduce fuel gathering and cooking duration.

In Chapter 4, we discussed the problems with stove design and dissemination strategies. If both problems were solved, efficient stoves could be provided to all the rural households in 5–10 years, at 20 to 10 million stoves/yr, respectively. The majority of this expenditure would be payment for skilled labour, possibly as part of various employment-generation programmes. The impact at national level on fuelwood consumption would be significant, with the potential for saving nearly 100 Mt of fuelwood annually.

6.9 COOKING FUEL OPTIONS AT THE NATIONAL LEVEL

The features of various technologies are summarized in Box 6.4 and the various options available are listed below.

Box 6.4 Comparison of features of energy options for cooking

Features of options	LPG	Electricity	Kerosene	Improved fuelwood stoves	Biogas
Availability	Limited supply and refining capacity, and partly imported	Shortages predicted into AD 2000 even to meet non-cooking needs	Partly imported, saving from current use in lighting could be used for cooking	Firewood scarce but could be grown sustainably	Needs of a third of population could be met with dung but combined with leaf litter near-full cooking needs could be met
Distribution system	Limited only to large towns, cities	84% of villages connected to grid, poor reliability	Good public distribution system exists	Locally obtained	Dung and leaf litter resource locally available
First cost of the device and fuel cost	High >Rs1500, low fuel cost	High investment for heating power connection (>Rs2000) + high cost of stove	Low cost of device (Rs100–150), no other investment, low fuel cost	Low cost of stove (≈ Rs150) fuel gathered	High cost at family level (>Rs8000) but no fuel costs, partly subsidized, community option feasible with economies of scale
Efficiency (%)	60	71	60	33	45
Control of heat supply	Very easy	Easy	Very easy	Difficult	Very easy
Carbon emission	CO_2 emission	Electricity from coal thermal source leads to CO_2 emission	CO_2 emission takes place but insignificant PICS (products of incomplete combustion)	CO_2 emission for wood from non-sustainable source; PICs are emitted	Methane is converted to CO_2, no net addition of GHG
Renewability	Non-renewable	If coal-based, non-renewable	Non-renewable	Currently part of wood comes from non-renewable sources	Renewable
Feasibility to meet all rural cooking needs	Difficult	Not feasible	Feasible	Most feasible	Feasible

Option 1 Use the potential of cattle-dung family biogas (cover 72–104 million people; Table 6.2) + provide improved stoves for the rest of the rural population.

Option 2 Use the potential of cattle dung through community biogas (cover 200 million people) + provide improved stoves to the rest of the rural population.

Option 3 Use the biogas potential of cattle dung + leaf biomass from dedicated energy plantations and crop residues, and meet the full requirement of biogas for cooking.

Option 4 Use the potential of cattle-dung community biogas option (cover 200 million people) + kerosene for the rest.

Option 5 Use cattle-dung biogas at community level + methanol (from lignocellulose) for the rest.

Option 6 Use the potential of family cattle-dung biogas (cover 72–104 million people) + kerosene for the rest.

Choice of options

To satisfy the goals of cooking energy, the ideal choice appears to be option 3, which has the following advantages:

1. It aims to provide gaseous fuel for all.
2. It is based on biomass feedstocks and is thus sustainable.
3. It leads to local control and self-reliance, as energy is produced within the village itself.

If option 3 is not feasible due to difficulties in operating a leaf biomass-based biogas plant, particularly in obtaining the feedstock, the second choice would be option 4, which entails a mix of biogas and kerosene. For the reasons explained earlier, kerosene is an easier and more manageable fuel than LPG.

Benefits of adopting the bioenergy-based option 3

1. Conservation of 181 Mt of fuelwood used for cooking. If a conservative market value of Rs500/t is assumed and if 50 per cent of the fuelwood is assumed to be high-value wood from trees, the market value of fuelwood saved would be at least Rs45 billion annually, since wood has alternative uses.
2. Removal of the drudgery for over 200 million women, as nearly all rural women are involved in gathering, preparation, and use of fuelwood, crop residues, and cattle dung.

3. Improvement of the health of women due to decreased exposure to pollutants associated with biomass combustion.
4. 32 Mt of dung cakes would be saved from burning, thus nutrients would be returned to the soil as enriched biogas digester effluent.
5. With elimination of the use of biomass for combustion, the softwoods currently burnt would be saved and could be used as raw material in the paper and the rayon industries. Currently India imports wood and wood products worth Rs15 billion annually (Ministry of Commerce 1992).
6. The fuelwood saved from cooking could be made available for power generation for lighting and shaft-power applications, leading to conservation of fossils fuels and reduced imports.

Problems associated with the biomass option

Implementation of option 3 involves the following:

(1) adoption of the community biogas option;
(2) raising energy plantations to provide leaf biomass as a by-product for the biogas plant (and the woody biomass for gasification);
(3) a community organization to motivate and manage the system through community participation; and
(4) investments at village level.

The problems, prospects, and approaches for adopting bioenergy options are discussed in Chapters 10 and 12. The investment required is justified due to the projected improvement in the quality of life of women in rural areas and the positive impact on the environment (see Chapter 8).

It should be noted here that for every 14.2 kg cylinder of LPG sold at the present time in India, there is a government subsidy of Rs71 and the retail price is Rs105. In other words, every affluent urban household using LPG (20 million, CMIE 1994) is provided with a minimum subsidy of over Rs71/month (*Economic Times* 1994). Thus the annual subsidy involved is about Rs12 billion. LPG use is growing at about 12 per cent annually (CMIE 1994). This is an example of the inequities in conventional energy planning. What would be more rational from the perspective of equitable development would be to provide better-quality fuels to the poor, and withdraw the subsidy currently provided to affluent urban households. The savings incurred could be diverted to bioenergy programmes to improve the quality of life of rural women and to reduce the pressure on trees and forests. Even if the subsidy on LPG is removed, the urban rich will continue to use LPG and are unlikely to switch to, say, fuelwood or cattle dung. Thus there is no rationale to continue the subsidy.

Economic analysis of choice of cooking energy options

Device and fuel costs are two of the crucial factors that determine the choice of fuel–device combination, apart from reliability of fuel supplies. Gupta and

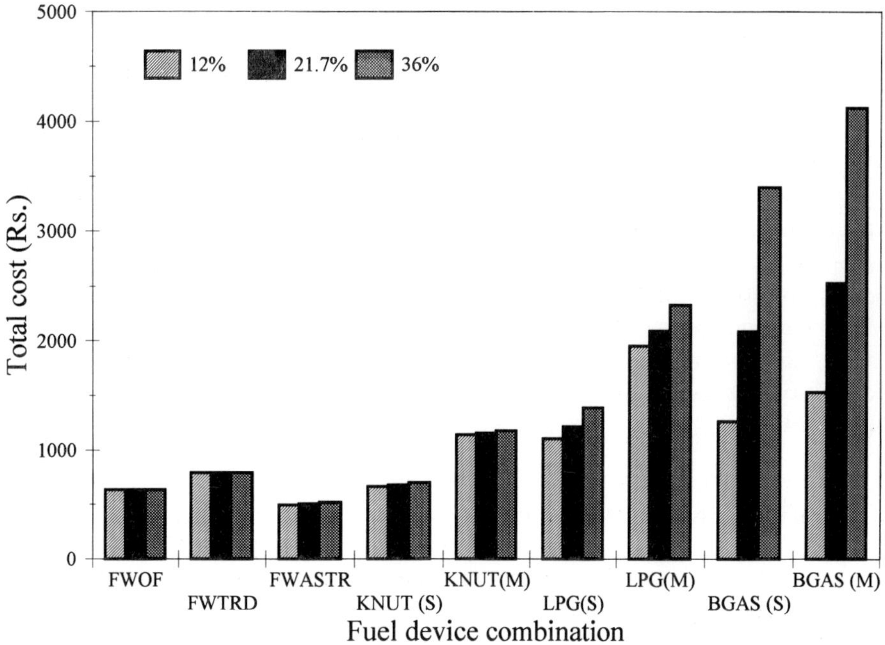

Fig. 6.4 Financial analysis of different fuel–device combinations using energy data from Dutt and Ravindranath (1993) and the market price for fuels and devices for mid-1994. Discount rates: 12, 21.7 and 36 per cent. The total cost was estimated using the annualized life cycle cost (ALC) (section 6.9). Notations are the same as in Fig. 6.3. S, subsidized price; M, market price. (Source: Gupta and Ravindranath 1994.)

Ravindranath (1994) have carried out a financial analysis of cooking energy options using the fuel requirements estimated by Dutt and Ravindranath (1993) and 1994 prices in South India. Both market and subsidized prices were used for LPG, kerosene, and biogas options. The three discount rates used were 12, 21.7, and 36 per cent. In rural areas with capital scarcity, interest rates are normally high. For details refer to the caption of Fig. 6.4 and the original studies (Dutt and Ravindranath 1993; Gupta and Ravindranath 1994). The total cost of cooking consists of the cost of the cooking device, operation and maintenance, and of fuel. The economic comparison is carried out using the annualized life cycle cost (ALC) method:

Total annual cost = capital cost × capital recovery factor + operation and maintenance costs + annual fuel costs.

The findings (Fig. 6.4) show that the efficient ASTRA stove is the least-cost option. Biogas is the most expensive option, even when a subsidy on investment is considered. The ranking of fuel options (solid to liquid to gaseous fuels to electricity) matches with the income-dependent energy ladder given in Fig. 6.1. Two points to be noted are: first, biogas, which is the only quality fuel option promoted in rural areas by government development programmes, is the most expensive option available in the market. Thus rural populations are presently given only the high cost option. Secondly, urban middle- and high-income groups benefit most from government subsidies for kerosene and LPG. It is interesting to note that only at the 21.7 per cent discount rate is LPG at the market rate comparable to the subsidized biogas option. Thus government subsidies distort the market in favour of the urban rich. The only fuel option that may be within the reach of the rural population is the subsidized kerosene option. The community biogas option with the attendant economics of scale may, however, make it more attractive to the rural population. Thus future energy planning for cooking for rural populations should centre around community-scale biogas and kerosene or methanol.

6.10 ENERGY OPTIONS FOR LIGHTING AND STATIONARY SHAFT POWER

Besides cooking, provision of quality lighting concerns all rural households and affects their quality of life. In rural India only 27 per cent of all houses are connected to the grid and use electricity for lighting, while the rest use kerosene for lighting. There is an overwhelming consensus that providing electricity for lighting for all should be one of the main goals of energy planning to improve the quality of life (Reddy *et al.* 1991).

Electricity is also required in villages for shaft-power applications, such as pumping drinking- and irrigation water, flour milling, and agro-processing. Meeting the energy needs of these activities is crucial for promoting development in rural areas. The possible options for power generation for lighting will be discussed in the next section, along with the electricity requirements for shaft-power applications in villages. This is necessary in order to consider all the electricity requirements of a village and to plan for power generation in an integrated manner.

6.10.1 Extent of rural electrification

Rural electrification programmes began in the 1950s as a social amenity, but they gathered importance in the mid-sixties as a source of energy for pumping irrigation water. The Rural Electrification Corporation was formed in 1969 with the task of electrifying all villages in India. Significant success in this effort has been achieved and 84 per cent of the villages were electrified by October 1993 (CMIE 1994). Many major states, such as Andhra Pradesh, Gujarat, Karnataka, Maharashtra, and Punjab, have the distinction of electrifying nearly 100 per cent of villages. However, in West Bengal only 73 per cent of the villages are electrified, while the lowest level of electrification is in Bihar where only 70 per cent of villages are electrified. Thus the achievement of the rural electrification programme has been excellent; for example between 1980 and 1991 the number of villages newly electrified was 212 634. However, there are many critics of the rural electrification programme, who argue that there are many villages (no data are available) where electricity transmission lines pass through the village and yet none of the households or farms has benefited from electrification. Such villages are, however, classified as 'electrified'. There could be some truth in these misgivings since the data from some states, such as Bihar and Uttar Pradesh show that only 4.4 and 9.3 per cent, respectively, of rural homes are actually electrified (NSS 1992). What is important is that at least 70 000 villages are yet to be connected to the grid, and even more important is that these villages could be remote, requiring very large investments to connect them to the grid. This assumption is based on the logic that the state electricity utilities, wanting to show progress and achieve physical targets, start with villages near existing grid transmission lines and leave all the remote and difficult villages to the last. These villages provide the first opportunity for alternative power generation systems.

6.10.2 Features of dependence on centralized electricity options

Conventional energy planners have so far relied on centralized electricity generation based on thermal and hydroelectric sources. In the light of the defined goals of medium- and long-term planning (Box 6.1), the following arguments have to be kept in mind when deciding on the options for power generation for meeting the electricity needs of all the half-million geographically dispersed villages. In the following analysis the power requirements of large urban and industrial sectors are not included.

Centralized power generation and dominance of coal

The dominance of centralized systems is evident from the allocation of funding during the Eighth (1992–97) development plan period. Out of the total national budget for commercial energy, the centralized power sector accounts for 69 per cent (PC 1992). To achieve the medium-term goals proposed, we have to shift away from overemphasis on centralized systems. The installed coal–thermal power generation capacity is projected to account for 70 per cent of the 100 000 MW capacity by 1997 (PC 1992). According to the business-as-usual projections, the coal requirement will increase to 626 Mt for power generation by 2009–10 (Table 6.1).

Environmental impacts

Coal mining, storage, transportation, and combustion lead to environmental degradation. Coal combustion for power generation is the dominant source of greenhouse gases, both in India and globally. With the global concern over greenhouse warming, India, like other developed and developing countries, needs to consider this aspect as one of its guiding principles in selecting energy options in the medium and long term. Coal combustion also leads to the emission of oxides of sulphur and nitrogen, and production of large amounts of ash. This results in environmental pollution for the population residing around the power plant as well as further afield. Currently more crucial for developing countries are local air pollution, land degradation, and waste disposal problems arising from coal use.

High transmission and distribution losses

The transmission and distribution losses in India are estimated to be around 23 per cent, with half of this being attributed to rural electrification (PC 1992). These losses for rural distribution in Karnataka, for example, are estimated to be 2.4 times the urban losses (Ramesh *et al.* 1988).

Low reliability and fluctuating voltage

The grid power supply was monitored at Ungra village (South India) for 12 randomly selected months during 1991 to 1994, representing different seasons and depending on the convenience of the authors. It can be observed from Fig. 6.5 that in nearly half of the months either there was no power supply or there was disrupted power supply during the crucial evening hours (6–10 p.m.) for over 50 per cent of the days. In 11 out of 12 months, there were days with no power in the evenings.

The power supply and voltage were monitored at Ungra village for 10 days during August and September 1994 at two-hourly intervals from 6 a.m. to 10 p.m. (Table 6.6). There were power failures on 9 out of 10 days, during the day as well as at night. The power supply was discontinuous on most days. An observation of 90 recordings made over 10 days showed that there was power supply for 72 per cent of the periods and, when present, the voltage was less than 150 V for 41 per cent of the time of observation.

This clearly establishes the poor reliability of supply and low and fluctuating voltage. This is also reflected by a study (by Bose 1993), which showed that, in rural areas, 84 per cent of the electrified households bought kerosene. The mean kerosene use in electrified households was 3.53 l/month compared to 4.86 l/month by non-electrified households. Fuelwood use for cooking among the electrified as well as non-electrified households was nearly the same. Thus it is assumed that electrified households used kerosene for lighting during power failures. This could be one of the demotivating factors for households seeking grid electricity. Even the Planning Commission recognizes poor power supply as a reason for the slow development of rural industries (PC 1992).

In rural India it is common to observe farmers waiting for hours by their

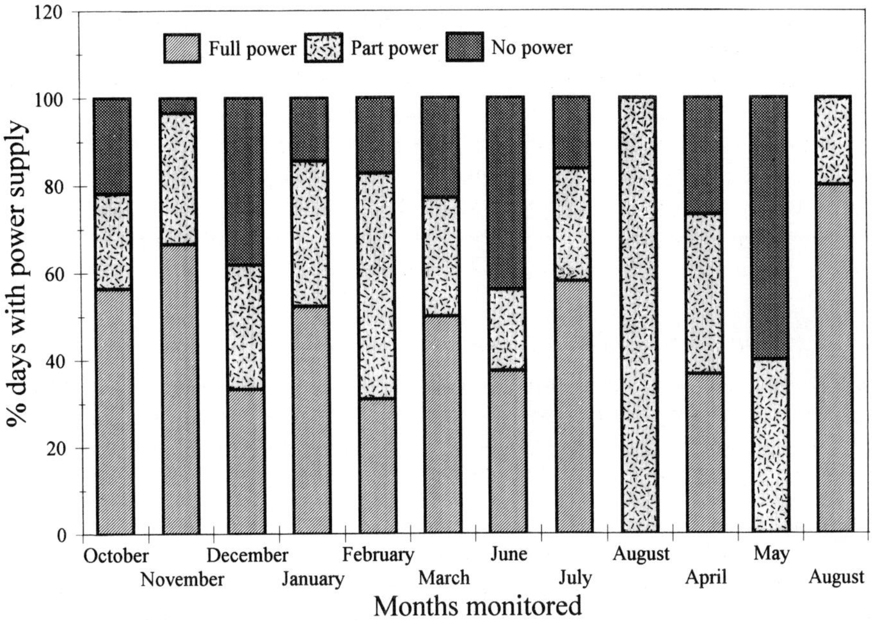

Fig. 6.5 Reliability of electricity supply for Ungra village in South India. Monitored for 12 months during 1991, 1992, 1993, and 1994, during the crucial 6–10 p.m. period. 'Part power' indicates that there were power failures during the 6–10 p.m. period. 'No power' indicates total power failure during the period (Ravindranath and Mukunda 1993).

irrigation pumps for the right voltage, and often they can pump only during the night after 10 p.m., when the load on the grid from urban and industrial sectors is low. In India, as a result of the burning out of irrigation motors resulting from abnormal voltages, incidents of irate farmers attacking and assaulting electricity utility officers are common. One reason why rural households do not use the efficient fluorescent tube lights is the low voltage in the evening hours. It is common to find households using high wattage incandescent bulbs, partly to overcome the low voltage problem. Thus proper voltages have to be ensured in order to deliver the benefits of electrification. In rural India, for every two electric motors there is one diesel engine (5.5 million) for irrigation-water pumping. The cost advantage of electric pumps over diesel engine pumps is well established (Rajagopalan and Demaine 1994). It is likely that farmers are forced to buy diesel motors due to low reliability and fluctuating voltage, or to difficulties in getting electrical connections.

Foreign-exchange requirements for importing oil

Diesel is a substitute for electricity for pumping irrigation water, and already 5.5 million diesel engines are used for this purpose. The business-as-usual

Table 6.6 Grid power supply and voltage situation in Ungra, a typical South Indian village, monitored during August and September 1994

Time of the day	Power supply and voltage situation									
	25 Aug	26 Aug	27 Aug	28 Aug	29 Aug	30 Aug	31 Aug	1 Sept	2 Sept	3 Sept
6 a.m.	110	100	no	200	100	100	190	60	220	no
8 a.m.	100	90	no	215	no	no	150	75	no	210
10 a.m.	100	100	100	200	100	no	170	180	no	200
12 a.m.	110	110	no	125	100	no	no	no	200	no
2 p.m.	120	110	no	no	100	210	no	210	180	no
4 p.m.	100	110	220	no	100	no	no	200	no	no
6 p.m.	105	no	200	210	180	185	170	160	200	210
8 p.m.	140	140	155	160	100	180	150	no	200	190
10 p.m.	215	160	220	210	100	225	180	no	200	190
Periods without power	0	1	4	2	1	4	3	3	3	4

no = no power supply or failure of grid supply.

projections show that oil demand will be 186 Mt by 2009–10 (Table 6.1). Such a large demand and the resulting imports could be disastrous to the foreign-exchange situation. Thus minimizing the use of petroleum fuels through efficiency improvements and finding sustainable alternatives is essential. The import of petroleum fuels is one of the dominant factors contributing to external debt, with India being one of the leading indebted countries in the world (WB 1994).

Electricity inaccessible to the rural poor

The low extent of household electrification in rural areas (only 27 per cent) is due to the high initial investment required and the administrative procedures involved. For example, installation of a meter for each house and the practice of each home being directly connected to transmission-line poles add considerably to the cost of the grid-connected systems. The cost reduction potential has been demonstrated in Hosahalli (a producer-gas system) and Pura (a community biogas system) villages where it has been shown that there is no need for meters if there is a community-managed, decentralized system providing certain basic services.

Projected power shortages

Even if all the planned additional capacity is installed, India is still projected to have power shortages in 1997 (PC 1992). The power shortage in 1993–94 is reported to be 11 per cent and the peak shortage is shown to be 32 per cent according to one estimate quoted in CMIE (1994).

Environmental implications of large hydroelectric power projects

In India today, unlike in the past, every hydroelectric power project, or even irrigation river valley project, is facing opposition from environmentalists and, more importantly, from local people. In many cases, the state governments have been forced to drop the power projects. Environmental problems, such as submergence of forest, and social problems, such as displacement of communities, are some of the major difficulties. The Narmada River Valley project, which is one of the world's largest irrigation and power projects, faced such severe local opposition that the World Bank withdrew from the project in 1993. The Gujarat State Government is still pushing ahead with the project, mainly due to the strong farmers' lobby, but more as an irrigation project than for electricity generation. We feel that if it was purely a hydroelectric power project it would possibly have been dropped long ago due to opposition from local people. Thus in future, hydroelectric power projects will be increasingly difficult to implement, due to environmental and social reasons.

Rural electrification uneconomic at current levels of use

The huge investments neccessary in rural electrification from a centralized generation system are not justified at the currently observed low levels of electricity use in farm households. The investment would be justified for a higher-capacity utilization (Rajagopalan and Demaine 1994).

6.10.3 Decentralized power generation options—why bioenergy?

The discussions of the features of rural electricity needs and problems of centralized generation and distribution in the previous section highlight the need to explore alternative decentralized options which appear environmentally sound and socially desirable. There are several renewable (biomass, wind, solar, hydro-power) and non-renewable (diesel generator) energy options for power generation at the decentralized level. It is necessary to understand why biomass-based energy options should receive priority over other options and to discuss the advantages to local and global communities as well as the environment.

Why renewable energy options? The factors that determine the choice of an energy source should be based on financial and environmental analysis, and, above all, local benefit considerations; how it will benefit the local community and the environment. An obvious option that needs to be considered is the diesel-based decentralized power generation system. Oil imports already account for nearly a quarter of all imports (*Economic Times*

1993) and India periodically faces foreign-exchange crises. One of the main goals of planning is also to reduce petroleum imports to promote self-reliance. Comparison of the economics of small-capacity gasifier-based power generation systems with diesel systems (5 kW and 96 kW) of identical capacity has shown that producer-gas systems are cheaper to operate (Rs/kWh) than diesel systems (Ravindranath *et al.* 1990; Mukunda *et al.* 1993). Another study also concluded that at the low loads operating in villages, biomass systems were financially cheaper than diesel generation systems (Sinha and Kandapal 1991). The economic analysis of decentralized systems is considered in detail in Chapter 9. A comprehensive study (Reddy *et al.* 1990) comparing all centralized and decentralized power generation options estimated the cost per kW of installed capacity as well as the cost of delivered electricity (per kWh) and concluded that decentralized options are cheaper than centralized options, particularly biomass-based options such as producer-gas and biogas systems, if a proper life-cycle costing procedure is adopted.

The next question is, which renewables are to be considered? The prominent renewable energy sources for power generation are solar, wind, micro-hydro, and biomass. India has development programmes for all these sources and has acquired some experience with them.

Solar energy

Solar photovoltaic (PV) technology has attracted much attention in the past 10 years and significant progress has been achieved in its development. PV systems have the potential for meeting the power requirements of remote locations for lighting and possibly small-scale water pumping. Globally available evidence suggests that the prices of PV systems are falling (Ahmed 1994). However, the prices are still too high in India to be taken as a serious alternative in the near or medium term (at over Rs29/kWh, TERI 1993). Keeping in mind such cost limitations, India has launched a small programme of field dissemination which is restricted to street lighting, operating community TV sets, domestic lighting of individual houses, vaccine refrigeration, and telecommunications. It is relevant to note that, other than electricity generation, these systems do not provide any social or economic benefit to the local community or environment, unlike, say, biogas or biomass gasifiers (Chapter 8). Thus even if we assume a situation where PV becomes competitive with other renewables such as biomass gasifiers and biogas dual-fuel systems, one has to give due consideration to the social, economic, and environmental benefits to local communities that accrue from bioenergy systems.

Wind energy

The potential for wind energy in India has been tentatively estimated to be about 20 000 MW (MNES 1993*a*). India has launched a moderate programme for tapping wind energy for power generation, with an installed capacity so far of 38 MW in 1992–93. India is concentrating on wind energy farms for feeding into the grid. Use of wind energy for pumping irrigation water or power generation is site-and season-specific, depending on wind patterns. Not all locations are suitable and, even in locations with moderate potential, the seasonality of wind patterns and variations within a day have to be kept in mind. Electricity is required in villages on all days in a year.

Wind energy systems in suitable locations cost around Rs30 000/kW and thus are competitive with grid electricity for large MW systems (TERI 1993). As in the case of solar PV, the social and economic benefits to the local community and environment are marginal. The concept of large, MW-size systems has a place in the national programme for feeding electricity to the grid, and this is the focus of the government programme (MNES 1993*a*). Wind energy may not be suitable for stand-alone, decentralized power generation at, say, the 10–200 kW level in all locations for year-round electricity supply.

Micro-hydroelectric

Micro-hydro is a highly location-and season-specific option, just like wind power. The total potential of micro-hydro in India is estimated to be 5000 MW (MNES 1993*a*). A large part of India is semi-arid, with rainfall in the range 50–100 cm, and some parts are also arid with annual rainfall less than 50 cm. In suitable locations micro-hydro systems can be financially very attractive (Reddy *et al.* 1990). Thus the micro-hydro systems are feasible only in some states, in some locations, and during certain periods of the year. The average cost of micro-hydro is around Rs28 000/kW installed capacity (TERI 1993). India has launched a moderate programme on micro-hydro, and so far 86 MW capacity has been installed and an additional 123 MW capacity is in the process of being installed. Most systems are in the range of a few hundred kilowatts to a few megawatts. Micro-hydro systems have a place in the national energy programme more to supplement the grid rather than to meet the power requirements of a village or a cluster of villages at a load of 15–200 kW.

The limitations of the above renewables are: first, location specificity and seasonality, and, secondly, the intermittent nature of wind, solar, and micro-hydro systems due to the fluctuating availability of sunlight, wind, and water flows. Thus they may require expensive electricity storage facilities to supply electricity demands at different periods (WEC 1993). Thus solar PV, wind, and micro-hydro systems have limited potential for large-scale decentralized applications. A summary of comparison of different renewable energy options for decentralized power generation within the context of overall goals of development is given in Box 6.5.

Box 6.5 Comparisons of features of renewable energy technologies for decentralized applications for loads up to 200 kW for village electrification

Features	Solar PV	Wind energy	Micro-hydro	Biogas (BG)	Biomass gasifiers
Suitability for loads up to 200 kW	Suitable	Suitable depending on wind pattern	Suitable depending on water flows	Capacity determined by availability of dung; inadequate in most locations	Suitable in most locations
Seasonality and fluctuation of supply	Seasonal in some situations	Seasonality of wind may not match with requirement	Seasonal	For BG plants, gas production stable/fixed; not suitable for varying loads	Varying capacity can be installed; any capacity could be operated, even at lower loads; no seasonality
Locational specifications	In most locations	Highly location specific, depending on wind potential	Location-specific, depending on water flow	Plants could be located in any village; not feasible to locate at all sites, such as for pumping tasks	Could be located in almost any village, farm, or elsewhere
Source of energy/ feedstock	Available	Available in potential locations	Only in some locations; water flow may have to be managed	Dung resource available sustainably but the quantities are fixed with some seasonal variation	Wood needs to be grown; can be grown sustainably; land available in most villages; some crop/forest residues could be used
Technology status	No indigenous technology; R & D on for cost reduction	MW level technology available but small rating (10–100 kW) not yet available	Technology available	Indigenous technology available, R & D and dissemination experience available, R & D required for cost reduction and alternative feedstock	Indigenous technology available; R & D, dissemination experience exists; R & D required for cost reduction, to make user friendly
Associated environmental benefits	Carbon and sulphur emissions avoided	C and S emission avoided	C and S emission avoided	C and S benefit + increase in available N of sludge manure, to a small extent can replace inorganic N	C and S benefit + if wood grown on degraded lands; soil and water conservation, biodiversity promotion, net emission reduction of CO_2
Need for storage of electricity	Storage required, high costs	Storage required, high cost	Storage required, high cost	Not necessary	Not necessary

Bioenergy options for decentralized power generation

With modernization, biomass is beginning to be used in various forms for various tasks: shaft-power applications, electricity generation, liquid fuel for transportation, and of course as a source of energy for heat. Biomass is a versatile form of energy and, unlike other renewables, offers a variety of possibilities of energy forms and processes and at different scales:

(1) variety of processes: combustion, thermochemical processing, and biochemical conversions;
(2) different scales: 5 kW/100 kW to several megawatts or even 50 MW and higher;
(3) various end uses: cooking, process heat, shaft power, electricity generation, liquid and gaseous fuels.

Furthermore, bioenergy systems do not require electricity storage facilities since there is little problem with respect to seasonality and fluctuations within the day. In particular, producer-gas electricity systems could be switched on or off at will. Bioenergy systems could be set up in any part of the world where biomass can be grown and harvested. Bioenergy options thus offer considerable advantages over other renewables. Biomass is used in all parts of the developing world, but usually in traditional, less efficient, and polluting ways, thus providing low-quality energy service. Modern bioenergy technologies will use the same biomass in the same locations to produce more acceptable forms of energy (electricity, liquid and gaseous fuels) and hold the potential to promote the quality of life. The positive and negative aspects of bioenergy and the barriers to their use are discussed in Chapters 8 and 12.

6.10.4 Electrical energy requirement for promoting development

The first step in analysing the bioenergy potential is to consider the present energy situation. Based on previous sections, the features are summarized here:

1. The electricity use levels are low in rural areas.
2. About 70 000 villages are yet to be electrified.
3. Only 27 per cent of rural households use electricity for lighting.
4. Some villages, such as Maulaganj and BNPura, do not have a flour mill, thus requiring women to walk to the next village several times a week to mill grains.
5. In villages such as Maulaganj and BNPura, water for domestic use is lifted and carried from water sources manually, by women.
6. As against a potential of 19 million irrigation pumps, 9.1 million pumps are presently in use. Many farmers still depend on diesel pumps for pumping water (5.5 million) and many cannot be connected to the grid, for several reasons, including shortages of power transmission-line poles and cables.

7. Villages such as BNPura and Ungra have inadequate or no street lighting.
8. The power supply is characterized by low reliability and fluctuating voltage.
9. Low loads and high transmission and distribution costs and losses.

The first question that arises is, what are the services that require electricity to promote the quality of life in rural areas? It is possible to define a set of services that most villages require irrespective of location, agro-climatic condition, resource endowment, or levels of development. An attempt is made to define a set of services requiring energy for promoting development and which is likely to from a part of any village energy plan. There is also need to ensure that the benefits reach women in the villages (Bose 1993).

Energy services for development

The services required in any typical village are listed:

1. Home lighting—every house should have electric lighting.
2. Appliances—every house should have provision for one or more appliances, such as radio, fans, TV, food mixer.
3. Community services—every street in every village should have street lighting at, say, one light for every 10 houses. Lighting in schools and hospitals, refrigeration in local hospitals, community TV, and so on.
4. Domestic water—every village house should have piped water at the doorstep, or every village street should have water taps.
5. Flour milling—every village should have a flour mill.
6. Agro-processing, including post-havest operations—Every village should have at least one agro-processing unit other than flour milling.
7. Agricultural pumping—the potential for pumping units is estimated at 19 million (NIRD 1991). Thus out of 104 million rural houses, one out of every five households is expected to have a pump for irrigation (data are not available on the number of households with farmland having potential for irrigation).

The above list includes not only activities using electricity at present but also activities or devices requiring electricity in the future, as incomes rise and the economy grows.

Electricity requirement

After defining the set of services and activities, the next step is to estimate the electricity required for each service and the total amount of electricity required daily and annually. This is estimated for two situations: currently used technologies (or devices), and efficient technologies (or devices) available. Pumping and lighting are the two dominant end uses of electricity and both offer potential for significant efficiency improvements, thereby leading to a reduction in electricity requirements. Any alternative plan for meeting electricity requirements must be based on efficient devices and processes, since efficient motors, efficient irrigation pumps, and efficient compact fluorescent lamps are some of the least-cost options (Reddy *et al.* 1990).

Lighting Fluorescent tubes and compact fluorescent lights (CFL) can be used for lighting in place of incandescent bulbs. The electricity required per household per year could be reduced to a quarter or less by shifting to CFLs. The life of a CFL is also about 8000 h compared to 1000 h for an incandescent bulb. Even a 40 W fluorescent tube gives as much light as a 100 W incandescent bulb. Financial analysis of CFLs has shown them to be financially attractive and feasible (Parikh *et al.* 1994).

Agricultural pumping Several studies have highlighted the potential for conserving electricity use in electric pumpsets (Reddy *et al.* 1991; ADB 1993). Simple replacement of the high-friction suction line and foot valve of the centrifugal pump by low-friction components is expected to result in a 22 per cent reduction in electricity use; a full retrofit could lead to a reduction of nearly 50 per cent. Thus an average 35 per cent saving in electricity use is assumed in estimating the electricity requirement in the efficient scenario (ADB 1993). Financial analysis has shown it to be highly attractive (ADB 1993). For other activities also there is potential for efficiency improvements. However, even in the efficient scenario, which is used for further analysis, the same level of requirement is assumed so that with efficiency improvements there would be scope for meeting the growing demand in the future for appliances and agro-processing.

The electricity requirement is given per household or per farm unit for lighting, appliances, and irrigation pumpsets, and estimates are made at village level for other end uses and community services (e.g. pumping drinking-water, flour milling). The standard norms of requirements are given in Table 6.7 and are used for estimating electricity requirements for the selected villages as well as at the national level.

Peak load and scheduling of activities

The peak load demand of a village has to be estimated to plan for the power generation capacity, especially with decentralized options. The peak load requirement can be calculated assuming the simultaneous operation of different activities. Activities requiring electricity could be scheduled in such a way as to uniformly distribute the load without making it inconvenient for the village community. For example, home lighting is definitely required in the evening hours, but flour milling and the domestic water pumping service are required for only 2–3 h/day and could be provided in the morning (6–8 a.m.) or evening hours (4–6 p.m. or 6–9 p.m.), as women would be away during

Table 6.7 Electricity requirement for promoting development in rural areas; norms for households (HH), farms and villages (kWh/yr)

Activities	Goals	Unit for the norm (village/HH/farm)	Electricity norm per unit (kWh/yr)	
			Existing technology	Efficient technology
Home lighting[a]	Electricity for all rural HH	Per HH	328	109
Drinking-water pumping[b]	Piped water to all villages	Village		
		Small and medium	3650	3650
		Large and very large	7300	7300
Street lighting[c]	Lights for all streets in villages	One light for every 10 HH	175	87
Flour milling[d]	Electricity for grain milling in every village	Village		
		Small and medium	7300	7300
		Large and very large	14600	14600
Agro-processing[e]	Electricity for agro-processing in all villages	Village		
		Small and medium	8000	8000
		Large and very large	16000	16000
Agricultural pumping[f]	Electricity for irrigation-water pumping	1 pump for every 5.5 rural HH	5362	3485
Appliances[g]	Electricity for radio, fan, others	Per HH	146	146

[a] Lighting: five lighting points for each home for 5 h/day.
Existing lights: 1 × 40 W fluorescent tube + 1 × 20 W fluorescent tube + 3 × 40 W incandescent bulb = 180 W/day = 328 kWh/yr.
Efficient lights: CFL (1 × 20 W + 1 × 13 W + 3 × 9 W) = 60 W/day = 109 kWh/yr.

[b] Domestic water pumping = 5 kW × 2 or 4 h × 365 days = 3650 kWh (small and medium), 7300 kWh (large and very large).

[c] Street lighting: existing lights = 40 W fluorescent tube × 12 h × 365 days = 175 kWh/yr; efficient lights = 20 W CFL × 12 h × 365 days = 87 kWh/yr.

[d] Flour milling: small and medium—10 kW × 2 h/day; large and very large—10 kW × 4 h/day.

[e] Agro-processing: small and medium—5 kW × 8 h/day × 200 days/yr = 8000 kWh/yr; large and very large—10 kW × 8 h/day × 200 days/yr = 16000 kWh/yr.

[f] Agricultural pumping: existing—5362 kWh/pump/yr (from CEA 1991). Assuming a 35% reduction due to efficiency improvements = 5362 kWh/yr × 0.65 = 3485 kWh/pump/yr (from ADB 1993).

[g] Appliances—assuming fan, or radio, or other non-heating appliances = 100 W × 4 h/day × 365 days = 146 kWh.

Village population size: small = <500, medium = 500–1000, large = 1000–2000, very large = 2000–5000.

day working hours (10 a.m. to 5 p.m.). Agricultural water pumping could be run during the day (10 a.m. to 4 p.m.) when the farmers are likely to be in the field. Obviously such timings would be a village option.

Given that power for agricultural pumping is the dominant activity, the requirements could be clustered into a minimum of two blocks:

(1) agricultural pumping;
(2) all other activities, such as lighting, flour milling, domestic water pumping, etc.

Even pumping irrigation water could be scheduled in such a way as to minimize the installed capacity. Not all farmers need to provide water at the same time to all croplands on all days. Even intensively irrigated crops, for example rice or sugarcane, do not require daily irrigation. Irrigation may be required once in 4–7 days for crops such as rice, and once in 15 days for other crops such as millets (sorghum, finger millet, *Panicum satium*), oil seeds (groundnut, sunflower), and fibre (cotton). The average size of farm holdings in India is 1.68 ha with 58 per cent of all holdings <1 ha; and only a part of the land owned by a farmer would be fit for irrigation (NIRD 1991). Thus the land area to be irrigated per farmer would be small (<1 ha). Even if the simultaneous load of, say, eight 5 kW irrigation systems is 40 kW, by arrangement among farmers on the appropriate scheduling of different pumps, the installed capacity could be reduced to say 10 kW, or at most 20 kW, such that each farmer could pump water every second day. Furthermore, every farmer may not have to pump for all the 8–12 hours every day, and water availability would be a major constraint. Thus it is possible to reduce the installed capacity to some optimum level and to increase the capacity utilization.

Energy for post-harvest operations such as threshing and winnowing is required when there is no demand for irrigation-water pumping. There is no clash between the two activities. Such a load sharing is feasible, since experience in Hosahalli village showed that, with a 5 kW system, the community agreed to stagger the services as follows:

(1) pumping domestic water during morning hours from 6 to 7 a.m. and evenings from 5 to 6 p.m.;
(2) flour milling during morning hours from 7 to 9 a.m.; and
(3) lighting during evening hours from 6 to 10 p.m.

Obviously there must be some flexibility for rescheduling or for emergency situations. Currently, it is common practice in India to find farmers irrigating their fields late at night, since the power supply is unreliable and the voltage is low during the day due to overloading of the grid. A decentralized system should nevertheless aim at providing electricity for irrigation during daylight hours. Even in large cities, such as Delhi or Bangalore, water is supplied for only about 2–4 h/day, and different parts of a city get water at different times of the day, including some at midnight.

Table 6.8 Electricity required (kWh/yr) to promote development and quality of life in the selected villages, based on norms for existing and efficient technology given in Table 6.7

End uses requiring electricity	Ungra[a] HH: 177, POP: 1047		Nanuda HH: 135, POP: 838		BNPura HH: 98, POP: 593		Maulaganj HH: 190, POP: 1143	
	Existing technology	Efficient technology	Existing technology	Efficient technology	Existing technology	Efficient technology	Existing technology	Efficient technology
1. Domestic lighting	58 056	19 293	44 280	14 715	32 144	10 682	62 320	20 710
2. Street lighting	3150	1566	2275	1131	1750	870	3325	1653
3. Domestic appliances	25 842	25 842	19 710	19 710	14 308	14 308	27 740	27 740
4. Domestic water pumping	3650	3650	3650	3650	3650	3650	7300	7300
5. Flour milling	14 600	14 600	7300	7300	7300	7300	14 600	14 600
6. Agricultural water pumping	193 032	125 460	128 688	83 640	96 516	62 730	182 308	118 490
7. Agro-processing	16 000	16 000	8000	8000	8000	16 000	16 000	16 000
8. Total electricity	314 330	206 411	213 903	138 146	163 668	115 540	313 593	206 493
9. Present use level[a]	77 310		185 555		8760		NO	
10. Electricity use/capita	300	197	255	165	276	181	274	180
11. Wood required (t/yr)		268		180	–	140	–	268
12. Land required (ha) at productivity				–				
4 t/ha/yr		67		45		35		67
8 t/ha/yr		34		23		18		34

[a] HH, Number of households in the village, wood required at 1.3 kg/kWh; POP, population of the village.

The crucial factor is empowering the village community to decide the scheduling of different activities in a manner most convenient to the majority in a pragmatic and democratic manner. However, the village community must be educated as to the advantages of staggering various activities and the need to distribute the load uniformly over the day and over the week. With proper information the village community is more likely to decide on a rational scheduling of activities. Any attempt to impose a schedule is likely to fail as villagers may not fully know the rational for spreading out the load and the scheduling may be inconvenient.

6.10.5 Case studies of four villages

For the sake of convenience, four of the six villages from Chapter 2 are selected for detailed consideration of bioenergy alternatives: Ungra (in the south), BNPura (in the east), Nanuda (in the west) and Maulaganj (in the north).

1. Maulaganj is a non-electrified village.
2. Nanuda is a progressive and developed village since the level of electricity and kerosene use is higher than in the other villages.
3. Ungra is a moderately developed village with good infrastructure such as pumps, flour mill, and agro-processing units.
4. BNPura is a backward village with the lowest level of home electrification and irrigation pumps.

Thus these four villages represent a gradient of levels of development, energy use, and geographical distribution. The current levels of electricity use for the four selected villages were discussed in Chapter 2 (Table 2.21).

Using the norms given in Table 6.7, the electricity requirement was estimated for the four selected villages and is given in Table 6.8.

1. Agricultural pumping is the dominant activity, accounting for 55–62 per cent of the total energy requirement in the existing, as well as the efficient, technology mix.
2. Lighting is the activity next in importance and it accounts for about 19 per cent in the existing scenario, declining to about 10 per cent in the efficient technology mix.
3. Agricultural pumping and lighting together dominate the total electricity required in the efficient technology mix, accounting for 65–70 per cent.
4. A switch to the efficient technology mix would lead to about a 35 per cent reduction in power requirement in the villages.
5. The per capita power requirement is in the range of 165–197 kWh/y in the efficient scenario. Electricity use in the development-oriented energy-efficient scenario is about 30–40 per cent lower than the present national average. It is thus theoretically possible to achieve significant improvements in the quality of life in rural areas with lower electricity use than the present national average.

Bioenergy options

The two bioenergy options considered for meeting the decentralized village electricity requirements are dual-fuel biogas and producer-gas systems. Both these technologies have been developed indigenously and are available off the shelf. Biogas dual-fuel systems based on dung as the feedstock for power generation, as well as energy-forest wood-gasifier-based power generation systems have already been implemented in villages. Their acceptability and operational feasibility has been demonstrated in Pura (Rajabapaiah *et al.* 1993) and Hosahalli (Srinivas *et al.* 1992) (Chapter 5).

Community rather than individual bioenergy systems

In order to estimate the potential of biogas and producer-gas, community-level systems are considered rather than individual family or farm-level systems for the following reasons.

1. To obtain economy of scale. To share the facilities such as sheds and civil structures, sharing peripherals like control panels, voltage stabilizers, circuit breakers, wood-cutting machines for producer-gas, water supply for biogas plants, etc.
2. Many services, such as the pumping of drinking-water, flour milling, and street lighting, will be village-level facilities.
3. To achieve uniform load distribution over 18 hours in a day, in order to achieve maximum capacity utilization and to reduce investment in installed capacity.

Cattle-dung biogas electricity

Biogas is a clean fuel and can be used in internal combustion engines directly or can be used to generate electricity by coupling a dual-fuel engine to a generator. Over 90 per cent diesel substitution is possible, or even full operation on biogas alone is feasible. Community biogas is the only option for power generation in a village, such as Pura, as it is not feasible to have several biogas-based power generation systems attached to several small biogas plants. The potential for power generation is given in Table 6.9.

Based on the cattle dung available at the community level, the electricity generated is adequate for lighting in all the four villages when the efficient scenario is adopted. However, in only two out of the four villages is the power generated using biogas adequate for meeting the minimal base load requirements, which include lighting, pumping drinking-water, and flour milling. Electricity needs of appliances and street lighting are not included.

Information from two additional villages in North India, three villages in a semi-arid region and three villages in a heavy-rainfall region of the Western Ghats in South India are also included in Table 6.9. In all three Western Ghat villages, as well as in two North Indian villages, the electricity requirement of lighting, pumping drinking-water, and flour milling could be met using

Table 6.9 Biogas for power generation at village level for a sample of villages and at national level, assuming a community biogas system with cattle dung

Village	Electricity generation potential[a]		Electricity requirement for lighting[b] (kWh/yr)	Electricity for lighting + water pumping + flour milling[b] (kWh/yr)	Installed capacity at 10 h/day of operation[c] (kW)
	kWh/day	kWh/yr			
Ungra	100	36 500	19 243	37 543	10
BNPura	97	35 405	10 682	21 632	10
Nanuda	60	21 900	14 715	25 665	6
Maulaganj	57	20 805	20 710	42 610	6
Dhanwas	84	30 660	12 644	23 594	8
Bishanpur	128	46 720	10 137	21 087	13
Pura	52	18 980	9 483	20 433	5
Hosahalli	26	9 490	4 687	15 637	3
Suggenahalli	68	24 820	17 004	27 954	7
Malenahalli	104	37 960	9 374	20 324	10
Sirsimakki	93	33 945	8 829	19 779	9
Tattikai	80	29 200	9 265	20 215	8
Rural India	53 GWh	19 345 GWh			5 300 MW

[a] Electricity generation potential estimated taking the biogas production/day (from Tables 6.2 and 6.3) and 0.75 m^3 of biogas/kWh.
[b] Electricity requirement estimated at village level using the norms given in Table 6.7
[c] Installed capacity = (kWh potential/day)/(no. of hours of electricity generation required/day = 10 h/day).

biogas. However, in the semi-arid villages of South India biogas power is inadequate to meet the power requirement of all the base-load activities. Based on the above examples from different parts of India, it is possible to draw the following conclusions on the biogas electricity potential:

1. The electricity needs of lighting (both home and street) could be met in all the villages.
2. The electricity needs of lighting, pumping drinking-water, and flour milling (the three services for which electricity is required on all days) could be met in some proportion of the villages.
3. The electricity requirements for appliances and irrigation water pumping and agro-processing cannot be met from the biogas option.

Installed capacity for biogas power

The amount of biogas available for a given quantity of dung feedstock is fixed (barring minor seasonal variations) for a given technology. Thus the generation capacity as well as the number of hours of generation will be determined by the biogas production/day. The first step would be to schedule the activities, for example:

1. lighting, 5 hours;
2. pumping water, 2 hours;
3. flour milling, 2 hours;
4. an extra hour for any additional activity.

Thus a village like Ungra would require an electricity supply for a minimum of 10 h/day for the critical base-load activities. With biogas production of 75 m^3/day (Table 6.3) and power generation of about 100 kWh/day, a 10 kW generation system could be installed. The potential for installed capacity, assuming 10 hours of electricity supply per day, is given in the last column of Table 6.9. In most villages the installed capacity potential is in the range of 5–10 kW. At the national level the potential for installed capacity for biogas electricity is crudely estimated to be 5300 MW, which was equivalent to 7 per cent of the installed capacity in 1991. Biogas electricity could only make a marginal contribution at the national level, but at the village level it can make a significant contribution to village requirements. However, this potential can be increased by using leaf biomass feedstock.

Producer-gas electricity option for the selected villages

Biogas is only adequate to meet a part of the total electricity requirement of any typical village. The dominant electricity-requiring service in villages is the pumping of irrigation water. But its needs cannot be met from biogas electricity. Electricity for pumping irrigation water is characterized by:

1. peaks and troughs;
2. dispersed water sources and pumping units, since farmers' holdings are small, fragmented, and geographically dispersed;
3. the capacity of the majority of the pumps is in the range of 3.7–5 kW and limited by water availability;
4. the need for a large number of pumping units for any village.

The producer-gas electricity option could be considered at two levels:

1. Use of biogas electricity for meeting the base-load requirement of services such as lighting, domestic water pumping, and, in some villages, flour milling; while the power requirement of the remaining services could be met from producer-gas.
2. Producer-gas electricity alone could be considered for meeting all the electricity requirements of all the services, thereby keeping the biogas option for other services, particularly cooking.

Biogas versus producer-gas for electricity generation

The biogas potential was suggested for use for meeting cooking rather than electricity needs (section 6.5.1). Furthermore, the full base-load of villages includes home lighting + appliances + street lighting + flour mill operation + pumping domestic water. Biogas electricity potential is inadequate to meet the full base-load requirement. It is best to depend on one mode of electricity generation for operational convenience at the village level. If all these electricity-requiring services cannot be provided by other bioenergy options (such as producer-gas electricity), then biogas could be used for electricity generation, as electricity has a higher preference value than heat energy. The option of producer-gas electricity alone for meeting all the electricity needs of villages is explored in the following sections.

6.11 PRODUCER-GAS ELECTRICITY

Producer-gas technology

Biomass can be converted into a combustible gas called producer-gas through a process of gasification, which is basically a two-stage process; in the first stage biomass undergoes partial combustion to generate gas and charcoal, and in the second stage charcoal reduces the product gas (mainly carbon dioxide and water vapour) to a combustible gas consisting of carbon monoxide and hydrogen (and other gases, such as nitrogen). The process also generates small amounts of methane and higher hydrocarbons, depending on the design and operating conditions (Mukunda *et al.* 1993).

Producer-gas can be used in either compression-ignition (diesel) engines or spark-ignition engines. Producer-gas–diesel engines operating on a dual-fuel mode can replace 75–85 per cent of diesel. They can be used for shaft-power generation for mechanical operations, such as pumping water or milling grains, and for electricity generation when coupled to generators. Gas engines are already available commercially in India in various kilowatt-scales, from 5 to 100 kW (MNES 1993*a*). As seen in Chapters 4 and 5, producer-gas systems of various capacities are already in use in India and other developing countries for small-scale applications.

Feedstocks for producer-gas systems

A range of crop residues and woody biomass from trees could be used as feedstocks for producer-gas systems. Currently, wood-based systems are available, and designs using other low-density biomass are under development and should soon be available in India (MNES 1993*a*). It can be inferred from Table 2.17 that crop residues with fuel potential are limited, since nearly all cereal and most pulse residues are used as fodder or manure and thus are not available as fuel (section 2.8). It is important to note that currently crop residues are used and have an opportunity cost. Rice husks are used in the cement industry, in rice mills and in the manufacture of bricks. Coconut leaves are used as thatch and the husk as fibre, and sugarcane bagasse is used in sugar mills. As observed in the case of Punjab, for rice-husk-based power generation systems, the price of residues such as rice husk could increase once new uses and demands are developed. Crop residues may continue to be used as a fuel in the domestic sector (at least for heating bath water) assuming that cooking-energy requirements are going to be met from bioenergy options. However, crop residues of mulberry, pigeon pea, and coconut shells could possibly be used directly in gasifiers, but their availability is limited and they are location specific (Table 2.17). In a given village, both cropping patterns and the area under crops could vary from year to year depending on rainfall distribution and market factors. Thus a constant supply of crop residues as feedstock cannot be assured over a long period on a continuous basis and the transportation of low-density residues is not feasible.

Woody biomass would be the dominant source of feedstock for gasification. The availability of woody biomass and production potentials are discussed in Chapter 7.

Steps in estimating the power generation capacity

The approach to the estimation of electricity requirements was explained in section 6.10.4, and estimation of generation capacity would involve the following steps:

1. Identify the services requiring electricity in the village, their seasonality, extent of requirement (kilowatts and hours/day), and the location of the requirements through surveys and discussions with the community.
2. Estimate the land available for an energy forest on private as well as common land, and estimate the crop residues available for gasification.
3. Discuss with the village community, explaining the costs, benefits, land requirement, organization, roles, and responsibilities involved in the producer-gas electricity option.
4. Discuss the benefits of scheduling operations to increase the capacity utilization so as to reduce investment costs and arrive at an optimum schedule.

The number of pumpsets estimated to be required according to the norms (Table 6.7) is 36 in Ungra, 34 in Maulaganj, 24 in Nanuda, and 19 in BNPura. The capacity of the pumps could be in the range 3.7–5 kW.

For the majority of locations, the following assumptions would hold good (at one pump for every 5–6 households) for each village:

(1) capacity of pump 3.76–5 kW;
(2) each pump is expected to pump for 4–5 h/day;

(3) each pump will be used on alternate days;
(4) half the number of pumps would be operated in the morning (7 a.m. to 12 noon) and the other half in the afternoon (12 to 5 p.m.);
(5) on any day at a given time only a quarter of the pumps in the village will be in operation;
(6) the village community should have the flexibility to make adjustments where necessary.

N (number of the pumps for the village)
- day 1 = 0.5 × *N* pumps to lift water
  - 0.5 × *N* 50% pumps to operate 7–12 a.m.
  - 0.5 × *N* × 50% pumps to operate 12–5 p.m.
- day 2 = 0.5 × *N* pumps to lift water

For example, in Ungra:

36 pumps
- day 1 = 18 pumps
  - 9 pumps operate 7 a.m. to 12 noon
  - 9 pumps operate 12 to 5 p.m.
- day 2 = 18 pumps

In Ungra, assuming a total requirement of 36 pumps, the installed power generation required would be 45 kW. Considering the power factor, the actual capacity to be installed is 56 kW (assuming 80 per cent of rated capacity). Similarly, the installed power generation capacity required for the four villages is:

Ungra	60 kW	(45 kW + 20 per cent of 45 kW—rounded to 60 kW)
Nanuda	40 kW	(30 kW + 20 per cent of 30 kW—rounded to 40 kW)
BNPura	30 kW	(25 kW + 20 per cent of 25 kW—rounded to 30 kW)
Maulaganj	60 kW	(45 kW + 20 per cent of 45 kW—rounded to 60 kW)

To operate all the other systems such as lighting, pumping of drinking-water, flour milling, and home appliances, the installed capacity required would be less than for pumping irrigation-water. Thus all other services could be provided simultaneously. Irrigation could be treated as one block and all other activities as another block, and could electricity be provided to these two blocks to avoid the clash of demand between them. For example,

Irrigation-water pumping	7 a.m.–5 p.m.
All other operations	6 a.m.–7 a.m.
	5 p.m.–11 p.m.

The decision to schedule the activities should be left to the local community, with the rational suggestion to avoid clashes between the two critical clusters of agricultural pumping and all the other operations. Thus the upper limit of installed capacity would be determined by the maximum number of irrigation pumps to be operated simultaneously and their capacity. This in turn would be determined by the availability of water.

How many units?

Should a village decide in favour of single or multiple units for reliability? It may be a good idea to have multiple units to ensure reliability of some critical services such as lighting and pumping domestic water on all days. For example, for Ungra 2 × 25 kW + 10 kW, or 50 kW + 10 kW, may be desirable. Street lighting is desired by village communities all through the night. With a load of less than 1 kW, one of the following options could be considered for street lighting after 10 p.m.:

(1) install solar photovoltaic lights;
(2) use a battery to store electricity for use between 11 p.m. and 6 a.m.

Thus it could be concluded that for villages in the population range of 593 (BNPura) to 1143 (Maulaganj), the electricity requirement for all the activities (according to the norms in Table 6.7) could be met with an installed capacity in the range 30 – 60 kW.

Thus a combination of biogas (for cooking) and producer gas (for electricity) has the potential to meet all the electricity and cooking energy requirements of selected villages.

6.12 BIOENERGY FOR ELECTRICITY AT THE NATIONAL LEVEL

In this section the potential of bioenergy is analysed at the national level, along with estimates of wood requirements.

6.12.1 Similarities in energy services at the national level

India has locations with diverse agro-climatic, socio-economic, and resource endowment situations with varying energy requirements. Any attempt to generalize at the national level will initially face the criticism of being unrealistic, naïve, and too extrapolatory.

Analysis of the energy consumption patterns, in Chapter 2, of villages belonging to a range of geographical locations interestingly showed similarity in activities requiring energy, the sources of energy used, and the share of different sources of energy, as well as per capita consumption. The predominance of biomass fuels, the large dependence on fuelwood followed by cattle dung and crop residues for cooking, the low efficiency of use of biomass for cooking, the use of kerosene for lighting, use of electricity for pumping irrigation water, and the human effort for gathering wood and fetching water are common to most villages. The services requiring energy are not likely to vary from region to region, at least for the basic activities or services. For example, no matter where the village is located or whatever the social or economic status, all village communities and homes are likely to aspire to piped safe water for residential use (requiring mechanical or electrical energy), electricity for lighting and flour milling, and quality fuel (liquid or gaseous fuel) for cooking, and so on. Currently, they may not be using these modern fuels due to various reasons such as non-availability, high costs, and low reliability. The challenge for the technical experts and energy planners is to provide reliable, efficient, convenient, and affordable energy options. In the name of conserving traditional practices, the rural or tribal population should not be condemned to traditional and low-quality fuels and less efficient devices. The ideal situation is to provide the opportunity of a choice of fuels and devices and let the housewife, farmer, or village community make decisions about their use.

The energy options for various activities may not vary. For example, biogas could be produced and used as a fuel for cooking in all village situations (except at very high altitudes), wood can be grown in any village, and producer-gas could be generated for operating dual-fuel engines for mechanical applications or for electricity generation in any village in any location.

The crucial aspects for national-level consideration of bioenergy potential require an understanding of:

(1) energy services required for different villages for different activities or processes;
(2) energy requirement per capita, per household, per farm, and per village;
(3) feasibility of different energy sources and technologies for different locations; and
(4) resources required; land, feedstock, capital, skilled manpower, etc.

The energy services and electricity required for typical villages are expected to be those based on the norms given in Table 6.7. The energy requirement may vary for agricultural pumping and agro-processing, depending on water availability, rainfall, and cropping patterns. However, the electricity requirement may not vary significantly for several other end uses such as lighting, pumping drinking-water, and flour milling. Thus to explore the potential of bioenergy a set of services, as described in Table 6.7, is considered for different situations in India and, if those power requirements can be met, some variants to the norms could surely be managed.

6.12.2 Bioenergy potential: size of village criteria

Different criteria could be used for analysis of the potential for bioenergy at the national level, based on agro-climatic factors, levels of socio-economic development, and size of the village. For the current analysis the village size is used as a criterion to assess the potential and impact of bioenergy, for the following reasons:

1. Data are available on the number of villages and population distribution for different village-size groups.
2. Size or population of the village determines the power requirement for most activities, particularly the installed capacity for power generation, or size of a community biogas plant for cooking.
3. Power requirements are generally estimated on a per capita or per household level for services such as lighting, appliances, and cooking, and at the village population level for services such as pumping drinking-water, flour milling, and street lighting. These may not vary much from location to location.
4. For a given size of village, the power requirement may not vary for most services, except for agricultural pumping and agro-processing. Monsoon rainfall distribution being highly seasonal and variable, irrigation is required to obtain high and stable yields. In every village water may have to be pumped from open wells, tube wells, rivers, streams, canals, or ponds. Thus it may be reasonable to assume that electricity is required in all villages for pumping irrigation water as every village will have some source of water, except in highly arid areas.

Indian census data are available giving the frequency distribution of number of villages and proportion of population residing in different size groups (Table 6.10). The villages are grouped into five classes, and it is interesting to note that 73 per cent of the villages or human settlements have less than 1000 population and account for 32 per cent of the rural population. Villages with populations up to 5000 are included in the analysis and they account for 85 per cent of total rural population (530 million) and 98.5 per cent of the villages. Villages with over 5000 population can be considered as small towns and thus are not considered here.

6.12.3 Electricity and installed capacity for typical villages

The average population and number of households per village (at a family size of six are given in Table 6.11 for different village size classes. Electricity is projected to be provided for all the stationary shaft-power requiring activities and lighting, assuming the norms as defined in Table 6.7. The electricity requirements for different services for the average village in each village size group, as well as the national total; are given in Table 6.11. Only the efficient scenario or technology package is considered for estimating the bioenergy requirement from Table 6.7. The two main efficient technologies considered are compact fluorescent tubes for lighting and retrofitting of irrigation pumps, as in the case of the sample villages.

Table 6.10 Distribution of Indian villages according to population and size of village, based on national census data (NIRD 1991)

	Population range of villages					Total
	<500	500–1000	1000–1999	2000–4999	>5000	
Number of villages	271 000	136 000	95 000	47 000	8000	557 000
% to total number of villages	48.7	24.4	16.9	8.4	1.6	100
Total population (million)	78.0	119.0	163.0	170	97	627
% of total rural population	12.5	19.1	25.9	27.1	15.4	100
Cumulative population (million)	78.0	197.0	360.0	530	627	627
Mean population of a village	287.0	875.0	1715.0	3617	12 125	1125
Mean number of households (at six/household)	48.0	146.0	286.0	603	2020	187.5

The total village-level requirement of electricity ranges from about 59 MWh/yr for villages with less than 500 persons to 618 MWh/yr for villages in the size group of 2000–5000 persons. As observed for the four selected villages, irrigation pumping accounts for 53–62 per cent of total village electricity requirements, followed by home lighting and appliances which account for 20–25 per cent of the village total. The national-level aggregate electricity requirement in rural areas under the efficient scenario is around 100 TWh/yr (rounded off from 94.5 TWh/yr) covering 98.5 per cent of the villages and 85 per cent of the rural population. This requirement of 100 TWh with an efficient technology scenario is equivalent to 150 TWh under the existing technology scenario. If the transmission and distribution losses are included, in order to deliver 150 TWh in villages, 250 TWh would have to be generated at the power stations, compared to the total of 287 TWh of electricity used nationally in 1991.

6.12.4 Per capita electricity requirement

Electricity requirements range from 170 kWh/capita/yr in large villages (2000–5000 population) to 206 kWh/capita/yr in smaller villages (<500 population). There is a small reduction in per capita requirement with increase in size of the village, due to reductions in per capita requirement for community

Table 6.11 Electricity requirement with efficient technologies for residential, community services, agricultural, and agro-processing sectors in rural India to promote development, using the norms given in Table 6.7 for electricity for village population data from Table 6.10

	Village population range (P)			
	<500	500–1000	1000–2000	2000–5000
Number of villages	271 000	136 000	95 000	47 000
Mean number of households/village	48	146	286	603
Mean population per village	287	875	1715	3617
Electricity requirement per village (kWh/yr)				
Home lighting	5232	15 914	31 174	65 727
Appliances	7008	21 316	41 756	88 038
Street lighting	435	1218	2436	5220
Domestic water[a] pumping	3650	3650	7300	14 600
Flour milling[b]	3650	7300	14 600	29 200
Agricultural pumping	31 365	90 610	181 220	383 350
Agro-processing[a]	8000	16 000	16 000	32 000
Total	59 340	156 008	294 486	618 135
National total TWh	16	21	28	29
National total				= 94 TWh/yr
Per capita kWh/yr	206	178	172	170

[a] Electricity requirement for water pumping for villages in P = 2000–5000 is assumed as double that of villages in P = 1000–2000. Similarly modified for agro-processing activity.

[b] Electricity for flour milling is taken as 7300 kWh for villages with P = 500–1000 (from Table 6.7); 50 per cent of this value for villages with P = <500; two times this value for villages with P = 1000–2000; and four times this value for villages with P = 2000–5000, i.e. it is taken as being proportional to mean village population size.

services. The national mean of all villages is 189 kWh/capita/yr, compared to the use of 340 kWh/capita/yr in India in 1991–92 (all sectors and all populations included). Assuming 189 kWh/capita/yr, the annual electricity requirement for a household would be 1134 kWh. However, it is necessary to stress that 189 kWh/capita/yr is with the efficient scenario, which is equivalent to 280 kWh/capita under the existing technology scenario. To compare with the existing electricity technology and use levels, it is also necessary to include transmission and distribution losses. If these factors are added, the estimated use is presently equivalent to 476 kWh/capita/yr at the electricity generation point (Ramesh *et al.* 1988). Therefore, the electricity requirement could be reduced to 40 per cent to deliver the same level of services by shifting to efficient devices and decentralized systems.

6.12.5 Power generation capacity for average villages

It is necessary to estimate the peak load to calculate the power generation capacity required for villages. This can be achieved using the assumptions made for the four selected villages (section 6.11). For example, scheduling of activities should be such that a near-uniform distribution of load within a day is achieved, keeping in mind the convenience of the local community. Assuming the scheduling procedure adopted for the sample villages in section 6.11, the installed capacity required ranges from 15 kW for villages with <500 population to 165 kW for villages with 2000–5000 population (Table 6.12).

The total decentralized power generation capacity to be installed at the national level (covering 98.5 per cent of villages and 85 per cent of rural population) would be about 26 000 MW (rounded from 25 540 MW). This value is estimated for the use of efficient options with marginal or no transmission and distribution losses.

6.12.6 Technology and wood required for producer-gas electricity

Three crucial issues in considering the producer-gas option are: the availability of technology, the economics, and the sustainable biomass supply, which depends mainly on the land availability and productivity.

Technology

India has several gasifier designs available in 5, 20, 40, and 100 kW capacities (MNES 1993*a*) and it is not difficult to tailor the designs to suit the requirements of individual villages. There are still, however, some minor problems with the designs as well as the dissemination system (Chapters 4 and 5). Once a demand or niche is created for such systems, and if 'a level playing field' is created for different energy options, such problems could be overcome. Thus the range of installed capacity required in villages is 15–165 kW. Any variations could easily be catered for by using the existing indigenous designs, either as a single generating unit or as multiple units. Multiple units would probably be more reliable for delivering services on an annual basis for some critical daily services, such as drinking-water supply and lighting. For details of the Indian designs of gasifiers refer to Mukunda *et al.* (1993) and Paul and Mukunda (1993).

Wood requirement

The amount of woody biomass feedstock required for power generation depends on the efficiency of the conversion system. The conversion efficiencies of current kilowatt-scale Indian designs are in the range of 17–18 per cent (Mukunda *et al.* 1993). The total wood required (at 1.3 kg/kWh) ranges from

Table 6.12 Installed capacity for power generation, wood and land required to meet all the electricity requirement of villages through producer-gas electricity with an 18% conversion efficiency system, for the estimates of electricity requirement from Table 6.11

	Village population-range class				Cumulative value
	<500	500–1000	1000–2000	2000–5000	
Power required kWh/village/yr	59 340	156 008	294 486	618 135	–
Capacity in kW to be installed[a]	13	35	65	137	–
Installed capacity required with power factor of 80% load (kW) at village level[b]	15	45	80	165	
Total installed capacity at national level for all villages (MW)	4065	6120	7600	7755	25 540
Wood and land required					
Wood required at 1.3 kg/kWh at 18% efficiency (t/yr)	77.0	203.0	383.0	803.0	
Land required (ha per village) at productivity:					
4 t/ha/yr	20.0	50.0	96.0	200.0	
8 t/ha/yr	10.0	25.0	48.0	100.0	
12 t/ha/yr	6.0	17.0	32.0	67.0	
Current wood use for cooking (t/yr) at 0.288 t/capita/yr (from Table 2.2)	82.0	252.0	494.0	1042.0	
Electricity required at national level (TWh/yr)	16.0	21.0	28.0	29.0	94.0
Wood required at national level (Mt/yr)	20.8	27.3	36.4	37.7	122.2
Land required at national-level at productivity:					
4 t/ha/yr (Mha)	5.2	6.8	9.1	9.4	30.5
8 t/ha/yr (Mha)	2.6	3.4	4.5	4.7	15.2
12 t/ha/yr (Mha)	1.7	2.3	3.0	3.1	10.1

[a] Capacity (kW) is based on the explanation in section 6.11. It depends on the number of irrigation pumps and the capacity of pumps (5 kW), and assuming only 1/4 of pumps have to be operated simultaneously.
Example: for a village of 500–1000 population (146 HH) = 26 pumps (@ 1 pump for 5.5 HH). Assuming 7 pumps (@ 1/4 of 26 pumps) × 5 kW capacity = 35 kW.

[b] Installed capacity considering a power factor of (0.8). For example, for a village of population 500–1000: (35 kW + 35 kW × 0.2) = 42 kW, rounded to 45 kW.

77 t for a village with a population of <500 to 803 t for a village with a population in the range 2000–5000 (Table 6.12). In per capita terms, typically about 0.25 t of wood is required annually to generate electricity at a decentralized level for all services. This level of wood requirement is lower than the current per capita fuelwood use of 0.29 t/capita/yr for cooking alone (Table 2.2). Nationally the total wood required for producing electricity for rural needs is 122 Mt which is equal to about 67 per cent of the fuelwood presently used in the rural domestic sector (mainly for cooking). The wood supply options are discussed in Chapter 7.

6.13 LARGE BIOENERGY ELECTRICITY GENERATION SYSTEMS

In earlier sections the electricity generation potential of small kilowatt-size bioenergy systems for decentralized power generation at the village level was explored. However, bioenergy also provides the potential for large-scale megawatt systems, as in the USA where over 8000 MW of installed capacity is in operation. The capacity of the plants ranges from a few megawatts to 50 MW (Williams and Larson 1993). In India, two options for large-scale biomass power are:

(1) wood from dedicated plantations and forest residues; and
(2) co-generation systems, such as sugarcane residues in sugar mills or agro-industry processing units.

The potential and prospects of both the options are considered here.

6.13.1 Biomass power from co-generation of sugarcane residues

Sugarcane is one of the dominant crops in the tropical world. In India about 3.8 Mha are under sugarcane. Sugarcane production and processing leads to two by-products; bagasse in the mills, and leaves and tops (tender stem parts) in the field. Typically, a third of the crushed cane ends up as bagasse, with a moisture content of about 50 per cent and with an energy value of 9.5 MJ/kg (fresh). The leaf and top residues account for 279 kg (dry)/t of fresh cane (Williams and Larson 1993). For every tonne of cane produced and processed, about 444 kg of dry residue is produced in the factories and farms. This large quantity of residue in sugarcane-producing countries is a potential energy source waiting to be converted efficiently to modern energy carriers. The Ministry of Non-Conventional Energy Sources set up a task force in 1993 to study the co-generation potential and prospects in India, and some of the findings from the report submitted to the ministry are considered (MNES 1993*c*). In India co-generation in sugarmills is presently limited to electricity production for captive consumption and steam requirements of the industrial processes at low pressures.

More than 95 per cent of the sugar mills produce their own electric power by utilizing the steam (22 atm, 330 °C) from bagasse-fired boilers in back-pressure turbo-generators, and the exhaust is used as process steam. Currently sugar mills are not generating surplus electricity in India. However, a few sugar mills are in the process of installing high-pressure boilers in order to generate surplus power, presumably at present to feed their ancillary units rather than for sale to the electricity grid.

Sugarcane and bagasse production in India

The area under sugarcane fluctuates slightly, but there is a general increasing trend (Fig. 6.6), particularly since 1980–81. The average productivity for 1990–91 was 66 t/ha of harvested cane, and the total production for that year was 241 Mt cane, out of which only 51 per cent was processed in sugar mills; about 39 per cent was used in the small gur and Khandasari units (raw and crude sugar), and about 10 per cent was used as seed material. There are 392 sugar mills in India, with capacities ranging from 1250 TCD (tonnes of cane per day) to over 5000 TCD. A standard size of 2500 TCD is assumed for estimation of the energy-producing potential. The quantity of cane crushed by the industry has increased from 87 Mt in 1981–82 to 133 Mt in 1991–92.

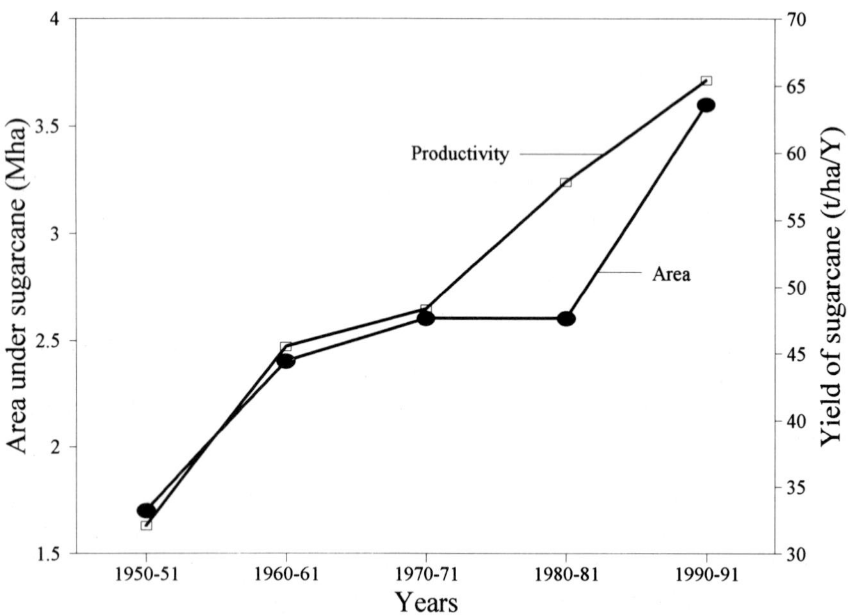

Fig. 6.6 Growth in area and productivity of sugarcane (MNES 1993*c*).

Potential for power generation

The potential for power generation from sugarcane bagasse depends on the conversion technology used. The task force mentioned above (MNES 1993*c*) suggested various technological options for the industry, depending on the availability of funds, to bring various levels of changes in the hardware:

1. One of the advanced options considered for the existing mills is to replace the boilers and turbo generator sets with new boilers of 65 atm and steam temperatures of 480 °C, with a turbo generator of 10.7 MW. Surplus power of about 7 MW could be generated for a 2500 TCD mill. This should be compared with the existing system, producing only 2.5 MW, which meets the captive power requirement, leaving no surplus.
2. For the new or retrofitted mills, the task force suggested an adaptation of the boiler and turbo generator parameters of 65 atm, 480 °C and 75 t/h of steam production. A surplus power of about 9 MW could be generated, through an extraction condensing turbo generator of capacity 12–13 MW.

The task force estimated the potential for surplus installed capacity to be about 3119 MW for the whole industry. Power generation is assumed for the working season only (165 days/yr). However, to operate all year round, sugar mills would have to either use coal and multifuel boilers or other agriculture wastes, such as sugarcane leaves and tops, rice husk, wood, etc.

Constraints in utilizing the co-generation potential

The task force (MNES 1993*c*) listed some of the constraints:

1. The incremental investment required for a typical 2500 TCD sugar mill for producing 9 MW of surplus electricity is about Rs230 million. The industry cannot presently generate such finances and the rate of interest is high.
2. The large power utilities are reluctant to buy small quantities of power from sugar mills that operate only seasonally.
3. There is no clear government policy to ensure a market for surplus power from sugar mills. Therefore the sugar mill managers cannot be expected to make large, long-term investments in a situation of uncertain market and prices.
4. Currently the power tariffs are not based on economic price or long-run marginal cost approaches but are actually subsidized. Thus the sugar industry cannot at present expect remunerative prices from the state utilities. This absence of a level playing field is unlikely to motivate the sugarcane industry to invest in new processes, equipment, and manpower training.
5. In addition to all these, there are some technical problems in upgrading

the existing mills, and in power generation systems, synchronization, and feeding electricity to the grid.

6. Opportunity cost for bagasse: bagasse has other attractive competitive end uses, for example as a raw material for the paper industry. Currently India imports pulp as well as newsprint. The Planning Commission (PC 1992) suggested the need for using sugarcane bagasse and for setting up integrated sugar–newsprint units to conserve forests as well as foreign exchange. Thus one has to consider comparative economics and the impact on external trade of bagasse use for power, or paper, or both.

Current plans for promoting co-generation in sugar mills

In India two pilot co-generation plants are being set up in South India at MRK (Sethiathope) and Cheyyar Cooperative Sugar Mills. The projects include the installation of high-pressure and high-efficiency boilers and turbo attenders (three 2.5 MW) for efficient utilization of bagasse and to produce about 4 MW of surplus power for export to the grid. A number of other sugar mills are contemplating co-generation systems for power generation.

Advanced bioenergy systems for co-generation in sugar mills

The surplus electricity generation potential of various technological options have been considered by Williams and Larson (1993) and their estimates are given below:

1. Existing sugar mills are designed to be self-sufficient with respect to energy requirements and not to have any surplus bagasse left at the mill, in order to avoid disposal problems. Electricity generation is in the range of 15–25 kWh/t of cane milled.
2. Advanced bioenergy systems such as the biomass-integrated gasification–steam-injected gas turbine (BIG/STIG), if operated year round, could produce a surplus of 595 kWh/t cane with a total production of 664 kWh/t.

Assuming the advanced BIG/ISTIG systems (Williams and Larson 1993) and the 1987 sugarcane production of 182 Mt, the potential for electricity generation is estimated to be 120 TWh (compared to the total of 287 TWh generated in India during 1991–92); for the year 2027 the electricity generation potential from the sugar industry is estimated to be 418 TWh. These estimates are probably on the high side as they do not consider the competitive uses of sugarcane bagasse in small-scale units and in the paper industry. But there is little doubt that if advanced technologies are adopted, the potential of sugarcane co-generated electricity will be very high and merits the serious attention of policy makers in order to create conditions for promoting co-generation systems in the sugar industry.

6.13.2 Megawatt-size biomass electricity systems

For a developing country with a high population density such as India the mention of moderate (say up to 10 MW) or large-scale (20–50 MW) biomass power generation systems or moderate-sized biomass-based liquid fuel (methanol) systems can bring an instant negative response from policy makers, energy experts, and also environmentalists. One tends to think of such systems as being suitable for countries with large amounts of surplus land, such as Brazil, Zaire, Canada, or Australia. However, even for a densely populated developing country like India, there could be locations where densities of human settlements are relatively low or where large areas of degraded land are available which could be good candidates for moderate-sized or large megawatt-scale bioenergy plants. What are the arguments for and against moderate-to large-scale bioenergy systems for electricity generation in developing countries such as India?

Arguments against large bioenergy systems

1. Population densities are high in many developing countries (over 200/km^2) and land could be a constraint even for food production; thus it is not feasible to dedicate large tracts of lands for energy.
2. Even if land is available in some locations, the next question would be 'energy for whom?' Who will benefit from the energy produced? If the beneficiaries of large systems are rural communities, it may be acceptable. If electricity is for urban areas and industries or liquid fuels for urban cars with relatively little rural benefits, the ethical question would arise as to why the land belonging to rural and tribal communities should be dedicated to the urban sectors, where it may be used largely for luxury consumption such as private transport and air-conditioners. This is a serious question which policy makers will have to face, particularly from environmental groups as well as local communities.
3. A large bioenergy plant may exert pressure on the existing forests and village trees, including primary forests. These outside vested interests (industrialists and wood contractors) who have no long-term stake in the forests, as do local communities, could find ways of exploiting the primary forests even if legally banned. Loss of primary forests for energy is not acceptable to local communities, environmentalists, and surely not even to bioenergy planners or promoters.
4. Even if degraded lands are available, there would be options such as promotion of natural regeneration and forest succession to enhance biodiversity and the ecological role of forests, or to promote grass production for cattle and not to use them for high-yielding, short-rotation energy plantations.

Arguments in favour of medium to large bioenergy systems

Two major categories of land are available apart from forest land: village commons and private lands. The decisions on village community land should be left to the communities, and decisions on the use of private land or its produce must remain with the farmers. The government and industry could, however, provide incentives or disincentives to growers to encourage various options. The ultimate decision makers should be the local communities or farmers, who must therefore be presented with all the options and implications of bioenergy and other options. Land owners and market factors will decide the end use of land, subject to appropriate controls. Land and production aspects are discussed in detail in Chapter 7.

1. Land reclamation projects. It is not difficult to find a number of locations where large tracts of degraded lands in fragments of various sizes may be available and need to be revegetated. These may have been left barren for years due to lack of incentives to revegetate. Megawatt systems could provide funds and technology to reclaim the land and also generate income as incentives.
2. Long-term employment for local people in growing energy forests, harvesting, transporting, processing the wood, power generation, and maintenance.
3. Part of the electricity generated will be available for use in local villages and the rest could be sold to nearby demand centres and the grid. A sugar factory will have several thousands of hectares of dedicated land and nearly all the sugar ends up away from the catchment area. An energy plantation could be considered in the same way as any cash crop, such as sugarcane, cotton, tea, or cocoa plantation. We are only contemplating the use of degraded lands and attempting to generate employment and incomes. After all, lands are dedicated for export products (cotton, jute, cocoa, spices, sisal, pulpwood) to earn foreign exchange to buy petroleum oil (Hall 1993).
4. Economies of scale will make large-scale power generation systems profitable. The local communities should share the profits.
5. With large-scale systems it may be possible to motivate industry to invest money in partnership with local communities to obtain high and sustainable yields. Land ownership should stay with farmers and the local community.

Technology

In a biomass gasifier–gas turbine system (BIG/GT), biomass can be gasified in a pressurized air-blown reactor. The products are cleared of particulates and other contaminates before being burned in an efficient power cycle, based on an aeroderivative gas turbine, such as steam-injected gas turbine

(STIG), intercooled STIG (ISTIG), or a combined cycle (Williams and Larson 1993). Biomass gasifier–gas turbines are characterized by high conversion efficiencies and lower expected unit capital costs in the 5–100 MW capacity range and are likely to become available commercially by the late 1990s (Larson 1993). BIG/GT systems have a number of attractive features for application in developing countries, such as low anticipated capital cost and potential for a high share of local manufacturing content (OTA 1992).

Land required per megawatt

Assuming a low conversion efficiency of 18 per cent and 1.3 t wood/MWh, the wood required for operating a 1 MW unit for 16 h/day would be 7592 t/yr. The land required per megawatt would be in the range of 1000–2000 ha (rounded from 949 to 1898 ha) at productivities of 8 to 4 t/ha/yr, respectively.

If the conversion efficiency is doubled to, say, 35–40 per cent using advanced systems, the land required would be halved to 500 and 1000 ha/MW at productivities of 8 and 4 t/ha/yr, respectively. A higher biomass productivity of 8 t/ha/yr, is assumed, since the power generation company managing the multi-mega watt-scale system is expected to provide technology packages for higher woody biomass productivity.

Size of units and possible number

The size and the number of units possible can be estimated depending on the availability of degraded land for dedicated energy plantations. If 1 MW systems are assumed, the catchment area required needs to be estimated (Fig. 6.7). In any given area or in a cluster of villages, only a fraction of the total area would probably be dedicated to energy plantations for supplying wood to large-scale systems. Assuming that only 5 per cent of the geographical area of a cluster of villages would be available for energy plantation and at 1000 ha/MW (at 8 t/ha/yr and 18 per cent conversion efficiency), the catchment area required would have a radius of about 8 km, which may be feasible. If 5 per cent of India's geographical area is assumed to be available for megawatt-level systems (amounting to 16 Mha) it would be possible to set up 16 000 units of 1 MW each, at one unit for every 35 villages. If the conversion efficiency is doubled to 36 per cent (which should be commercially feasible, Larson 1993) it would be possible to install 32 000 MW by dedicating only 5 per cent of India's geographical area. It does seem feasible for an entrepreneur to set up a 1 MW unit for power generation by having an agreement with farmers of 35 villages within a radius of 8 km, similar to a sugar mill, cotton ginning mill, or jute mill. Thus megawatt-scale systems seem feasible and demand the serious attention of energy planners. The size of megawatt-scale systems could vary from 1 MW to tens of megawatts, depending on land availability in different locations.

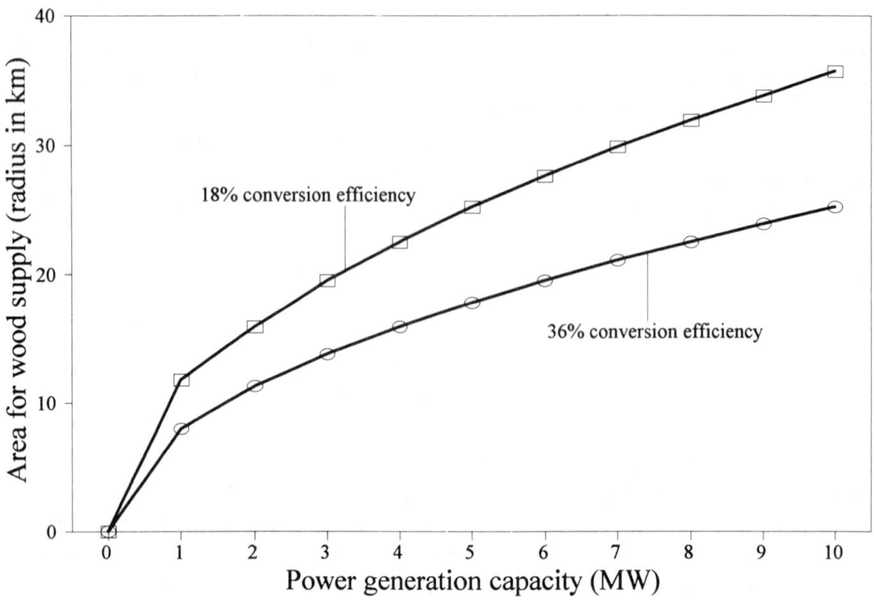

Fig. 6.7 Catchment area expressed as radius (km) required to supply woody biomass for producer-gas electricity systems of varying megawatt capacity with a conversion efficiency of 18 per cent and 36 per cent, assuming only 5 per cent of the total catchment area is dedicated for energy plantations.

6.14 THERMAL APPLICATIONS OF BIOENERGY

Apart from cooking, biomass energy is currently used as a source of heat energy in a range of rural establishments and industries. For example, rice husks are used as a fuel in rice mills for parboiling of rice, in firing bricks, in crop drying, and in establishments such as restaurants. A study of efficiency in rice mills in Madurai district in South India shows that the efficiency of rice husk use is only 13 per cent and concluded that there is considerable potential for increasing the efficiency using the gasifier option (Muthurasu *et al.* 1993).

There is a large potential for using bioenergy efficiently for thermal applications such as rice parboiling, firing brick and roofing tiles, tobacco curing, crop drying, and in establishments such as hostels and restaurants. Not much information is available on the type or extent of biomass used, or the efficiency of its use and the features and extent of demand for process heat. Thus no detailed analysis is attempted in this study.

6.15 CONCLUSIONS

Conventional energy planning based on commercial energy (oil and electricity) is shown to be non-sustainable and leading to an inequitable distribution of benefits, particularly neglecting the needs of rural areas. We consider that promotion of renewables is essential to achieve the joint goals of sustainable and equitable development. A set of minimum services required by rural communities for promoting development is defined. The energy end use approach is adopted to meet the energy needs of the defined services. Decentralized bioenergy systems for producing biogas and electricity, using local biomass resources, are shown to be more appropriate to promote development compared to other renewables. This is because, apart from meeting energy needs, multiple goals could be achieved, such as self-reliance, local employment, and land reclamation.

Analysis of the bioenergy potential shows that it is possible to meet the cooking energy needs using biogas from cattle dung and leaf litter, and the electricity needs of lighting and shaft power using the 'energy forest–producer-gas' option. The potential for biomass (lignocellulose)-based methanol production and megawatt-scale power generating systems was explored, and the feasibility and large potential were shown, especially for electricity generation. Land requirements were estimated for decentralized and megawatt-scale electricity production and large-scale liquid fuel systems. The critical issue of land availability for producing woody biomass for decentralized, as well as megawatt-scale, systems in a sustainable way has to be analysed in order to understand the feasibility of bioenergy systems. This is discussed in the next chapter.

Photo 6.1 Village electrification for street lighting, households, and water pumping.

Photo 6.2 A farm pump house with an electric pump connected to grid electricity for water lifting for irrigation.

Photo 6.3 A new 300 kW gasifier using agricultural residues for rural industrial applications (courtesy NARI, Phaltan).

Photo 6.4 Large-scale (50 MW) wood-waste power plant in Anderson, California, owned by Wheelabrator Shasta Energy Co. (Courtesy of NREL, photographer W. Gretz. Printed in *Biomass* and *Bioenergy*, Vol. 3, 1992).

7

Land and biomass availability for sustainable bioenergy

In previous chapters the potential and the technical and operational feasibility of biomass electricity systems for both decentralized village-level and megawatt-level systems were shown. The utilization of this large bioenergy potential requires sustainable supplies of biomass, particularly woody biomass. Thus the woody biomass requirement, the current biomass availability, and options for sustainable supply have to be considered. In a developing country such as India the crucial question always asked is whether it would be able to simultaneously meet the food requirements of a growing population and supply biomass for energy sustainably. Food production is always a priority, and thus it is necessary to analyse the food–fuel issue and the current woody biomass production programmes with reference to India.

The biomass requirements for the decentralized village kilowatt-scale biomass electricity option (122 Mt/yr) as well as megawatt-scale systems (128 Mt/yr) were estimated in Chapter 6. If India were to develop fully the biomass electricity option, it would need to ensure a sustainable supply of about 250 Mt of woody biomass annually. The features of such a wood requirement are that it is required in an air-dry state, all the year round, in the form of thin twigs and branches and not necessarily tree trunks or large branches. Between the decentralized village and megawatt options, the former gets priority over the latter as it pertains to the self-reliance of half a million villages for all their electricity requirements. Several opportunities exist for supplying wood on a sustainable basis and these are discussed in the following section.

7.1 OPTIONS FOR WOOD SUPPLY

Option 1 Conservation potential of wood used in cooking. The total fuelwood use for cooking in rural India is 181 Mt (Table 2.2). The fuelwood currently used for cooking is projected to be replaced by liquid and gaseous fuels, e.g. biogas, methanol, and kerosene. It was also shown in section 2.7 that a large part of the fuelwood used in rural India was obtained in a sustainable way. Substitution of fuelwood by liquid or gaseous fuels would lead to the saving of all the fuelwood used. Thus only 67 per cent of the fuelwood currently used for cooking is adequate as feedstock for biomass electricity generation to meet all rural electricity requirements. There would still be some surplus fuelwood for use in large megawatt-scale electricity systems.

Option 2 Producing wood on community, government, or degraded forest land.

Option 3 Producing wood on degraded private or farm land.

Option 4 Sustainable harvest from existing forests. In India there is a ban on felling trees in reserve forest areas (primary forest controlled by the Forest Department). This source is not considered due to the existing ban as well as the following reasons:

1. These primary forests are maintained for their ecological roles, such as conservation of biodiversity and their watershed role.
2. Periodic harvesting of wood, even if done sustainably, could damage biodiversity and forest succession due to the mode of cutting, transportation, and trampling. Furthermore, it may not be feasible to enter dense forests for periodic felling of branches without damaging the ground vegetation. The total annual woody biomass productivity of the forests is estimated to be 138 Mt (section 2.7.1).
3. Even if a sustainable rate of harvesting is determined and suggested, contractors or managers of utilities may overharvest, leading to degradation. Hall *et al.* (1993) also acknowledged the existence of potential for sustainable harvesting of wood from existing forests, but due to social and environmental concerns, particularly biodiversity, did not suggest the use of primary forests as sources of biomass for energy.

Option 5 Logging waste. The potential of logging wastes, estimated to be 9.5 Mt annually (Table 2.11), is too small to make any significant contribution to wood requirements of the biomass electricity option.

Consideration of options 2 and 3 involves a range of related issues, such as land availability, land quality, competitive uses of land, and sustainability of wood production. These options can be considered at two levels:

1. Some proportion of wood currently burnt as cooking fuel would become available for the producer-gas electricity option. Thus only a part of the 122 Mt is required to be produced annually. Tree plantations, farm trees, homestead gardens, and degraded lands are the various sources of fuelwood used for cooking (Table 2.11). Among these sources, only wood from tree plantations amounting to 40 Mt could be considered as easily available as feedstock for power generation. These plantations are grown to produce industrial wood and poles for the urban market and thus are harvested. This implies that 90 Mt of woody biomass needs to be sustainably produced for kilowatt systems and 120 Mt for megawatt systems.

2. All the 122 + 128 Mt is expected to be produced sustainably from dedicated energy plantations on community and farm land, so that the existing village tree or forest resources are not used.

7.2 TRENDS IN NON-CROPLAND AREA

There are three major categories of land, namely cropland, forests, and non-cropland (which is also non-forest land). Let us compare the population growth and changes in area under the cropland and non-cropland categories. Only some categories of degraded land are included here; detailed estimates are given in Table 7.1 and discussed in section 7.5.1. The population of India grew at an annual rate of 2.47 per cent during the 1970s and 2.36 per cent

Table 7.1 Estimates of degraded land availability in India (Mha)

SPWD (1984); degraded (waste) land, quoted in PC (1992)	Degraded forest	36
	Degraded non-forest	94
	Total degraded land	130
Chambers *et al.* (1989); land available for tree planting	Cultivated land	13
	Strips and boundaries	2
	Uncultivated, degraded land	33
	Degraded forest land	36
	Land for tree planting	84
Kapoor (1992); land available for tree planting	Agricultural land (143)[a]	45
	Forest land (75)	28
	Pasture land (13)	7
	Fallow (long) (17)	10
	Fallow (short) (23)	15
	Urban land (18)	1
	Total land for tree planting	106
Ministry of Agriculture (1992)[b]; land use statistics	Forest land with <10% tree crown cover	11
	Grazing land (pasture land)	12
	Tree groves (degraded)	3
	Culturable waste	15
	Old fallow	11
	Total	52
	Current fallow	14
	Total degraded land	66

[a] Values in parentheses are total area under the category.
[b] Definitions of land categories are the same as in Table 2.11 and an explanation of these is given in section 7.5.3.

during the 1980s (TSL 1992). Thus one would expect the area under crops (known as 'net sown area' in India) to grow during the 1970s and 1980s and the area under non-cropland to decline. Surprisingly, the non-crop area has stabilized at around 40 Mha between 1970–71 and 1988–89 (Fig. 7.1). The area under food grains has remained nearly stable at around 124–127 Mha between 1970 and 1990 (Fig. 7.2). Again, one would have expected that the non-crop area would have been converted to cropland as a result of population pressure. The minor variations in crop and non-cropland areas from year to year are probably due to variation in rainfall.

This observation at national level could be further supported by considering the land-use changes in Karnataka state (Ravindranath and Gadgil 1991). In Karnataka a similar trend was observed, where the cropped area stabilized at around 10 Mha (10.0 Mha in 1965–66 and 10.5 Mha in 1988–89) and the non-crop (and non-forest) land area also stabilized around 5.5–5.8 Mha between 1965–66 and 1988–89. A partial explanation for the observed trend is presented in the next section. Two conclusions are possible: first, the cropped area has remained constant over a period of 20 years at both the national and Karnataka state level, and may continue to stabilize; and, secondly, over 40 Mha of non-cropland could be available for energy plantations or wood production.

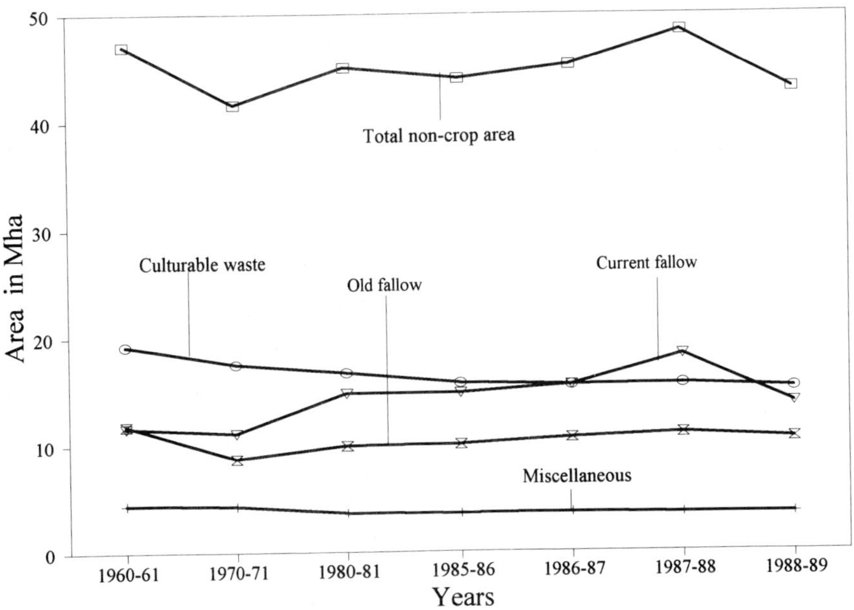

Fig. 7.1 Area under non-cropland categories over the years (1960–89). For an explanation of the land categories refer to section 7.5.3 and Table 2.11 (Ministry of Agriculture 1992).

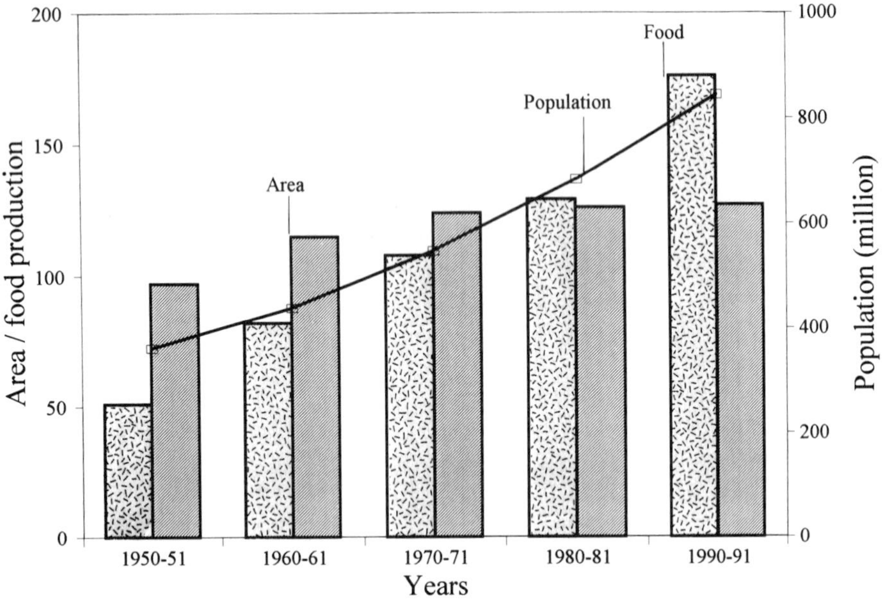

Fig. 7.2 Growth in population, area under food crops, and food production; area in Mha, food production in Mt (CMIE 1992; TSL 1992).

7.3 FOOD VERSUS FUEL

The food and fuel issue has been debated on many occasions. A general feeling is that land availability is not a constraint for biomass production for energy except in populous Asia (Woods and Hall 1994). Predictions like the one by the IPCC's Response Strategies Working Group (IPCC 1991) show a 50 per cent increase in cropland by 2025, in parallel with the world's rising population. Another study for major countries, including India, also predicts increases in cropland for food production (Bekkering 1992).

Population and food production

The population of India nearly doubled in the 30 years between 1960–61 and 1990–91 (Fig. 7.2). However, the area under food crops increased by only 10 per cent, while food grain production more than doubled. The per capita food grain availability (cereals + pulses) increased from 469 g/capita/day in 1961 to 510 g/capita/day in 1991 (TSL 1992). Thus the rate of growth of food grain output has been higher than the population growth.

Potential for increased food production

The potential for increasing food production without increasing the area under food crops is assessed here by considering the current and the potential

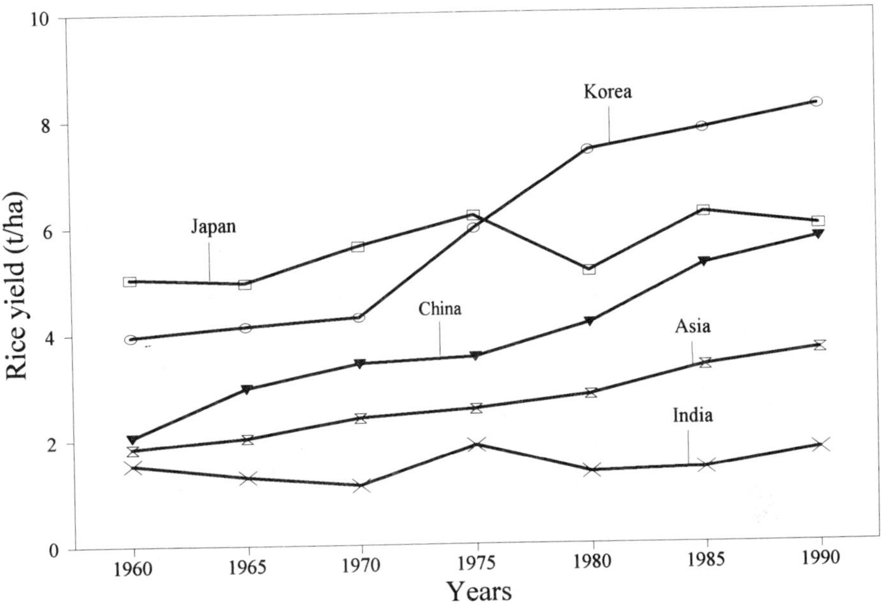

Fig. 7.3 Trends in rice productivity in some Asian countries.

productivity as well as irrigation levels. A comparison of productivities can be made at two levels; first, between Asian countries (Fig. 7.3) and, secondly, within India between different regions.

Rice is the most important crop with respect to area and food production in India as well as the world. The productivity of rice in India in 1990 was 1.8 t/ha, compared to 5.7 t/ha in China and 8.2 t/ha in Korea (Fig. 7.3). Even the Asian mean yield for rice is double that of India. Rice productivity is generally increasing in all the Asian countries. If India could achieve the rice yields of China, rice production in India would increase from 74 Mt (in 1990–91) to 252 Mt, which is higher than India's total food grain production of 176 Mt for 1990–91 (TSL 1992). This clearly shows the large potential for increasing food production in India by intensive cultivation rather than expansion of area.

The major crops in India are rice (42 Mha), wheat (24 Mha), jowar or sorghum (14 Mha) and pulses or legumes (24 Mha) (for the year 1990–91) (TSL 1992). The productivities of these crops and their production trends are given in Fig. 7.4. The productivities of all the crops are low, particularly those of sorghum, bajra (pearl millet), and pulses (less than 1 t/ha) and these three account for 49 Mha, which is more than the area under rice (42 Mha). The productivities of food grains, as well as the major cereals, millets, and pulses, have been increasing since 1970–71, but they are still very low. Overall food grain productivity is 1.38 t/ha/yr (1990–91). If the overall food grain

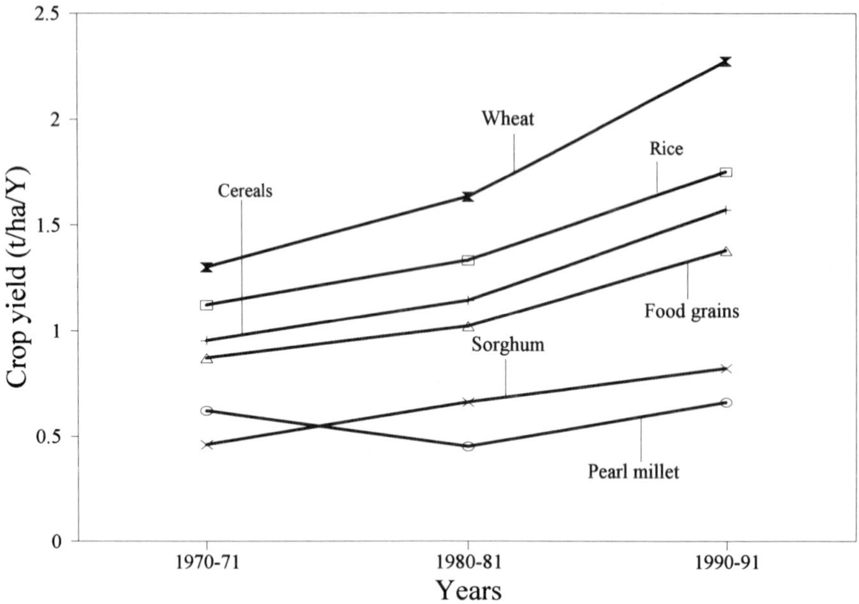

Fig. 7.4 Productivity trends for rice, wheat, sorghum, pearl millet, cereals and food grains in India (CMIE 1992; TSL 1992).

production were to be increased to a moderate 3 t/ha/yr, the total food production would increase from 176 Mt to 423 Mt (for 1990–91).

Within India the productivities of major crops, such as rice, wheat, and sorghum, vary from state to state (Fig. 7.5). Rice productivity in Punjab and Tamil Nadu states is nearly double that of Bihar and the national average. Similarly, wheat yields in the Punjab are double that of Bihar. If the national average yields could be enhanced to the level of yields in states like Tamil Nadu and Punjab, there would obviously be a significant impact on food production.

Barriers to crop productivity

Detailed analysis of the barriers contributing to low crop productivity or the factors required to achieve high crop productivity, is beyond the scope of this book. The main aim is to understand the availability of land for sustainable woody biomass production. The major factors contributing to productivity are simply listed here.

1. Irrigation water In India 77 per cent of the cropped area is under rainfed conditions. Apart from contributing to increased yield, irrigation also increases the cropping intensity. In India, under monsoon-dependent rainfed agriculture only one crop is grown, but with irrigation it would be possible

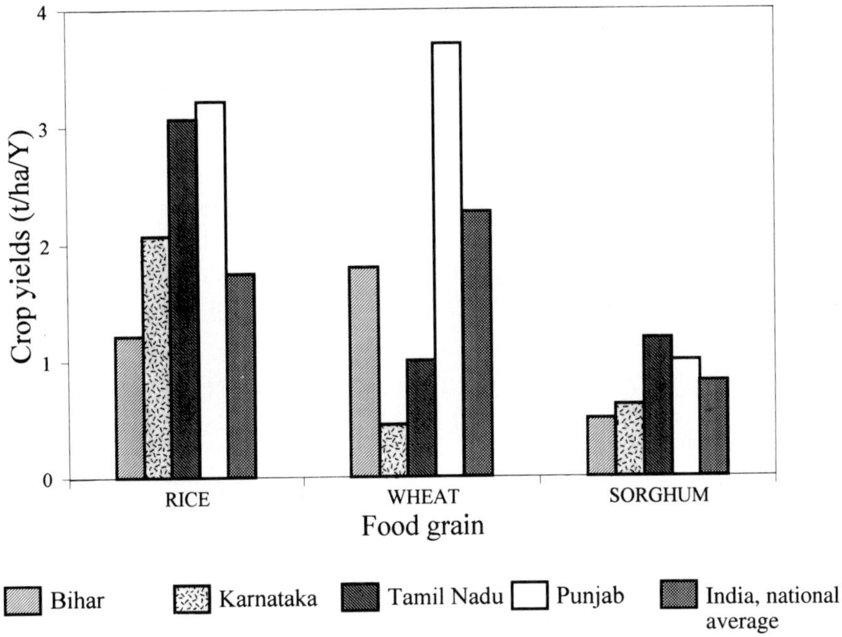

Fig. 7.5 Comparison of rice, wheat, and sorghum yields in different states of India for 1990–91 (TSL 1992).

to grow at least two crops. Currently only 38 Mha out of 143 Mha are cropped more than once. The total irrigation potential in India is estimated to be 113 Mha, which is 80 per cent of the currently cropped area (TSL 1992). Therefore, there is a large potential for increase in the cropped area under irrigation, thereby increasing cropping intensity as well as crop productivity. To enhance irrigation water supplies, energy to lift water is critical since nearly half of the potential is in the form of minor irrigation systems, which include open wells, tube wells, and lift irrigation from streams and rivers. It is thus considered possible to double or triple crop productivity and cropping intensity with provision of irrigation without increasing the area cropped.

2. Fertilizer or nutrient supply In conjunction with water supply, crop productivity is highly sensitive to nutrient supply. The mean fertilizer application rate in India is 68 kg/ha, compared to 106 kg/ha in Sri Lanka, 262 kg/ha in China, 452 kg/ha in Korea, and 417 kg/ha in Japan (TSL 1992). Sustainability of the high rate of application of fertilizers as observed in Japan or Korea is questionable, but in India there is certainly potential for increasing crop productivity through increased rates of organic as well as inorganic nutrient application.

3. *High-yielding varieties (HYV)* In India 66 per cent of the area under rice, 46 per cent of that under sorghum, and 49 per cent of the area under pearl millet are under high-yielding varieties. There is still potential for increasing crop productivity by increasing the area under HYV in India.

Crop productivity has to be increased, irrespective of bioenergy considerations, in order to increase the productivity of land, labour, and capital. The main conclusion from the above discussions is that current crop productivities in India are low and there is large potential for doubling or tripling crop yields and total food production without expansion of the area under crops.

7.4 LAND REQUIREMENT FOR FOOD, FODDER, FUEL, AND TIMBER

A large, populous developing country such as India must adopt an integrated strategy to meet the growing food, fodder, fuel, and timber requirement. The main reason is that these dominant biomass needs should not compete for land, given the high human and livestock population densities. Traditionally, fodder is a by-product of cereal crops, which account for 72 per cent of the total cropped area (TSL 1992). The biomass requirement for energy should not compete for land with food or fodder requirements. A detailed analysis of strategies for food, fuel, fodder, and timber for India is beyond the scope of this book. However, based on the analysis in earlier sections, we attempt to summarize the arguments to show that land is not a constraint for bioenergy options. The area under crops has stabilized at around 140–143 Mha from 1970 to 1991 (Fig. 7.2), and the degraded land estimates vary from 66 Mha to 130 Mha (Table 7.1).

Food production

The food grain productivity is low, for example in 1990–91 it was 1.38 t/ha/yr (TSL 1992). According to a projection by a working group set up by the Ministry of Agriculture, the demand for food grains for a population of 1 billion is projected to be 180 Mt by 2000 (*Deccan Herald* 1994). An estimate of the food grain requirement (not demand) made by the International Food Policy Research Institute (IFPRI), quoted in *The Observer* (1994), is 210–220 Mt by 2000. The projected food grain requirement of 210 Mt can be achieved by increasing the food grain yield from the current 1.38 t/ha (in 1990–91) to 1.64 t/ha by 2000 using the 127.5 Mha (1991) of land currently under food grains. In fact there is potential to at least double food grain productivity to the range of 2–3 t/ha with expansion of irrigation, multiple cropping, and modern cultivation practices. Thus it is possible to meet the food grain requirement of the growing population without increasing the cropped area.

Fodder requirement

Increased irrigation and cropping intensity, leading to an increased area under cereals, will lead to a corresponding increase in fodder production as a by-product. The grass productivity of pasture or village commons or degraded forest lands is very low (section 7.5.2). Shailaja *et al.* (1994) have shown that through mixed forestry of appropriate density, for bioenergy, the grass production from the currently degraded lands could be increased.

Timber requirement

Timber is required for industrial (softwood) and structural (hardwood) uses. It is shown (section 7.5.2) that only 3 Mha for industrial timber plantations and 4 Mha for hardwood plantations are adequate to meet the projected demands for the year 2000. This shows that only a fraction of currently degraded land is adequate to meet the timber needs. Bringing more area under timber would lead to a market surplus and decline in price.

Biomass fuel

Taking the lowest estimate of degraded land, 66 Mha (Table 7.1), about two-thirds of this is adequate for decentralized rural bioenergy centres for 0.5 million villages as well as 16 000–32 000 MW of megawatt-scale biomass electricity (depending on woody biomass productivity and efficiency of conversion). Even after dedicating land for producing woody biomass for energy and timber, surplus degraded land would be available.

It may be necessary to invent some incentive schemes to reclaim and revegetate surplus degraded lands. It is assumed that cooking-energy requirements would be met through biogas potential (Chapter 6) and in all the estimations the timber and fuel contribution of existing tree plantations (*Eucalyptus*, teak, pines) and village/farm trees and shrubs are not included. Thus an integrated approach to food, fodder, timber, and fuel shows that all these requirements could be met for the growing population without any competition for land.

7.5 LAND AVAILABILITY FOR BIOENERGY AND OPPORTUNITY COST

Having shown that there is a potential to double or triple India's food production even with the current cropped area, it is necessary to analyse the actual area available for fuelwood production, its opportunity cost, and the wood production potential.

7.5.1 Total degraded land available

Available estimates of the extent of degraded land are contentious, and the subject is complicated by the definitions used. Degraded land could be defined simply as land whose current biological production is lower than its potential and is declining. The various estimates available are given in Table 7.1.

SPWD estimate (1984) The total degraded land is estimated to be 130 Mha, with non-forest land accounting for 72 per cent.

Chambers et al. *(1989)* They estimated the total land available for tree farming as 84 Mha, which includes degraded forest, crop- and non-cropland.

Land use statistics Excluding degraded cropland, the other categories of degraded land were estimated to be 66 Mha (Ministry of Agriculture 1992).

Kapoor (1992) Based on various estimates of the National Wasteland Development Board, Kapoor (1992) gave a detailed estimate of the total area of degraded land as 106 Mha.

Kapoor (1992) and Chambers *et al.* (1989) also gave estimates of degraded land under crops. Farmers may be cropping such degraded lands for the lack of any other options. Lands on steep slopes, saline soils, and highly eroded soils may not be fit for profitable crop production. Alternative options, such as tree plantations (e.g. *Eucalyptus*) or horticultural gardens (e.g. mango) involve long gestation periods and large initial investments and thus are often unattractive, especially for poor farmers. With appropriate incentives even a part of the currently degraded cropped area, not fit for cropping, could be shifted to tree farming. In India, farm forestry occurs on cropped land; for example, part of the *Eucalyptus* plantations in Gujarat, Karnataka, and Haryana states are on previously cropped lands (Saxena 1992). Notwithstanding these considerations, in our analysis tree farming on current cropland is not necessarily required for woody biomass production for energy (irrespective of the status of land or crop productivity) since large areas of non-crop land appear to be available.

7.5.2 Opportunity cost of land

Unfortunately there are no studies to show the vegetation status of these non-cropland categories. India has a cattle and sheep + goat population density of 1.4 and 0.8/ha of non-cropland, respectively. This implies that every hectare of non-cropland (including dense forests and deserts) supports more than two livestock. If dense forests and desert areas are excluded, the grazing pressure on the remaining lands would be even higher. Cattle only partly meet their feed requirements from grazing, but sheep and goats meet their feed requirements fully. Thus the opportunity cost of the land considered for energy plantations is the grass production grazed by livestock.

Case studies of grass productivity of grazing lands in Rajastan (Kumar and Joshi 1972), Gujarat (Pandya and Sidha 1987), and Haryana (Panda *et al.* 1992) indicate that the grass productivity is low, in the range 0.40–1.6 t/ha/yr. A study of mixed tree plantations has shown that grass productivity can be higher in mixed tree plantations compared to monocultures or even open, degraded lands (Shailaja *et al.* 1994). Thus with silvipasture plantations, it may not be difficult to ensure the current levels of grass production. In fact with proper mixing of tree species and densities the grass productivity may increase as the soil quality improves. If the cropping intensity and productivity of cereals and millets are increased, more crop residues will be available as fodder. Thus there need not be any conflict between energy plantations and meeting the grass requirement of livestock.

The potential alternative uses for the degraded lands are crop production timber plantations, and forest succession.

1. The land categories under consideration are degraded so that profitable cropping may be ruled out.
2. The timber (hardwood) and industrial softwood requirements are projected to be 8 Mt and 12 Mt annually by 2000 (Kadekodi and Ravindranath 1994). In India since 1980 about 17 Mha have been put under timber and softwood plantations. Even if all timber and softwood requirements have to come from new dedicated plantations, the land required is 3 Mha for softwood (at a productivity of 4 t/ha/yr) and 4 Mha for structural timber (at 2 t/ha/yr productivity). Thus only 7 Mha are adequate to meet the projected demand for timber for the year 2000. If large areas of degraded land were used for timber and industrial softwood plantations, there could be a surplus and market collapse. We conclude that there is only a limited potential for using the degraded land for timber or industrial softwood options.
3. The option of promoting forest succession on degraded lands may not provide sufficient incentives to farmers or local communities, particularly for degraded farmlands. This is also probably the case for fallow lands due to the absence of any short-term financial gain.
4. In order to reclaim farmers' (private) and local communities' (degraded forest and village commons) degraded land, financial incentives are required. Energy plantations for bioenergy could be one of the motivating factors to reclaim degraded lands profitably.

7.5.3 Land categories for forestry

The three main categories of lands based on tenureship are degraded forest land, village commons, and degraded farmland with potential for fuelwood production (Table 7.1).

Forest Department

Of the three categories of land based on tree crown cover:

(1) forests with >40 per cent crown cover (37.8 Mha) could be considered as dense forests requiring protection;
(2) forests with 10–40 per cent cover (25.7 Mha) could be termed partially degraded forest, with potential for enhancing standing biomass as well as productivity;
(3) forest land with less than 10 per cent crown cover (11 Mha) is fully degraded and requires revegetation.

Community or local government land

Traditionally, villages had some 'common land', called by various names, mainly for meeting the diverse biomass needs of villages. These village commons or pastures (12 Mha) are often in a degraded state and are used largely as grazing lands. The goal of forestry on these lands should be to meet the diverse biomass needs of the local community, such as providing timber and non-timber tree products. The non-trunk woody parts of trees, such as branches and twigs, can be sustainably harvested as feedstock for biomass electricity systems. Demand for timber for structural purposes in villages is relatively small, since the number of new houses built annually is not large; even at the national level the structural timber required is only about 8 Mt annually. Thus only a fraction of common land has to be dedicated for structural timber products and the remaining land could be dedicated for growing wood for energy. As demonstrated in Hosahalli village (Chapter 5), the common land can be used for wood production for biomass electricity if it is acceptable to the village community.

Adequacy of village commons It is very difficult to assess the distribution of village commons among villages. At the national level the average area of village commons is 22 ha/village (12 Mha for 0.55 million villages). The village commons category by itself may not be adequate in most villages, thus use of degraded private land is necessary to provide wood as feedstock for decentralized power generation. But it is necessary to make full use of all surplus village commons for energy as a mode of reclaiming the lands.

Farmers' lands

There are several categories of land, depending on usage and frequency of cropping (Table 7.1):

(1) miscellaneous tree crops and groves which are currently in a degraded state with respect to tree vegetation;
(2) culturable waste land which is fallow (not cropped) for long periods (over 5 years);

(3) other fallow land, left fallow for periods of 1–5 years, could be considered as available for forestry;
(4) current fallow land which is left fallow during the reference year and which is cropped depending on rainfall could also be available.

However, degraded land under this last category is not considered as available for forestry in this study for several reasons. First, if large timber, softwood or fuelwood surpluses are generated, the markets and prices are likely to decline. Secondly, it may not be desirable to divert cropland from food production to wood production, even if degraded.

As seen in Fig. 7.1 the degraded land under various categories of non-cropland, excluding current fallow, has remained stable at around 40 Mha between 1970 and 1990. The question that arises is why these land categories exist, or why farmers do not cultivate them.

Highly degraded These land categories are in a highly degraded state and are thus not suitable for cropping.

Absentee landlords The owner of the land may live far away in a town or city, and the land is not cultivated for several years.

Large holdings Farmers with large holdings are unable to cultivate all the land they own due to labour scarcities or inability to supervise cropping adequately. Thus they leave part of the land fallow, particularly fragments away from settlements or degraded patches.

Although the authors feel that energy plantations should not be grown on regularly cropped land, even if it is in a highly degraded state, it may not be possible to stop farmers growing energy plantations on currently cropped land. However, market factors are expected to prevent any large-scale shift to energy plantations. If considered necessary, it is possible for the government, banks, or industry not to extend the incentives for raising energy plantations on regularly cropped land, particularly irrigated cropland.

7.5.4 Land for megawatt-scale producer-gas systems

It was shown in Chapter 6 that by dedicating only 5 per cent of geographical area or 16 Mha of degraded land it would be possible to install a generation capacity of 16 000–32 000 MW, depending on the efficiency of conversion. Demand for wood for megawatt-scale biomass electricity systems would provide the needed incentive to revegetate the surplus degraded lands after meeting the wood requirements of decentralized kilowatt systems. Thus land is not a constraint in adopting the megawatt-scale option, given that only 5 per cent of the total land is adequate for making a significant impact on the power supply.

7.5.5 Leaf litter for biogas production

It was shown in Table 6.4 that the leaf litter from land dedicated for energy plantations for producer-gas electricity, along with cattle dung, is nearly adequate to meet the cooking energy requirement through biogas. Using leaf litter for biogas production improves nutrient cycling through a reduction of the loss of nutrient value due to anaerobic digestion of leaf biomass. If megawatt plants are also set up, this would lead to enhancement of leaf litter availability for biogas production, due to an increased energy plantation area.

7.6 AFFORESTATION AND BIOMASS PRODUCTION PROGRAMMES

In India there is a long history of forestry programmes aimed at producing biomass for various needs, including fuel. The two major approaches to biomass production are promotion of natural regeneration and artificial plantation forestry. India has implemented one of the largest afforestation programmes in the world, as is evident from the 14.4 Mha afforested during the period 1981–90 (FAO 1993*a*). Thus it may be important to understand and learn from the world's largest forest regeneration or biomass production programme.

Form 1950 to the early 1970s the revegetation programmes were on a small scale, dominated by promotion of revegetation on partially degraded forest lands through coppice regeneration. Since 1978–79 artificial regeneration replaced natural regeneration and the scale of afforestation was significantly increased. The two major factors that changed the afforestation drive towards the end of 1970s and early 1980s were, first, the report of the NCA (1976) which recommended the social forestry concept, and secondly, funding from various external agencies, such as the World Bank (in Gujarat and Uttar Pradesh), SIDA (in Tamil Nadu), and USAID (in Madhya Pradesh and Maharashtra). Social forestry, when launched in India during the early 1980s, was a novel programme of afforestation as it included the term 'social' and caught the imagination of the world. The main goals of social forestry were and continue to be:

(1) increasing fuelwood, fodder, small timber, and minor forest product supplies for the rural population;
(2) restoring a proper ecological balance and optimum utilization of land, water, livestock, and human resources.

Although the goals contain many broader desirable objectives, including ecological ones, the real purpose of social forestry was to meet the large fuelwood demand projected by NCA (1976) and ABE (1985) and the possible impact of such a large demand for fuelwood on deforestation. Thus afforestation programmes are directly relevant to any discussion on bioenergy. In the following sections the achievements of biomass production programmes, biomass productivity, impact on biomass fuel supply, and socio-economic aspects are discussed.

7.6.1 Area afforested since 1980

In the 30 year period 1950–80, the area afforested was on a small scale. However, during the 5 year period 1980–85, the area afforested increased significantly, and it peaked during 1985–90, when 1.5–2.1 Mha were afforested annually (Fig 7.6). Even in the post-1990 period about 2 Mha are afforested annually. The afforestation programme includes:

(1) community forestry on village commons and territorial forestry on degraded forest land;
(2) farm forestry on farmers' lands; and
(3) strip or row plantations along roads, streams, and irrigation tank beds.

According to a report of the Ministry of the Environment (1990) nearly half of the afforestation or social forestry programme is farm forestry on private lands and the rest is on degraded forest and village commons.

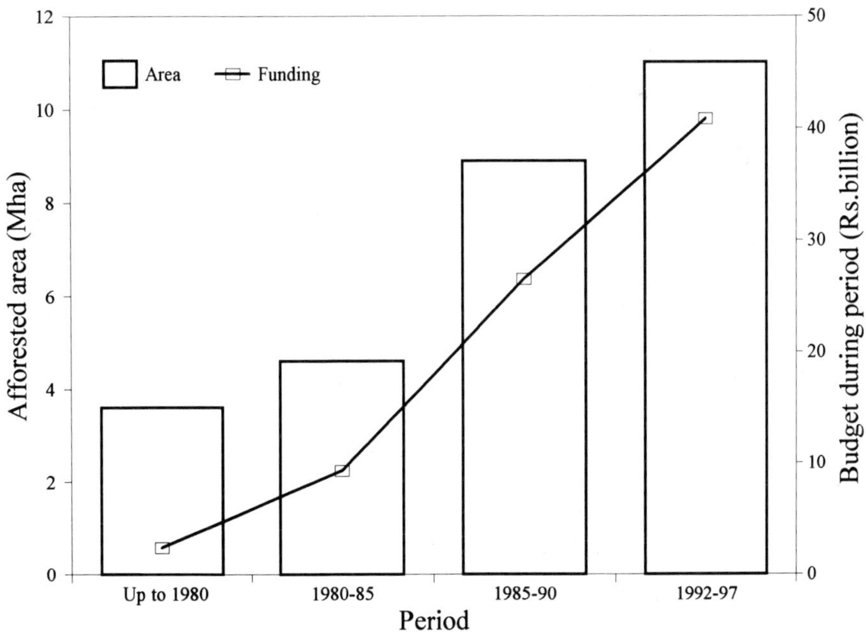

Fig. 7.6 Area afforested and budget for forestry over the years. Proposed area to be afforested and budget allocated for the period 1992–97 (Kapoor 1992; PC 1992).

Community forestry: a case study of Sirsi

Sirsi is in the Western Ghat hilly region of Karnataka with an annual rainfall of about 300 cm. The area brought under afforestation in Sirsi administrative block from 1983 to 1991 was 6370 ha or 66 ha/1000 rural population, compared to 21 ha/1000 rural population at the national level. A study was conducted in a cluster of 45 villages in a contiguous area around Sirsi during 1993 (Bhat, D. M. and Ravindranath 1994).

Species composition Monitoring of tree plantations in 45 villages established during the period 1983–91 showed that *Acacia auriculiformis* accounted for 85 per cent of seedlings planted, followed by *Casuarina equisetifolia* (10 per cent). These two exotic species dominate even in an area with high rainfall and where moist evergreen and moist deciduous forests dominate.

Density of plantations The density of trees in about half the plots was in the range 1500–3000 plants/ha and the rest of the plots had a density of less than 1500 plants/ha.

Productivity of plantations The productivity of plantations is in the range 5.3–7.9 t/ha/yr (Fig. 7.7). In Sirsi, one peculiarity is that the area

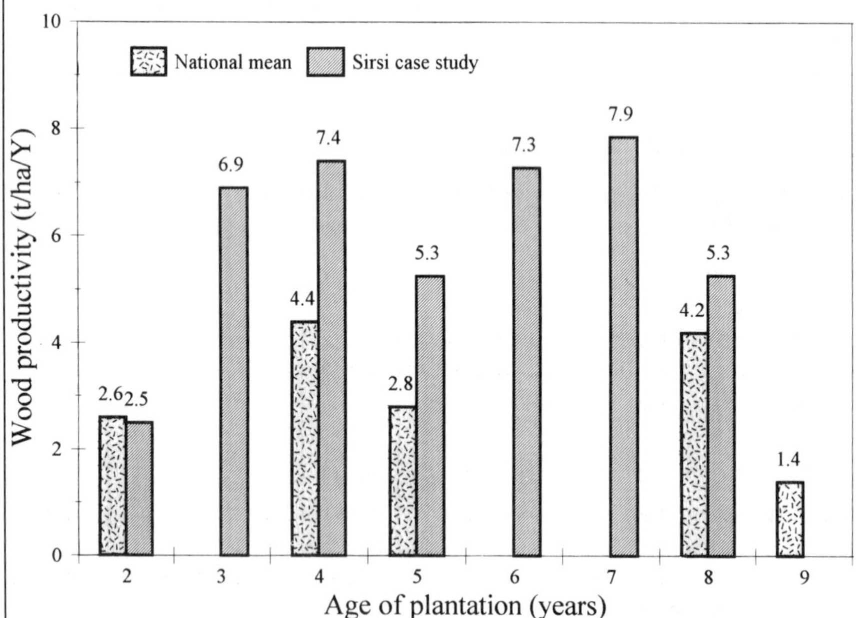

Fig. 7.7 Mean annual increment in woody biomass in social forestry plantations at different ages in the Sirsi area (Bhat, D. M. and Ravindranath 1994) and at national level (Seebauer 1992).

under tree plantations is all on village commons (minor forest) and on Forest Department land. In the selected villages there were no plantations on private lands. The productivity in Sirsi is nearly double that of the national average for non-farm forestry plantations. The high rainfall could be the main contributing factor.

Contribution of forestry plantations The adequacy of wood production available from the afforested areas was estimated using data on areas under plantations, annual productivity, age of plantations, per capita fuelwood requirement, and village population. In the Sirsi block the annual woody biomass production available from tree plantations is adequate to meet 100 per cent of fuelwood requirement in 31 per cent of the villages, 50–100 per cent of the requirement in 33 per cent of the villages, and in 36 per cent of the villages it was less than 50 per cent of the annual fuelwood requirement. Thus it is concluded that the afforestation programme has the potential to meet biomass fuel requirements to a significant level. However, what fraction of the biomass production is actually accessible to the local communities is not known. The second question is the dominance of *Acacia auriculiformis*, which is known to be a poor coppicing or non-coppicing species and once harvested will require replanting. Afforestation in the Sirsi area is a success story with respect to area afforested and woody biomass productivity.

7.6.2 Evaluation of afforestation programmes

The criteria and parameters used to evaluate afforestation programmes are along the following lines:

(1) species diversity of forestry plantations;
(2) survival rates;
(3) biomass productivity and production;
(4) contribution to fuelwood requirements; and
(5) access and socio-economic impacts.

Species compositions of social forestry plantations

One of the main goals of social forestry is to meet the diverse biomass needs of rural communities, such as fuelwood, fodder, small timber, and minor forest products (fruits, leaves, flowers, etc.). To meet such a diverse biomass requirement an appropriate species mix is required.

The species composition of social forestry plantations at national level is given in Table 7.2. The exotic species *Eucalyptus* hybrid and *Acacia auriculiformis* together accounted for 80 per cent of mixed tree plantations at national

Table 7.2 Species composition of farm forestry and community forestry under an overall social forestry programme (% to total trees)

Species	Community forestry (national survey)[a]	Farm forestry (national survey)[a]	Tumkur[b] community forestry	Sirsi block in western Ghats community forestry[c]
Eucalyptus sp.	30.0	71.6	55.0	
Azadirachta indica	2.1	1.4		
Casuarina equisetifolia	6.6	6.5	4.0	10.0
Dalbergia sisso	2.6	0.2		
Tamarindus indica	1.3	1.7		
Acacia nilotica	4.8	0.4		
Leuceana leucocephala	1.7	3.9		
Tectona grandis	12.0	0.3		2.0
Terminalia arjuna	1.5	1.5		
Bamboo	1.9	0.1		
Acacia auriculiformis			25.0	85.0
Others	35.5	12.4	16.0	3.0
Total	100.0	100.0	100.0	100.0

Sources:
[a] IIPO (1990).
[b] Ravindranath *et al.* (1992*c*).
[c] Bhat, D. M. and Ravindranath (1994).

level (Seebauer 1992). A study in the semi-arid Tumkur district (Karnataka), has shown that *Eucalyptus, Acacia auriculiformis*, and *Casuarina equisetifolia* accounted for 84 per cent of tree seedlings actually planted. In the heavy-rainfall Uttara Kannada district (Karnataka) *Acacia auriculiformis* accounted for 85 per cent of the seedlings in social forestry plantations (Bhat and Ravindranath 1994*a*). This preference for exotics whose leaves are not palatable to cattle and which do not yield any fruits or flowers, tends to reflect decisions of the Forest Department, rather than local preferences or priorities (Arnold and Stewart 1991). When the species diversity under farm forestry was examined, a study in 10 states showed that 70–100 per cent of farmers planted *Eucalyptus* in block plantations (PC 1987). Another study in Ungra area showed that all farm forestry plantations were that of *Eucalyptus*. Some of the reasons according to the study (Ravindranath *et al.* 1992*c*) are:

(1) it is a fast growing, quick coppicing, and hardy species;
(2) there is a large urban market for poles, which fetch a higher price than fuelwood or softwood supplied to industry;
(3) it is not grazed, since it is not palatable to livestock;
(4) seedlings are easily available from the Forest Departments, as well as

from private nurseries, and it is easy to propagate with a high survival rate;

(5) there is an absence of information on the performance of other species and species mixes.

The species choice of *Eucalyptus* and *Casuarina* really shows the preference for trees yielding poles (for construction) and fuelwood, rather than promotion of biodiversity or multiple benefit features. In fact, *Acacia auriculiformis* cannot be used for poles like *Eucalyptus*, because of its branching features, and it is thus only useful for fuelwood or pulpwood. None of the three dominant species provide fodder for cattle, any fruits, flowers, seeds, or even leaf manure to the local community.

Survival rates

In the afforestation programme it is not enough just to plant seedlings; their survival rate and subsequent productivity are important. There have been two national-level studies (IIPO 1990; Seebauer 1992) and a number of location-specific studies (Ravindranath *et al.* 1992*c*; Bhat, D. M. and Ravindranath 1994). It can be seen from Table 7.3 that the national mean survival rate for all plantations is 77 per cent, on farm forestry plantations it is 82 per cent, and on government plantations it is 75 per cent (Seebauer 1992). Thus we

Table 7.3 Survival rates of tree plantations under social forestry

	Type of plantation	Survival of seedlings planted (%)	Reference
Ownership of plantations	1. Farm forestry + community forestry	82	Seebauer (1992)
	2. Community forestry	75	
Plantation type; species	1. *Eucalyptus* sp.	73	
	2. *Acacia* sp.	78	
	3. Mixed species plantation	74	
	4. Other timber	91	
National mean	All plantations	77	
Karnataka	1. Community forestry	82	IIPO (1990)
	2. Farm forestry	67	
Sirsi	Community forestry	78–100	Bhat, D. M. and Ravindranath (1994)
Tumkur (around Ungra)	Farm forestry + community forestry	90	Ravindranath *et al.* (1992*c*)

conclude that the survival rates of plantations on farms as well as in Forest Department and village common lands are reasonably high, and should dispel any doubts on the establishment phase of tree plantations. One explanation for the high survival is that the species chosen are exotics, hardy, and non-palatable to livestock. The private plantations are protected by farmers, and plantations on village commons are protected by fences and/or forest guards.

Productivity

The annual above-ground woody biomass productivity for different locations and types of plantations is given in Table 7.4. The productivity of social forestry plantations on farmers' lands (farm forestry) is in the range of 4.2 t/ha/yr (at national level, according to Seebauer 1992) to 8.2 t/ha/yr (in Karnataka, Ravindranath *et al.* 1992*c*); these productivities are higher than that of Forest Department-promoted plantations, which are in the range 1.5–2.6 t/ha/yr. The productivity on Forest Department lands in heavy-rainfall areas such as Sirsi is 5.3–7.9 t/ha/yr (Bhat, D. M. and Ravindranath 1994).

Table 7.4 Annual above-ground woody biomass productivity of forestry plantations (air-dry t/ha/yr) in different locations and different categories of lands

	Type of plantation/ location	Productivity (t/ha/yr)	Source	Location
Karnataka	Farm forestry	7.2	DES (1985)	Kolar and Bangalore districts—semi-arid
	Farm forestry	8.2	Ravindranath *et al.* (1992*c*)	Tumkur district—semi-arid
	Forest Department	1.5	Ravindranath *et al.* (1992*c*)	Tumkur district—semi-arid
National	Farm forestry	4.2	Seebauer (1992)	National
	Forest Department	2.6	Seebauer (1992)	National
	Eucalyptus plantations	6.6		National
States	West Bengal	6.4	Seebauer (1992)	West Bengal
	Gujarat	4.1	Seebauer (1992)	Gujarat
	Tamil Nadu	3.8	Seebauer (1992)	Tamil Nadu
	Himachal Pradesh	1.6	Seebauer (1992)	Himachal Pradesh
	Andhra Pradesh	1.2	Seebauer (1992)	Andhra Pradesh
National mean	All plantations	3.2	Seebauer (1992)	India
West Bengal	Regenerating *Shorea*	6.6	Ravindranath *et al.* (1994*a*)	West Bengal
Karnataka	Forest Department	5.3–7.9	Bhat, D. M. and Ravindranath (1994)	Western Ghats heavy rainfall

Data on productivities according to age are available from two studies (Fig. 7.7). In Sirsi block (annual rainfall 300 cm) the productivity is high and is in the range of 5.3 t/ha/yr at age 5 years to 7.9 t/ha/yr in 7-year-old plantations, but declined thereafter. When the national-level mean productivity at different ages is taken, it seems to increase up to 4 years and then declines, but, as in Sirsi, productivity seems to peak again at about 8 years and then declines. It appears that 7–8 years is the optimum period for harvest for short-rotation plantations. The productivity of farm forestry plantations (largely *Eucalyptus*), both according to national-level as well as location-specific studies is in the range 4.2–8.2 t/ha/yr. There is an expected variation in productivity according to location, land category, and type of plantation. The productivity of farm forestry plantations could be termed moderate, and on village commons and degraded forest lands as low to moderate, depending on local rainfall. It seems reasonable to aim at a productivity of 2–4 t/ha/yr on degraded village commons and 4–8 t/ha/yr on farm plantations, without any major changes in production practices or genetic improvements.

Social aspects and access to woody biomass production

Social forestry programmes are aimed at meeting the biomass needs, particularly fuelwood, of local rural communities and especially the rural poor. When a tree is harvested, branches and twigs are obtained in addition to the main trunk. Even if a village tree plantation is sold as industrial wood or poles to an urban contractor, the twigs, branches, and tree tops are left behind in the village, due to their high cost of transportation, since it is easier to load straight poles than to load the tree tops or crooked branches with multiple twigs. The proportion of tree tops, twigs, and branches is estimated to be 32 per cent of the total standing biomass (Seebauer 1992). Thus even if tree plantations are contracted out for sale to urban and industrial sectors, at least 32 per cent of the standing biomass would be left behind in the village, estimated to be 40 Mt annually (Table 2.11). It is also common to lop the branches of standing trees and to collect fallen twigs, thus even before cutting down the trees, local communities benefit, to a certain extent, from the by-products of social forestry programmes. The rural poor are likely to get at least a part of the 40 Mt of lops and tops left behind in the village. Even the wood sold in urban areas will end up benefiting the urban poor, since they predominantly use fuelwood for cooking and even the poles used for scaffolding purposes will end up as fuelwood if they are damaged. Thus it is reasonable to conclude that the urban poor have also benefited from social forestry programmes, even though they were mainly intended for the rural poor.

7.6.3 Potential available, achievement, and barriers

Here we make a comparison of the area afforested with the extent of land potentially available for afforestation. If we take an estimate of degraded land available for tree plantations to be in the range 66–106 Mha (Table 7.1), at the current rate of afforestation of about 2 Mha/yr, 30–50 years will be required to plant trees on all the degraded land. The afforestation rate of 2 Mha/yr, although the largest globally, is far from adequate in India, given the large extent of degraded land available and the increasing biomass needs. What are the constraints in expanding the rate of afforestation?

Financial constraints

The establishment costs for social forestry programmes are about Rs10 000–15 000/ha (NWDB 1991). At such a cost the funding required is minimally Rs660 billion, if only 66 Mha of degraded land is considered. If a ten-year phased programme is planned, an annual expenditure of Rs66 billion would be required, compared to current funding levels of Rs8 billion/yr (PC 1992). Thus it seems impossible to revegetate all the lands at current establishment costs using the Forest Department funded and promoted conventional block plantations approach. The cost per hectare is high with the current approach since the programme is largely managed by the Forest Department without involving the local communities, NGOs, or industry. Some of the factors that contribute to cost escalation are:

1. The forestry plantations have to be protected from grazing and for this Forest Departments adopt an expensive approach of barbed wire fence + cattle-proof trenches + fully paid guards.
2. The organizational cost of the Forest Department staff is also partly included in the established costs. According to an estimate, the organizational costs (salaries and infrastructure) account for a third of the budget (Ministry of the Environment 1990).
3. Inefficiencies of functioning of government institutions also contribute to higher costs; for example, delays in decision-making and implementation, leakages of funds, no accountability, no genuine stake, use of poor-quality inputs and seedlings etc.

It thus makes eminent sense to look for alternative approaches to revegetate degraded lands.

Technical barriers

These are of two types: those that contribute to the high cost of raising the plantations and those that contribute to low biomass productivity. Afforestation programmes in India have suffered from both types of technological limitations. The important gaps are: poor-quality planting materials, lack of methods for revegetation of highly degraded lands, inappropriate planting and silvicultural practices for commercial and farm forestry (Tiwari 1994).

The productivity of forestry plantations, especially those on Forest Department and common lands, is low. Low productivities give little incentive for farmers or local communities to take the initiative to raise, protect, and manage tree plantations. Relatively long growth periods are also a barrier to large-scale implementation as farmers or local communities have to wait for 7–8 years before getting any significant benefits. Currently there is no mechanism to provide incentives such that they can obtain some biomass products or financial returns starting from the second or third year.

Institutional barriers

These are of two types: first, those concerning empowerment and involvement of local communities, and secondly, those involved in harvesting, transportation, marketing, and pricing of forest products. Currently there are no institutional arrangements to ensure proper land and product tenure to local communities or to meaningfully involve them in decision-making processes regarding species choice, mode of protection, harvesting, and sharing benefits in most locations. In the absence of such institutional arrangements, local communities are unlikely to commit themselves to the protection or management of degraded forests or village commons. There are also controls on the harvest and transportation of wood, including *Eucalyptus*, which deters private growers. Without assured markets, farmers are unlikely to invest significantly in fuelwood or softwood plantations, as currently the market for wood or timber is poorly developed, leading often to distress sales at low prices by farmers (Saxena 1992).

7.7 PRODUCTIVITY OF ENERGY PLANTATIONS

High woody biomass productivity is needed for the success of the energy plantation programme for bioenergy. This is because a large proportion of the wood required for bioenergy options is expected to come from farmers' lands where the incentive for farmers should be employment and profits. The productivity of energy plantations will depend on a number of factors, such as site quality, species choice, density of planting, soil and water conservation measures, and silvicultural practices.

Site quality

For a given site, the productivity will be determined by other factors. The land categories considered in Table 7.1 are all in a degraded state, which means their productive potential is presently low, due to factors such as overharvesting, overgrazing, soil compaction, soil erosion, and loss of plant diversity over the years as a result of high livestock and human population pressures. No specific information is available regarding the soil or vegetation status of the different categories of land.

In order to obtain high biomass yields, initial supplements of nutrients and organic matter may be necessary. At the present, the afforestation programme in India does not involve any significant levels of fertilizer application. In fact, all forestry on degraded forest land, village commons, and even farm lands, according to our observations, are raised without any fertilizer application or with small applications at the initial stages, unlike in Brazil and the USA where fertilizer application rates are high (Betters *et al.* 1991).

Even in the absence of fertilizer application, a mean productivity of 4.2 t/ha/yr at the national level, and 7.2–8.2 t/ha/yr in Karnataka, have been recorded for farm forestry plantations (Table 7.4). It is expected that the practice of application of little or no fertilizer may continue, given the fact that the present rate of application of fertilizer to cropland itself is also very low (68 kg/ha, TSL 1992). Thus proper land management, including multi-use crops, agro-forestry, water retention, and use of nitrogen-fixing species, will be necessary to reduce or avoid requirements for fertilizer (Hall 1993).

Species choice

The Indian afforestation programme is dominated by a few exotic species, such as *Eucalyptus* hybrid, *Acacia auriculiformis*, and *Casuarina equisetifolia*. On farmlands *Eucalyptus* seems to be a popular species in all regions of India (PC 1987). At present there has been no research on the adaptability of different clones, species, and species mixes to different soils and climatic situations in India. It is obvious from several studies that one or two tree species dominate in a region irrespective of soil quality or rainfall variations; for example, *Acacia auriculiformis* in the Western Ghats of Karnataka state and *Eucalyptus* + *Acacia auriculiformis* in southern Karnataka (Ravindranath *et al.* 1992*c*; Bhat, D. M. and Ravindranath 1994).

Major plantation programmes in many countries have embarked on tree-breeding strategies to increase the yield and quality of plantations, including introduction of new species, selection of varieties, formation of hybrids, clonal propagation, and ensuring seed and seedling quality (Evans and Hibberd 1993). For example, at Aracruz in Brazil, where vigorous testing of *Eucalyptus grandis* and *E. urophylla* seed origins, followed by clonal propagation of outstanding trees, has raised productivity from 22 to 55 m^3/ha/yr and has greatly increased resistance to cryphonectria canker (Campinhos and Ikemori 1989).

Realizing the importance of producing biomass for energy, the ministry (MNES) set up biomass research centres in 11 agro-climatic regions in India. One of the main goals is to identify productive tree species and to develop a package of practices for high yields from short-rotation tree crops. A summary of preliminary experimental results on productivity from various biomass research centres is given in Table 7.5, derived from Chaturvedi (1993). No information is available on irrigation or fertilizer application practices for the species and locations. The annual increment of woody

Table 7.5 Biomass productivity of selected plant species obtained in biomass research centres

Location	Name of species	Age years	Total yield (t/ha)	Yield (t/ha/yr)
Bhubaneshwar	*Cassia siamea*	3.5	59.2	16.9
	Leucaena leucocephala	3.5	39.5	11.3
	Casuarina equisetifolia	3.5	22.1	6.3
	Acacia auriculiformis	3.5	37.7	10.8
	Gliricidia sepium	3.5	25.1	7.2
Srinagar (UP)	*Prunus cerasoides*	7.0	126.0	18.0
	Alnus nepalensis	5.0	70.0	14.0
	Albizia stipulata	6.0	120.0	20.0
	Leucaena leucocephala	3.0	110.4	36.8
	Albizia lebbek	3.0	60.0	20.0
	Hardwickia binnata	3.0	100.5	33.5
NBRI, Lucknow	*Samania saman*	5.0	80.1	26.7
	Cassia siamea	3.0	62.1	20.7
	Erythrina indica	3.0	79.8	26.7
	Prosopis juliflora	8.0	96.5	12.1
NARI, Phaltan	*Terminalia arjuna*	6.0	42.0	7.0
	Acacia nilotica	8.0	59.5	7.4
	Acacia auriculiformis	8.0	54.5	6.8
	Prosopis juliflora	8.5	187.0	22.0

Source: Chaturvedi (1993).

biomass of all the tree species in the Madurai centre (semi-arid region) is very high, with 37 t/ha/yr for *Leucaena leucocephala* to 20 t/ha/yr for *Albizia lebbek*. Legumes tend to dominate the species selected for trials in India. *Prosopis juliflora* is a very hardy legume with profuse coppicing capacity and has recorded an annual increment of 12 t/ha/yr in Lucknow (North India) to 22 t/ha/yr in Phaltan (western India), both in semi-arid regions. However, it is now recognized that these experimental trials were restricted to monocultures.

Currently farmers or village communities have little or no access to any information on the relative performance of different (exotic or local) tree species, monocultures, and mixed cultures, or various packages of practices. There are very few studies reporting the performance of local tree species and mixed forest plantations in India or elsewhere (FAO 1993*d*). A study in South India has shown that under identical soil, rainfall, and tree density conditions, mixed tree cultures give higher productivities than monoculture plantations of *Eucalyptus* hybrid and *Leucaena leucocephala* (Shailaja *et al.*

1994). The benefit of mixtures of tree species is also shown by the *Eucalyptus saligna* and *Albizia falcataria* experiments in Hawaii. A 40:60 *Eucalyptus: Albizia* mixture resulted in a 14 per cent increase in total wood productivity over pure *Eucalyptus* at the age of 65 months. Even when only the *Eucalyptus* component of the mixture was considered, its yield was 60 per cent higher than the *Eucalyptus* in pure plantations (Debell *et al.* 1989).

Soil and water conservation

The moisture-holding capacity of soil subjected to erosion would be low due to the low organic matter status and loss of silt and clay. In monsoon regions, rainfall distribution is highly seasonal, with dry spells during monsoon and long dry summer months. There is a need to enhance the moisture retention in soil and to halt soil erosion wherever possible. Irrigation is not usually feasible due to the non-availability of water, undulating topography, and high costs. Inadequate information on the efficiency of different techniques in conserving soil and moisture is presently a constraint.

7.8 PRODUCTIVITY OF ENERGY PLANTATIONS FOR LAND ESTIMATIONS AND PROFITABILITY

The plantations on farmers' lands in several locations have shown annual productivities of 7.2–8.2 t/ha/yr, with a national-level mean productivity of 4.2 t/ha/yr (Table 7.4). There are two examples of community forestry: first, in the Western Ghats where 5.3–7.9 t/ha/yr productivity has been obtained; and secondly, in West Bengal state where about 6.6 t/ha/yr is obtained in the naturally regenerating *Shorea robusta* forests under community management (Ravindranath *et al.*, 1994*a*). The productivity of Forest Department or community forestry plantations is thus not necessarily low, though it is generally lower than that of farm forestry (Table 7.4).

The afforestation programme in India by farmers as well as the Forest Department is presently not based on any package of practices to encourage high yields. There is no programme to upgrade the seed quality based on breeding, selection, or propagation methods. No selection process is involved, and all seedlings that germinate and survive in the nursery are planted. There is no attempt to match the species to the site, as is evident from a number of studies covering many locations which show similar tree species and planting densities irrespective of site characteristics (Ravindranath *et al.*, 1992*c*; Bhat, D. M. and Ravindranath 1994). Appropriate genetic resources are fundamental to successful plantation forestry. Brazil has achieved significant increases in *Eucalyptus* productivity through genetic improvements coupled with other practices, such as vegetative propagation. Forestry research in India receives insignificant funding (<$US3 million annually during 1985–90, which is less than 2 per cent of the total research budget or less than

10 per cent of funding alllocated to agriculture) (PC 1992). There has also been little attempt to understand and incorporate traditional knowledge and practices of raising trees. There is thus an urgent need for tree improvement, including selection of species, variety and provenance, formation of hybrids, and clonal propagation for high and sustainable yields (Evans 1992).

An important reason for discussing these issues here is to show that even with little or no scientific input to plantation forestry, moderate yields (4–8 t/ha/yr) have been obtained in India. Thus the premise is that with the development and adoption of appropriate tree improvement and plantation forestry practices it should be possible to achieve higher productivity levels than currently obtained on both farmers' and other land categories. In this book we have assumed an annual increment of 4–8 t/ha/yr for all estimations of woody biomass productivity; in the long run this could be termed as conservative, given the potential for improvement.

In India nearly half of the afforested area is on farmers' land (Ministry of the Environment 1990). Obviously farmers must be growing forests for profit. The profitability of biomass feedstock production could be enhanced through a number of measures:

1. By mixing highly profitable fruit- or nut-yielding tree species, such as mango, coconut, and tamarind (*Tamarindus indica*), along with fast growing and coppicing tree species.
2. Similarly, high-value timber species (such as teak and mahogany) could be mixed in the energy plantation.
3. Crop residues of highly profitable crops, such as mulberry, could be used as feedstock.

The potential of such measures is limited by market factors. For example, large-scale mixing of, say, mango trees may lead to a glut of fruits in the market and a sharp decline in prices. The same will hold good for mulberry and coconut. What is desirable is to provide the option of mixing economically valuable tree species and leave the choice to farmers.

7.9 LONG-TERM SUSTAINABILITY OF BIOMASS PRODUCTION

7.9.1 Ecological aspects of long-term productivity

Some of the common perceptions, or rather apprehensions, of biomass use for energy are non-sustainability of productivity due to regular harvesting and removal of biomass, decline of soil fertility, loss of biodiversity, and high susceptibility to pests and diseases. The maintenance of long-term sustainability and site fertility is all too often given insufficient emphasis in plantation schemes (Lundgren 1980). Furthermore, many of the soils available for energy plantations in the tropics and sub-tropics are inherently infertile and are easily degraded through loss of nutrients as a result of harvesting and removal of biomass (FAO 1993*d*).

To generate woody biomass an option is to adopt short-rotation forestry by (1) replanting or reseeding, (2) promotion of coppice practices, and (3) mixed species forestry with sustainable logging. The crucial factor is to maintain high productivity from a given area in a sustainable manner. This issue gains further importance when promoting energy forests on degraded soils. Some information is available on the sustainability aspects for some species, such as *Eucalyptus*, pine, and teak, particularly under coppicing. Sustainable coppice yields are crucial for energy plantations subjected to periodic harvests.

Eucalyptus The commercial plantation experience with *Eucalyptus* in Brazil has reported reductions in coppice yields (Betters *et al.* 1991). In contrast to this, the experience in Australia suggests that *Eucalyptus* yields increased by 20 per cent in the first coppice, then declined by 0–20 per cent for the second, and by more than 20 per cent for the third coppice (Betters *et al.* 1991). In the US, studies suggested that coppice yields would be 20–100 per cent higher than the first harvest (Betters *et al.* 1991). However, in case of successive seedling crops of *Eucalyptus* there are no long-term declines in productivity. It should be noted that seedling replanting permits genetic changes through new species, variety and superior seed (Evans 1992).

Pines Evans (1992) concluded from studies in Swaziland and South Africa that there are no reports of declining yields of *Pinus patula* during second and third rotations.

Acacia The results of a study on *Acacia mearnsii* from an area subjected to eight rotations reported no apparent growth decline (Lewis 1967).

According to Evans (1992) no certain answer can be given about the yields of later rotations as there are little data available. Thus it is possible to suggest that with appropriate management practices it may be possible to obtain sustainable yields from a given site, particularly through replanting if necessary. The next question would be to identify management practices that ensure sustainability of productivity. The crucial practices for sustainable yields include species choice, quality clones and seed material, silvicultural practices including harvesting practices, and site quality maintenance.

Species choice for sustainability

This is relevant to ensure long-term productivity, reduction of soil erosion, restoration of soil nutrient status, and reduction of the risk of pests and diseases. Some measures are listed from FAO (1993*d*):

1. Mult-tier forests could provide better protection to the soil at the establishment stage as well as at harvest time, as all tree species do not come to the harvest stage at the same time. At the establishment stage some species could grow quickly and give protection to the soil.
2. On sites where there is a risk of litter build-up, there is a case for growing many tree species, such that the litter of some tree species will decompose faster than others, leading to quicker litter breakdown and nutrient mineralization; this being necessary for the release of nutrients, promotion of natural regeneration, and enhancement of soil organic matter.
3. Inclusion of nitrogen-fixing tree species can benefit overall yields on nitrogen-deficient sites.
4. Inclusion of shallow- and deep-rooted species can be beneficial due to the 'nutrient pump effect' where deep-rooting species tap nutrients in the subsoil and deposit them at the surface through litter.
5. Multispecies plantations are likely to use soil nutrients from different depths and soil profiles, leading to optimum use of soil nutrients.

FAO (1993*d*) concluded that due to the unpredictable nature of epidemics and climatic change, the maintenance of genetic diversity in plantations is essential. Diversity is necessary at different levels, both within and between species. Large diversity gives a better chance to evolve, to cope with new conditions, and also provides insurance against total loss.

Soil fertility management for sustainability

In the case of energy plantations, the woody biomass is harvested periodically, leading to depletion of soil nutrients. However, there is also a view that declining yields in successive forest crops due to nutrient depletion are very much the exception (Evans 1990). The depletion of nutrients could be minimized through a range of measures such as appropriate species choice, silvicultural practices, and harvesting practices. Litter accumulation takes place in plantations, leading to immobilization of nutrients, and during harvesting significant loss of nutrients may take place (Evans 1992). Mixtures of species will alleviate this problem to some extent by improving the carbon : nitrogen ratio.

Nutrient loss due to harvesting of woody biomass could be minimized by leaving the thin branches and twigs, leaves, reproductive parts and tree tops (not required for gasification) in the plantation itself, as nitrogen tends to concentrate in the leaves, twigs, and small branches. Leaving the unused parts on the soil surface leads to the reduction of soil erosion and enhancement of organic matter content. In addition:

1. Appropriate soil and water conservation measures should be used to reduce topsoil and litter erosion and to increase soil moisture availability due to a higher organic matter content.

2. If leaf litter is used for anaerobic digestion for biogas production, the nutrient-rich digested slurry must be returned to the same site.
3. The ash obtained from combustion of wood should be returned to the site as it is rich in some mineral nutrients. It is easy to transport ash and spread it on the field.
4. Wherever possible, at the beginning of the rotation inorganic fertilizer could be added to supplement the soil nutrients.
5. Soil disturbance should be minimal to reduce compaction, erosion, and leaching.
6. The existing vegetation should be retained, wherever it is advantageous; for example, as buffer zones to prevent soil erosion or as shelter to newly planted seedlings and indigenous flora and fauna.
7. Soil organic carbon is a good indicator of ecosystem recovery and long-term sustainability, especially as we are considering annual harvests from originally degraded lands. Thus the biomass production systems must be designed to constantly add substantial amounts of organic matter to the soil. Ground cover, root growth, leaf litter, and harvest residues, which will end up as organic matter, should usually be left on site (Ranney 1992).

7.9.2 Biomass production and biodiversity

Biomass production for energy immediately brings to most people a picture of high-density monoculture plantations of exotic species. This need not be so, and much has been written about this. Unlike the pulp or fibre-processing industry, for combustion in gasifier reactors the woody feedstock could come from any tree species of moderate or high density. Even low-density wood could be used, except that larger volumes would have to be handled. Wood chips of specific size from a mixture of species are also acceptable as feedstock, permitting the use of many tree and shrub species yielding woody biomass.

An important issue is how to ensure promotion of biodiversity in programmes of biomass energy plantations. Appropriate guidelines should be promoted for biomass energy plantations to enhance the biodiversity in currently degraded lands. The necessity of diversity of plant types and species has been explained in earlier sections. In Brazil, for example, companies have found that by leaving one-quarter to one-third of the estate area in its natural state (for forest succession to take place from a degraded site through natural regeneration) helps to minimize pesticide use as a result of the activity of diverse predators that reside in the naturally vegetated areas (Carpentieri *et al.* 1993). A study of plant diversity in the Western Ghat region (South India), in *Eucalyptus* and teak plantations and primary forests, concluded that in *Eucalyptus* and teak plantations there do not appear to be any serious constraints to the growth of other native species, provided there is adequate space for other species to grow (Bhat *et al.* 1984). Some of the management practices/guidelines to promote biodiversity in energy plantations are:

1. Use many species rather than monocultures, and where two tree species grow equally well, one native and another exotic, for ecological benefit plant the native species (Evans 1992).
2. Depending on the land category, leave a fraction of the total site for forest succession to progress through natural regeneration.
3. Leaving a few gaps or islands for natural revegetation in plantations would lead to colonization by a large number of species.
4. Many clones for a given species, variation in age of plants, firebreaks, protection from grazing, controlled weeding, optimum spacing, thinning, etc. will all contribute to the promotion of biodiversity (Evans 1992).

Thus there need not be any conflict between promoting biodiversity and plantation forestry for energy. In fact, energy plantations could provide an incentive to reclaim degraded lands and thereby promote biodiversity.

7.9.3 Biomass production and the risk of pests and diseases

Plantations of only one or two species with plants of comparable age can be more vulnerable to pest and disease attack than natural forests. Plantations in the tropics and sub-tropics tend to be more affected than those in temperate regions. Proper species selection and good plantation design and management can be helpful in controlling pests and diseases (Hall *et al.* 1993). Many of the species selection, silvicultural, and management practices suggested above for obtaining sustainable productivity and for promoting biodiversity also contribute to a reduction in pest and disease occurrence and damage. Tree improvement programmes offer a long-term strategy for upgrading plantation quality and disease resistance. In planning tree improvement programmes, short-term gains in yield must be balanced against the long-term need to maintain genetic diversity as an insurance against new pests, diseases, and uncertainties due to climate change scenarios (Evans and Hibberd 1993).

7.10 IMPACT OF CLIMATE CHANGE ON BIOMASS PRODUCTION

One of the environmental factors that is likely to have a significant impact on forests and plantations in the medium and long term is CO_2 build-up, global warming, and the resulting climate change.

The IPCC (1992) projections regarding climate change include:

(1) effective doubling of the CO_2 in the atmosphere between 1990 and 2025 to 2050 for a 'business-as-usual' scenario;

(2) increase of global mean temperature in the range 1.5–4.5 °C and an unequal global distribution of this temperature change; and
(3) changes in precipitation.

Several positive and negative impacts have been projected. If the projected climate change materializes, natural or artificial regeneration of forests with local seed sources will become increasingly difficult (Ledig and Kitzmiller 1992). Insects and diseases can be expected to cause significant losses to forests, and in addition fire severity may also increase. It is also expected that the climate will change faster than tree species are able to respond (IPCC 1990). Climate change may also lead to forest dieback. Thus socio-economic impacts of climate change could be significant in regions dependent on forests. Climate change could have positive as well as negative impacts on woody biomass production for energy.

Climate changes of the predicted magnitude are expected to affect the net primary production (NPP) of the world's land ecosystems (IPCC 1990; Melillo *et al.* 1993). The factors that are most likely to affect biomass for energy productivity are an increase in CO_2 concentration in the atmosphere, an increase in temperature, and a change in precipitation. Projections of precipitation changes suggest enhanced precipitation and cloud cover throughout the tropics (Mitchell *et al.* 1990). Although there are uncertainties about the impact of these factors on biomass production, there are some indications of the possible effects from several studies:

1. The effects of elevated CO_2 levels on plants have shown increased rates of photosynthesis, lowered plant water use requirements, increased carbon sequestration, and increased soil microbial activity resulting in nitrogen fixation, and thus ultimately stimulating growth.
2. Water-use efficiency of both C_3 and C_4 plants will increase with rising CO_2 levels. Water-use efficiency could increase by about 35 per cent for C_4 plants and by 75 per cent for C_3 plants (Nobel 1991). What is even more important for regions where moisture stress is a major limiting factor is the finding that increases in water-use efficiency with elevated CO_2 are generally greatest in water-stressed systems (Mooney *et al.* 1991).
3. Impact of elevated CO_2: according to a recent modelling exercise, under doubled atmospheric CO_2 levels, tropical forests are predicted to incorporate a substantial proportion of the elevated CO_2 levels into production. For example, increases in NPP of 22 per cent and 18 per cent are predicted for tropical evergreen forests and tropical deciduous forests, respectively (Melillo *et al.* 1993).
4. Impact of changes in climate and CO_2: for tropical and dry temperate ecosystems the increase in NPP may be dominated by the effects of elevated CO_2 (Melillo *et al.* 1993).

Although there are uncertainties, the general conclusion is that rising CO_2 levels may be beneficial to growing biomass for energy purposes (IPCC 1992; Hall *et al.* 1993; Melillo *et al.* 1993). However, if the climate changes rapidly (as is expected) and if the current trends in energy production and use and deforestation persist, it may be difficult for many individual species and ecosystems to adapt or migrate, and pest and disease problems may be aggravated (IPCC 1990). Thus bioenergy planners as well as forest managers may have to include climate considerations into their planning and adopt some strategies to reduce potential adverse impacts as a precautionary measure. Some of the adaptation measures could be:

1. Reforestation strategies should emphasize conservation, diversification, and broader deployment of species, seed sources, and families. Planting a diverse array of species and seed sources is a hedge against the uncertainties associated with projections on climate change (Ledig and Kitzmiller 1992).
2. Shorter rotations of energy plantations will enable quick responses to climate change.
3. Management practices such as thinning and sanitation harvests should be employed.

7.11 ENERGY BALANCE OF BIOMASS PRODUCTION

In India the afforestation or tree plantation programme involves marginal use of fossil fuels, as land preparation, planting, after care, and harvesting operations are generally carried out using human labour. In fact, afforestation is considered as an employment-generation programme. The fertilizer input is marginal or absent, and even for crop production it is low. The tree plantations are raised under rainfed conditions and thus do not use fossil fuels for irrigation. Thus the fossil fuel energy content of biomass feedstock is not a serious issue in the Indian context as the fossil fuel use is small in biomass production.

7.12 CONCLUSIONS

The critical issue in pursuing bioenergy options for gaseous and liquid fuels or electricity is the sustainability of supply of biomass. This issue is crucial as the populations in developing countries are growing at a rate of 2–3 per cent annually. Thus the first question to be answered in a densely populated country such as India is whether enough land is available for biomass production. Interestingly, if all the wood presently used in cooking were to be saved by shifting to biogas or liquid fuels, it would be adequate to meet the woody biomass requirement of electricity generation for meeting the electricity needs of 98.5 per cent of villages and 85 per cent of the rural population, plus feeding to the grid. However, we strongly suggest that none of the existing tree biomass resources be used for electricity generation, except possibly the harvest residues of industrial plantations. We propose dedicated energy plantations in degraded lands. Surprisingly, analysis of land-use patterns over the decades shows that the area under cropping, as well as the area of degraded land, has stabilized since 1970. However, food production has increased at a higher rate than population growth, as food grain productivity has continuously increased. Crop productivity of all crops is low. Rice productivity is particularly low compared to other Asian countries. Thus there is potential to double or even treble food production without increasing the area on which it is grown.

One-third to half of the degraded village commons and land owned by farmers would be adequate for growing woody biomass in a sustainable way for decentralized kilowatt-scale electricity generation. Thus the large megawatt-scale option also seems to be feasible. In fact, the land categories considered are in a degraded state with no incentive to revegetate them. Thus the bioenergy option could provide an incentive to reclaim degraded lands. An integrated land-use strategy for producing food, fodder, timber, and fuel to meet the needs of a growing population is highlighted. Such an analysis for India showed that all these requirements can be met using the currently cropped land and a part of the currently degraded land.

The discussion on sustainability of woody biomass production shows that with appropriate mixes of species, varieties and clones, silvicultural practices, harvesting practices (leaving the unused plant parts in the plantations), returning ash and biogas sludge to the plantation, and soil and water conservation practices it would be possible to ensure higher and sustainable yields. Energy plantation managers should also recognize the positive as well as the negative impacts of climate changes, although on balance it looks as if climate change will provide positive benefits to plantation forestry. In fact, short-rotation energy plantation forestry may offer opportunities to adapt to changes in climate. The energy balance or fossil fuel content of biomass feedstock is not a serious issue in India as fossil fuel use is insignificant in biomass production.

The ultimate decision on land use choice rests with the village community for use of village commons and with farmers for private lands, although governments and industry could provide incentives or disincentives. Land does not seem to be a constraint for the bioenergy option, and in fact bioenergy has the potential to provide incentives to reclaim degraded lands and opportunities for generating employment and income to farmers and village communities.

Photo 7.1 Monoculture plantation of *Eucalyptus* in semi-arid region of South India.

Photo 7.2 Mixed species plantation grown for producing woody biomass as feedstock for gasification in Hosahalli village.

8
Environmental and socio-economic benefits of bioenergy

The current trends of energy provision and use based on commercial energy were shown to be non-sustainable, environmentally damaging, and leading to inequities. It will be useful first to understand the socio-economic and environmental implications of continuing with the existing patterns (business as usual, BAU) and the possible implications of adopting the proposed bioenergy options. Major environmental problems such as deforestation, submergence of forest (from hydroelectric projects), acid rain, global warming, and local air pollution are all linked in various ways to the extraction, processing, and use of energy. Environmental considerations have, however, rarely determined decisions on energy in the past. Now, due to local and global pressure, there is a change in the attitude of developing as well as industrialized countries, and environmental considerations are frequently included in energy planning. India, for example, has a clear policy to promote renewables on a medium- and long-term basis (Box 6.1). The need for and the potential of bioenergy options in meeting rural energy needs were presented in earlier chapters. In this chapter the social, economic, and environmental implications and opportunities created by adoption of bioenergy options in India are discussed.

8.1 IMPLICATIONS OF THE BUSINESS-AS-USUAL SCENARIO

To understand the implications of continuing the current patterns of energy, first the projected impact on consumption levels of various fuels is estimated; secondly, the impact on availability, adequacy, and access to fuels is analysed; and, finally, socio-economic and environmental impacts are assessed. It is also important to compare the projected impacts of the BAU projection on the quality of life and development compared with the planned goals of development.

8.1.1 Consumption of fuels and spread of electricity under BAU

Several projections of energy use in the future are given in Table 6.1. According to projections based on estimates made by TERI (1992*a*) for the year

2009–10, the commercial energy consumption is expected to increase from 200 Mtoe in 1990–91 to 600 Mtoe by 2009–10. The consumption of non-cooking coal, electricity, and petroleum oil is expected to at least treble by 2009–10.

Rural electrification

During the development plan period of 1992–97, an additional 50 000 villages are proposed to be electrified, so 92 per cent of villages will be electrified by 1997 (PC 1992). Furthermore, 2.5 million irrigation pumpsets will be electrified, taking the total number of irrigation pumpsets using grid electricity to 11–12 million. India would have made significant strides with respect to the spread of rural electrification with nearly 0.5 million rural settlements being connected to grid electricity by 1997. This will be a phenomenal achievement compared with other developing countries, where the percentage of villages electrified is still small (Chapter 11).

8.1.2 Renewable energy programme

The national energy planners have a medium- and long-term commitment to renewable energy (Box 6.1). However, the financial allocation for the period 1992–97 is only 1.2 per cent (Rs14 billion) of the national energy budget. Assuming that the targets set for the plan period (1992–97) are achieved, the following renewable programmes will be implemented during 1992–97 (PC 1992):

1. Biogas plants:
   (a) 0.75 million family plants;
   (b) development of low-cost designs;
   (c) multifeed biogas plants;
   (d) community biogas programmes in a small way.
2. Improved stoves:
   (a) 10 million stoves to be built;
   (b) high-efficiency stoves to be developed.
3. Biomass electricity:
   (a) several hundred megawatts of electricity from biomass including co-generation units in sugar mills.
4. Other renewables:
   (a) 100 MW of wind power;
   (b) 150 MW of small hydroelectric.

The targets for renewable energy have been enhanced recently to 2000 MW for the period 1992–97 (Cherail 1993). Since the total installed capacity by 1997 is projected to be about 100 000 MW, the contribution of all renewables to electricity generation would still be very small even if all the planned renewable energy projects are implemented (PC 1992).

8.1.3 Demand versus supply of energy

Electricity According to the projections, the shortfall in power generation during 1997 is estimated to be 18.4 per cent of peak demand (CMIE 1992). The four South Indian states are projected to have a deficit in the range of 18 per cent (in Tamil Nadu) to 25 per cent (in Karnataka). This clearly shows that India will continue to experience power shortages.

Oil The projected demand for petroleum products for the year 1996–97 is estimated to lead to the import of 37 Mt, aggravating the foreign-exchange problem. Continuation of this trend beyond 1997 will only aggravate oil imports. It is difficult to visualize how India will be able to finance the projected oil demands (186 Mt) for 2009–10.

8.1.4 Impact of the BAU demand scenario on the provision of energy services

Domestic lighting

In India, although 84 per cent of villages are already supplied with electricity (in 1992) only 27 per cent of village households have electricity supplies in rural areas. Under the BAU scenario there could be a marginal increase in the percentage of households connected to grid electricity. The majority of rural households would continue to depend on low-quality kerosene for lighting, as currently there is no plan to reduce the first cost or mode of providing electricity services to households. In urban areas probably all residential buildings will soon have electricity.

Irrigation pumping

By 1997, nearly 12 million pumps and, by 2000, probably all the potential 15 million minor irrigation water-lifting systems (NIRD 1991) should be electrified. Thus the share of total electricity used in agriculture will increase. The requirement for agricultural pumping (at the current rate of 5362 kWh/pump/yr) will be 80 TWh, up from 49 TWh in 1991, thus greatly increasing the load on the system.

Quality of electricity services

Currently 30 000 MW of the connected power load is for agriculture. This is a substantial load on the rather weak electricity distribution network, which is currently overloaded. Furthermore, overloading could lead to continued low voltage, and unreliable and erratic supplies to rural consumers. In a situation of power shortages and mounting losses, utilities adopt measures to minimize loss by introducing power cuts to rural areas, maintaining a highly unreliable power supply, charging high initial costs, and causing long delays in providing connections (Ranganathan 1993). All such measures lead to a low-quality service and discourage rural consumers. If it were not for the highly subsidized electricity provided to farmers in India, farmers probably would have opted for alternatives, including bioenergy options.

Shortage of fuelwood

Dependence on low-quality crop residues and cattle dung is an indication of the shortage of fuelwood. With the current afforestation programmes being inadequate to meet current demands, along with the absence of any significant programme for conservation or substitution of fuelwood in rural cooking, the growing population and the resulting increased demand could lead to severe scarcities of wood, escalation of prices, diversion of fuelwood to urban areas, and shifts to lower-quality solid fuels. All this will lead to a lower quality of life for rural women.

Escalation of demand for kerosene and LPG

The kerosene demand is likely to escalate due to increases in rural population and the increase in the absolute number of households requiring kerosene for lighting in rural areas and for cooking in larger villages and urban areas. The urban population is projected to account for 37 per cent of the total population by 2010 (TSL 1992). Thus, the growing urban population will generate higher demands for LPG and kerosene, because with a rise in household incomes there will be a shift from solid to liquid to gaseous fuels along the energy ladder (Fig. 6.1).

8.1.5 Social problems resulting from the BAU scenario

The discussion in section 6.1 led to the inference that the present supply-oriented policies and the present pattern of reliance on commercial fuels has accentuated inequities between and within rural and urban areas. The poor in villages are likely to continue to depend on kerosene for lighting and biomass for cooking. Even when the family biogas option is considered, poorer households with the potential for biogas plants are unlikely to benefit from such bioenergy programmes, because of the high initial cost. Use of LPG will be restricted to urban areas and to the middle- and upper-income classes due to the high initial cost and non-availability to rural users. Any scarce resource (e.g. LPG, kerosene, electricity) will be accessible only to the rich. The current programmes and consumption patterns of energy, if continued, will only accentuate the inequity, particularly between rural and urban societies.

8.1.6 Escalation of the foreign-exchange problem and external debt

The domestic production of oil seems to have peaked during the late 1980s and has declined from a high of 34 Mt in 1989–90 to 26 Mt in 1993–94 (CMIE 1994), resulting in an escalation of oil imports. If oil prices increase by the late 1990s, India is likely to have a severe foreign-exchange problem, as experienced during the 1990 oil crisis. The oil import bill for the year 1991–92 was Rs5.3 billion, accounting for 29.5 per cent of total export earnings and a quarter of the total import bill (*Economic Times* 1993). Unless major oil reserves are discovered soon, projected levels of consumption under the BAU scenario for year 2000, 2005, or 2010 may push India into a foreign-exchange crisis, leading to increased debt. Faced with severe financial restraints to fund power projects, India is resorting to borrowing large amounts from the World Bank and other sources, pushing India further into debt.

8.1.7 Environmental problems

Deforestation and forest degradation The projected fuelwood demand of over 350 Mt (by 2005) is expected to increase the pressure on forests and village trees, leading to the degradation of both.

Burning of cattle dung Cattle dung is a source of nutrients to crops. The demand for dung as a cooking fuel will increase as fuelwood becomes scarce. According to predictions by ABE (1985), by 2005 all the cattle dung available could possibly be burnt as fuel, with loss of its manurial value.

Local air pollution and greenhouse gas emissions Taking the BAU projection for coal and petroleum fuels, the total annual energy-related carbon emissions during 2010 are projected to be 422 Mt annually (TERI 1992*a*) compared to 146 Mt in 1989–90 (Mitra 1992). India has large coal resources. With the growing demand for electricity, India may be forced to enhance its dependence on coal in the absence of any serious carbon-neutral options. This would lead to large emissions of greenhouse gases, oxides of nitrogen and sulphur, and waste (ash), leading to local pollution and contributing significantly to global warming. From this, it is evident that both the local and global community have to worry about such energy-related emissions.

8.1.8 Financial problems

The Indian energy scene is dominated by state controls and high subsidies. The total financial loss to the state electrical utilities due to subsidies for the year 1993–94 is Rs95 billion (Rs74 billion due to subsidies for the agricultural sector and the rest due to the domestic sector (CMIE 1994)). In addition to this, the petroleum subsidy projected for the year 1993–94 was Rs75 billion (*Indian Express* 1993). Thus the total direct subsidy to the energy sector is over Rs170 billion. This excludes a range of indirect subsidies involved in coal mining, coal transportation (railways), displacement costs in hydroelectric schemes, and so on.

In the BAU scenario, energy consumption is projected to nearly treble by 2010 and thus the subsidies are also likely to increase proportionately. The electrical utilities are already facing financial crises and cannot stand any more losses. The current pattern of consumption and subsidy cannot persist, but the question is who is going to sound the alarm? The *status quo* could soon make the electricity and petroleum utilities non-viable, if they are not already, and the quality of services would suffer, particularly for the rural areas.

8.2 ENVIRONMENTAL IMPACTS OF BIOMASS PRODUCTION

Environmental impact analysis of conventional energy is generally concerned with the negative effects of its extraction, processing, and use. When considering bioenergy, one is really talking about the positive impacts on the environment, even the potential problems (such as loss of biodiversity due to monoculture plantations) are amenable to solution. Environmental implications at both the local and global level are considered in this chapter.

8.2.1 Land reclamation

The extent of degraded land availability was discussed in Chapter 7 and various estimates show that large areas (66–106 Mha) are available. As seen earlier, excessive livestock grazing pressures (an average of 2 livestock/ha of vegetated area, excluding cropland) have led to losses of vegetation cover and exposure of the bare soil to tropical rainfall, leading to soil erosion. One of the options to reclaim degraded lands is revegetation through afforestation.

The task of revegetating 66–106 Mha of degraded land at the current rate of afforestation of 2 Mha/yr (Chapter 7) is considerable and it would take at least 33 years to afforest 66 Mha. The constraints to afforestation of degraded land (financial and organizational) were discussed in Chapter 7.

Bioenergy—an incentive to reclaim degraded lands

Assuming that liquid and gaseous fuels would be provided for cooking, the other two major biomass demands are industrial softwood and structural timber. Analyses in section 7.4 showed that only 7 Mha of degraded lands will be sufficient to meet the estimated structural timber and industrial softwood requirements for the year 2000. Thus bioenergy, with a large woody biomass requirement of 122 and 128 Mt annually for decentralized power generation and large megawatt systems, respectively, is one of the attractive options available to provide the opportunity for commercial demand and profitable wood production on farmers' degraded lands (Chapter 6). A continued demand for sustainable supplies of wood would act as an incentive to reclaim degraded land and maintain productivity.

Promotion of biodiversity

The need for and measures to achieve biodiversity in energy plantations were presented in Chapter 7. The demand for sustainable supplies of wood for commercial purposes with the appropriate guidelines would act as an incentive to promote biodiversity in degraded lands, as practised in Brazil where a third of the plantation area is frequently left for natural regeneration (Carpentieri *et al.* 1993).

Reduction of pressure on primary forests

With the implementation of biogas programmes based on dung and leaf litter, it would be possible to conserve the fuelwood used for cooking (181 Mt annually) and reduce, or eliminate, the pressure on forests and village trees, thereby conserving them.

8.2.2 Reduction in CO_2 emissions

There is an increased awareness of the global warming potential of energy-related emissions. In this context it is necessary to understand how bioenergy can help mitigate this problem, by considering its potential for carbon-emission reduction and sequestration of carbon through forestry on degraded land. Analyses on the use of biomass as a source of energy conclude that bioenergy offers important opportunities to reduce greenhouse gases through displacement of fossil fuels (Hall *et al.* 1990; Marland and Marland 1992). It is noted that CO_2 emissions from renewable biomass do not add to the overall atmospheric burden of carbon on a life-cycle basis. Greenhouse gas emissions can be reduced both by sequestration of carbon and production of bioenergy (IPCC 1992).

Bioenergy can play a larger role in reducing global warming when used to displace fossil fuels rather than when used for carbon sequestration. This is partly because land can be used indefinitely in displacing fossil CO_2 emissions, whereas CO_2 removal ceases at forest maturity in the sequestration strategy (Hall *et al.* 1991). A modelling exercise of carbon uptake during forest growth and the fate of this carbon when forests are harvested and used as fuel to replace fossil fuels has led to the following conclusions by Marland and Marland (1992). The most effective strategy for using forest land as a mitigation option will depend on the current status of the land, its productivity, and the efficiency of biofuel conversion to substitute fossil fuels. Three options have been visualized:

Option 1 For forests with large standing biomass and low productivity—protect the existing forests.

Option 2 For forests with little standing biomass and low productivity—reforest or manage the land for carbon storage.

Option 3 For lands with high productivity potential—manage the forests for use as long-lived products or as substitutes for fossil fuels.

In India the situation is complicated by the Forest Acts, ownership and tenurial status of different categories of land, and the current and proposed end uses. For example, the reserve forest (primary forest covered under the Forest Conservation Act 1980) area (irrespective of the vegetation or soil status) is not available for harvesting due to preventive legislation. Furthermore, as suggested in Chapter 7, the 64 Mha of forest area (excluding monoculture plantations) should be left for the promotion of natural regeneration and forest succession for ecological goals as well as for storing carbon (option 1). Village commons must be used to benefit the local community (e.g. energy as well as other end uses, such as timber and non-timber products) and cannot be used solely for storing carbon (option 2) irrespective of productivity levels. In the case of various categories of farmers' degraded lands where their productivities are low, option 2 should be adopted. However, this option will not be acceptable to farmers unless they have a guaranteed flow of income as an incentive to not cut down the trees at any stage in their carbon-storage lifetime; this may not be politically and financially feasible at both local and national levels. The profit motive alone will motivate farmers and village communities to adopt tree farming or forestry on their lands.

In the case of option 3, India, with its high population densities, would have little or no land available with high productivity potential for energy plantations or carbon forestry, as such land would most likely be used for food production. Thus option 3 is also not considered feasible in India.

Potential for carbon benefit

The estimated carbon emission reduction from the substitution of fossil fuels, if all the lighting and shaft-power needs of 98.5 per cent of Indian villages (equal to 100 TWh/yr) are met using the bioenergy option (Table 6.11), is given in Table 8.1. In addition, diesel is used for operating irrigation pumpsets (4.5 Mt annually). If the 100 TWh/year electricity for meeting rural energy needs were to come from coal-thermal sources, 40 Mt of carbon would

Table 8.1 Emissions of carbon and sulphur avoided by substituting coal-thermal power generation with biomass electricity systems, assuming a conversion efficiency of 18 per cent

	Net electricity of 100 TWh to be generated[a]	Including T & D losses, 167 TWh to be generated[b]
Coal combustion required to generate electricity to meet projected rural needs at 0.62 kg coal/kWh[c]	62 Mt	103 Mt
Carbon emission avoided through bioenergy technology at 65% C in coal or at 0.4 kg C emission/kWh of electricity generation[d]	40 Mt C	67 Mt C
Sulphur emission avoided through bioenergy technology (S content of bituminous coal is 4.55% of coal by weight)	2.8 Mt S	4.6 Mt S

[a] Electricity to be generated under efficient technology mix (in Table 6.11: 94 TWh/yr rounded to 100 TWh/yr).
[b] T & D loss for delivered electricity: according to Sinha and Khandpal (1991) for every kWh of electricity used in rural areas 1.67 kWh needs to be generated at the busbar.
[c] Total coal used in power generation in India during 1990–91 = 116 Mt, and power generated = 186 TWh = 0.62 kg/kWh (CMIE 1992).
Coal used in 1991–92 = 135 Mt to generate 208 TWh = 0.65 kg coal/kWh (CMIE 1992).
[d] Assuming C content of coal as 65 per cent, C/kWh = 0.65 × 0.62 kg coal/kWh = 0.4 kg C emissions/kWh.
It should be noted that coal use/kWh of electricity generation does not include several factors that contribute to further C and S emissions: mining, transporting coal from pits to generation plants, and auxillary power consumption in the plant.
There will be net C sequestration in soils due to energy plantations but it is not considered here.

be emitted annually as CO_2. When the high transmission and distribution (T & D) losses are incorporated, the total carbon emissions avoided by shifting to bioenergy would be about 67 Mt annually. In addition to the carbon emissions avoided, carbon would also be fixed in soil and standing vegetation, since degraded lands are expected to be used. At a productivity of 4 t/ha/yr, 30 Mha would need to be used as energy plantations in order to supply wood for biomass electricity at the village level. The carbon in standing vegetation at any time would be 210 Mt (mean standing stock), assuming the mid-rotation-period standing biomass (30 Mha × 4 t/ha/yr × 3.5 years in a 7 year rotation). When biomass is produced for energy markets, producers will seek to maximize the harvestable annual yield rather than the total amount of carbon that can be sequestered at forest maturity (Hall *et al.* 1990) and such a shift will eventually lead to a higher carbon benefit annually. In addition to the 100 TWh for village-scale biomass electricity systems, the potential for at least 16 000 MW–32 000 MW of large-scale biomass electricity systems has been shown to be feasible (Chapter 6). This provides an additional opportunity to reduce carbon emissions.

Sequestration versus bioenergy for India

It is necessary to consider the relative carbon benefits with respect to these two options in India. Several options are available for a given hectare of degraded land for mitigating the effects of climate change:

(1) creating carbon pools—to grow timber forests, harvest sustainably, and store carbon in long-term products (structural timber);
(2) carbon sequestration through forestry; and
(3) bioenergy from energy plantations and use the wood sustainably for bioenergy to substitute fossil fuels, thereby reducing net carbon emissions.

Analysis of the three options should be based on factors such as the land tenure and demands for timber and bioenergy.

Creating carbon pools Wood used for structural timber will lead to long-term storage of carbon, since structural timber used in housing and furniture has a life of about 50 years. The timber demand in developing countries will increase due to population growth, and every year new timber could be converted into a long-term carbon storage pool. Large tracts are already dedicated to hardwood timber tree plantations of *Tectona, Dalbergia*, and *Pinus*. Given the current relatively small demand for structural timber (Chapter 7), the timber forest option has a limited potential for converting degraded lands to timber forest, since surplus timber would lead to price destabilization. It should also be kept in mind that any timber used 50 years ago will be decomposed or burnt in any given year, leading to carbon emissions. Therefore the net emission avoided would be less than the carbon incorporated and stored in structural timber in a given year. Using wood for building materials results in far less energy use and fewer carbon emissions by replacing steel, aluminium, and cement (Sampson 1993).

Carbon sequestration This option involves promoting forest succession (natural regeneration) or growing a forest plantation and using the site as a carbon sink, with no harvest of timber. Only reserve forest land controlled by the state and covered under the Forest Conservation Act 1980 (41.5 Mha, FSI, 1992) is suitable. There is certainly the potential for enhancing the standing carbon stock by increasing the tree crown cover of open forests (26 Mha, Fig. 3.1) from the present 10–40 per cent tree crown cover to over 40 per cent or even close to 100 per cent. The non-availability of village commons and farmers' lands for the carbon sequestration option was discussed earlier in the section. However, afforestation and sustainable harvesting and management could lead to carbon sequestration indirectly, by reducing pressures on primary forests and fixing carbon in soils and other long-term carbon pools. Thus in countries such as India there appears to be little scope for using private or village common lands for significantly expanding carbon sinks.

Degraded lands for bioenergy to reduce C emission Fossil fuel substitution using various bioenergy options provides an opportunity to use different categories of private or farmers' degraded lands for growing biomass for energy with a profit motive. Even part of the protected lands and village common land could be used to meet the electricity needs of village communities.

The following options are suggested for using non-cropland in India:

1. Reserve forest or primary forest land (41.5 Mha) to remain as carbon sinks and for expanding carbon storage by enhancing tree crown cover and standing stock.
2. Protected forest plus village commons to grow biomass for meeting the structural timber and non-timber needs of local communities, and any such surplus land could be dedicated for bioenergy options.
3. Degraded land owned by farmers should be used for bioenergy options and industrial softwood plantations.

Comparison of carbon benefit from sequestration versus bioenergy option Land is a limited resource and therefore should be used to derive maximum climate mitigation benefits, apart, of course, from providing local benefits. The potential for reducing atmospheric carbon is estimated for both the options cumulatively for 50 and 100 years for a hectare of degraded land.

Carbon sequestration Assuming a maximum standing woody biomass potential of 270 dry t/ha at maturity (Table 8.2, footnote [a]) and an average woody biomass productivity of 4 t/ha/yr, the peak standing woody biomass would be achieved in 67 years (on simple assumptions). Beyond this period (67 years) there would be marginal or no increase in standing woody biomass. Thus even in 200 years there will be only marginal change or some loss in carbon storage (at 135 t/ha).

Fossil fuel substitution by bioenergy When land is used for bioenergy, more intensive management would be adopted and thus moderate productivity levels of 4–8 t/ha/yr of woody biomass could be expected.

The comparison between the two options (Table 8.2) clearly shows that the sequestration option has a very limited potential. In comparison, the bioenergy option could lead to 244 t of carbon emission reduction/ha in 100 years, even when an energy conversion efficiency of only 18 per cent is used. Thus if wood productivity could be sustainably managed, the bioenergy option (through biomass electricity) should be the clear choice.

Assuming the estimated rates of carbon benefit from Table 8.2 for the 15 Mha proposed to be dedicated for bioenergy plantations to generate electricity for 98.5 per cent of villages,

Table 8.2 Comparison of carbon benefit via sequestration from afforestation and emission reduction via bioenergy substituting for coal in electricity generation

	C benefit/ha at woody biomass productivity of[a]				
	2 t/ha/yr	4 t/ha/yr		8 t/ha/yr[b]	
C-sequestration via afforestation					
in 50 years	50	100			
in 100 years	100	135			
C emission reduction via biomass electricity at conversion efficiencies of 18 and 36%		18%	36%	18%	36%
in 50 years		61[c]	122	122[c]	244
in 100 years		122	244	244	488

[a] Standing biomass/ha under forest succession or afforestation programme (estimated from Ravindranath *et al.* 1992*b*):
Mean standing biomass in Indian forests according to Table 2.10 (footnote [a]) = 126 t biomass/ha = 63 t C/ha.
Mean standing biomass in Indian tropical moist deciduous forest = 409 t/ha (23 Mha), in tropical dry deciduous forest = 94 t/ha (18 Mha). These two account for 65% of total forest area in India.
Mean woody biomass/ha of forest = (409 t/ha × 23 Mha + 94 t/ha × 18 Mha)/41 Mha = 270 t wood or 135 t C/ha.

[b] A productivity of 8 t/ha/yr is assumed for the bioenergy option as farmers are expected to adopt practices for high yields for profits.

[c] C emission reduction in 50 years = (4 t/ha/yr of wood/1.3 t wood/MWh of electricity) × 0.4 kg C emission/kWh electricity generation (Table 8.1) in coal thermal plants × 50 years = 61.5 t C at 18% conversion efficiency.
C emission reduction is similarly estimated for 100 years at 18% and 36% conversion efficiency.

(1) via the sequestration option a total of 2 Gt carbon could be sequestered in 100 years;
(2) through the bioenergy option a total of 3.7 Gt carbon emissions could be avoided in 100 years at 8 t/ha/yr, even at an energy conversion efficiency of only 18 per cent; and
(3) at a conversion efficiency of say 36 per cent, 7.4 Gt carbon emission could be avoided.

Some additional features of bioenergy options

1. A higher atmospheric carbon benefit could be obtained per hectare at higher productivity levels and at higher energy conversion efficiency.
2. Income flow would be generated from year 2 or 3 continuously in the bioenergy option, by selling wood as feedstock to electricity generators.
3. Local employment would result every year from planting, harvesting, transportation, and decentralized electricity generation.
4. The sequestration only option may not lead to any significant income

flows, except for some non-timber forest product flows, depending on plant species and the product market.

5. The timber and energy options will have negative costs per tonne of carbon sequestered or carbon emission avoided due to income generated by selling wood feedstock, timber, or even non-timber forest products.
6. Farmers and communities (for village commons) can be motivated to grow wood for energy for profit, whereas it is not possible to persuade them to dedicate their lands for creating carbon sinks with uncertain flow of financial benefits.

In spite of all these and other known benefits, most proposals for reducing global warming through forestry have focused on the need to plant more trees and/or protect forest reserves to sequester carbon. Relatively little attention is paid to the option of using bioenergy options for fossil fuel substitution. This is now, however, becoming a serious energy option (Johansson *et al.* 1993; Nakicenovic *et al.* 1993). Even at the global level the potential of forest conservation and sequestration options to offset current fossil fuel emissions is estimated to be 5–15 per cent; in comparison, fossil fuel substitution with bioenergy could offset about 20 per cent of current fossil fuel emissions (Dixon *et al.* 1994).

8.2.3 Air pollution

Coal-fired thermal power is likely to dominate the electricity scene in India. Air pollution due to emissions of sulphur and ash around the power plants and further afield is currently a severe environmental hazard and ash disposal is also a serious problem. The advantage of biomass as a feedstock compared to coal is seen from the low concentration of sulphur and ash on a dry basis;

	Sulphur (%)	*Ash (%)*
Coal	4.55–4.74	18.3–18.7
Woody biomass	0.01–0.07	0.41–1.59

By shifting to bioenergy, emissions of sulphur and ash could be greatly reduced. The ash obtained after combustion in biomass gasifiers could be returned to the energy plantation as a mineral fertilizer.

Decreased domestic air pollution It was shown in Chapter 3 that combustion of biomass (wood, crop residues, and cattle dung) for cooking emits smoke into the kitchen and is a health hazard, apart from being inconvenient for cooking. A shift to liquid and gaseous fuels (Chapter 6) would lead to elimination of the smoke problems, thus resulting in improved health and a decrease in drudgery for women in rural areas.

8.3 SOCIAL IMPLICATIONS OF A BIOENERGY OPTION

Self-reliance

As against the current dependence on external sources of energy, the local communities could become self-reliant with respect to electricity and cooking energy (with biogas). A shift to bioenergy will lead to conservation of diesel and kerosene used in rural areas and thus leading to a reduction in oil imports and promotion of self-reliance at national level.

Power to the people

One of the essential principles of the Indian development plan is the recognition of the need to involve people in the process of development (PC 1992). One of the necessary conditions for the implementation of the bioenergy option is community control and management; whether it is growing and protecting energy plantations, operating community biogas or producer gas systems, or sharing benefits. The community should obtain control over their resources and affairs through controlling a key aspect of their development, viz. energy supply. It is essentially transferring power (or control) over power (electricity) to the people. One of the development goals is to promote decentralization of political power to people in village and tribal communities through the 'Panchayat Raj System' (local self-government).

There are several examples in India where the local communities have demonstrated their willingness and capacity to manage energy and forestry systems, such as community biogas electricity in Pura (Rajabapaiah *et al.* 1993), community forest-wood gasifier for decentralized power generation in Hosahalli (Srinivas *et al.* 1992), community biogas for cooking at Methan (Venkataramana 1992), and joint forest management in promoting natural regeneration in degraded *Shorea robusta* forests in West Bengal (Poffenberger and Singh 1992).

If sustainable development is crucial for long-term development, sustainable management of natural resources is essential. As Gadgil and Guha (1992) emphasize, promotion of conservation and sustainable management of resources require the following conditions: communities living close to a resource, who have a long-term stake and dependence on the resource, and who must have full control over the management of that resource. Sustainable and equitable development also requires (Agarwal and Narain 1989):

(1) a participatory approach to planning and development; and
(2) decentralization of political power to the people.

Thus decentralization of energy production and its distribution could provide the impetus for a genuine transfer of power to people.

Building local institutions

One of the critical conditions necessary for promoting local control and management is establishment of appropriate village-level institutions. The Rural Bioenergy Centre concept based on biomass requires decision-making on a continuous basis for the protection of energy forests or for sharing water, electricity, and biogas. Thus a village-level institution is a precondition for the success of bioenergy systems. There are examples, such as a Village Development Society in Pura village, Village Energy Forest-Wood Gas Management Committee of Hosahalli, and the Forest Protection Committee of West Bengal. The village-level institution should have a broader mandate than just the distribution of wood, electricity, or protection of energy forests. It should include management of a wider range of resources: common land, water, and communal vegetation of the village ecosystems.

Local skill and employment generation

In centralized power generation or petroleum refining all employment generated is outside the village, with the exception of marginal employment in maintenance of the distribution system. Even in the case of renewable energies, such as wind or solar PV systems, all the machinery is manufactured in a centralized facility, installed by professionals, and its operation requires only small amounts of local labour. All wealth generation takes place outside the village. In contrast, bioenergy systems create employment within the villages in growing energy forests, preparing wood feedstock, collecting cattle dung or leaves, operating biogas- and producer-gas-based systems, etc. For development to spread uniformly over the country and not just promote islands of urban affluence, employment has to be generated within villages. Currently, village communities depend on nearby urban centres for all repairs and servicing of electrical and mechanical devices. With the implementation of bioenergy systems, local youth would be trained in the operation, repair, and maintenance of generators, biogas plants, gasifiers, pumps, flour mills, diesel engines, and domestic appliances. The creation of jobs in rural areas is essential to check migration to urban centres and to achieve equitable development between rural and urban areas.

Benefit to women

Provision of liquid or gaseous fuels for cooking, pumping domestic water, domestic lighting, and flour milling will all directly benefit women more than men. The above options will decrease drudgery (gathering wood), remove health hazards (freedom from smoke), save time (in gathering fuelwood and carrying water), and result in an improved quality of life. Compared with this, the current rural electrification programme mainly stresses the provision of electricity for irrigation-water pumping, and social forestry schemes perpetuate dependence on low-quality solid fuels for cooking. Expansion of irrigation and agro-processing will also create employment for women. Women could be employed in raising energy plantations, preparation of feedstock, and possibly the operation of biogas and gasifier systems.

Promotion of equity

Provision of lighting for all households, piped water for all, street lighting, and biogas for cooking for all will help promote equity. The success of community management systems depend on equitable distribution of benefits (Agarwal and Narain 1989). This interdependence of equity and community management, and equity and sustainable use of resources, is crucial for the development of rural areas. The decentralization of political power to village communities, establishment of village-level institutions, and decision-making through the concept of 'gramasabha' (village councils) are all likely to promote equity. For example, a good discussion of the choices of species for the energy plantations in an open village council meeting is more likely to take into consideration the diverse biomass needs of the poor, women, artisans, and the landless. In contrast, the species choice made by the Forest Department, as well as in some cases, the panchayat (local government) committees, very often does not reflect the needs of the wider community (PC 1992). We conclude that adoption of bioenergy options could catalyse social and institutional development and equitable distribution of the benefits of energy resources.

8.4 LABOUR EMPLOYMENT IN WOODY BIOMASS PRODUCTION

Forestry in India and elsewhere is considered a means of creating employment in rural areas. Energy plantations in currently degraded lands in India are expected to create significant employment opportunities. Labour requirements depend on the density of trees, species choice, mode of land preparation (e.g. human, animal, mechanical power), intensity of aftercare (e.g. weeding, watering, fertilizer application), and productivity (for harvesting). Labour data from field studies are not readily available in the literature. Based on a study by Mohan (1984) and our investigations, we have estimated the labour required for a coppicing stand of monoculture plantations for three rotations (Table 8.3) with a density of 1500–2000 trees/ha and wood productivity of 4 t/ha/yr. Harvesting is the dominant operation as it involves felling trees, lopping off tops and branches, transporting to the road, and loading and unloading to tractor or truck. The labour required for harvesting is about 168 person days/ha, accounting for 68 per cent of the total labour for the first rotation, and its share increases in the second and third rotation as the labour

Table 8.3 Labour employment in wood feedstock production for biomass electricity, assuming three rotations per crop of 18–21 years

	Farmers' plantations[a]		Forest Department plantations[b] (human labour)
	Bullock	Tractor	
First-rotation crop (person days/ha)[c]	255.0	245.0	285.0
Second-rotation crop (person days/ha)	196.0	196.0	196.0
Third-rotation crop (person days/ha)	198.0	198.0	198.0
Total (person days/ha)	649.0	639.0	679.0
Wood production[d] (t/ha/3 rotations)	84.0	84.0	84.0
Person days/t wood[e]	7.7	7.6	8.0
Person days/GJ wood[f]	0.5	0.5	0.5
Person days/MWh[g]			
conversion efficiency, 18%	10.0	10.0	10.0
conversion efficiency, 36%	5.0	5.0	5.0

[a] Includes plantations raised on farmers' land using bullocks or tractor power for land preparation.
[b] Forest Department plantations are largely planted using human labour, as the land is not suitable for using bullocks or tractor power.
[c] Labour employment is expressed as person days (of about 6–8 hours' work/day) by men and women. Labour employment for nursery, land preparation, planting, aftercare, harvesting, and loading and unloading for transport are included.
[d] Wood production = (4 t/ha/yr) × 21 yr/crop with three rotations.
[e] Person days/t = (649 days/ha) / (84 t/ha/crop in 21 years).
[f] Person days/GJ = (649 days/ha/crop) / (84 t/ha/crop) × 15 GJ/t.
[g] Person days/MWh = (649 days/ha) / ((84 t/ha/yr) / (1.3 t/MWh)) at 18 per cent conversion.

required for land preparation declines for the coppiced crops. The labour employment potential per t of wood is about 7–8 person days, and it ranges from 5 to 10 person days per MWh, depending on the efficiency of conversion.

The employment generation in conventional plantation forestry is characterized by peaks at the time of planting (first year) and harvesting (last year of rotation). Harvesting the feedstock for gasification does not require 6–7 years for the stems to develop. After the second or third year, depending on the growth rate, thin stems that are suitable for gasification need to be harvested and transported. This would be the dominant labour-employing activity. Wood needs to be harvested daily or weekly for gasification, resulting in a good spread of employment all year round. Wood production as feedstock would lead to full employment for seven persons in a moderate-sized village such as Ungra (274 t of wood × 8 days/t of wood)/(300 working days/year), to four persons in small villages such as BNPura, based on the electricity and wood requirements given in Table 6.8. Thus every village stands to gain from employment opportunities, not only in wood production and harvesting but also in processing and conversion.

8.5 CONCLUSIONS

In this chapter the current pattern of energy consumption is shown to be non-sustainable, since it leads to environmental degradation, social inequity, and financial and foreign change problems. Modern bioenergy options, on the other hand, are shown to lead to reclamation of degraded lands, promotion of biodiversity, reduction in net carbon emission by substituting for fossil fuels, and reduction of indoor air pollution. Bioenergy-based energy systems are shown to be sustainable and equitable, with rural women benefiting the most. The need for local community control and management is highlighted. In addition, bioenergy programme implementation would lead to significant local employment generation. Bioenergy options will become important in future energy programmes due to their environmental and local benefits. In spite of all these and other known environmental and social benefits, modern bioenergy is yet to make any significant impact on the global energy scene. The first condition for bioenergy to become a serious option is its financial feasibility, which is analysed in the next chapter.

9

Economic analysis of bioenergy options

In previous chapters the potential and need for modern bioenergy options, the potential beneficial environmental and socio-economic impacts, and the feasibility of the technologies have been discussed. However, for any technology to find acceptance and spread it must be economically viable apart from being technically sound. There are image problems for all alternative energy technologies, including bioenergy, such as high capital and operational costs, limited potential, and the idea that they are not yet ready for large-scale dissemination.

The costs of all commercial forms of renewable energy have declined markedly over the past two decades (WB 1992). Much evidence has emerged from various countries to show that bioenergy options are becoming commercially attractive compared to conventional energy options, and provide significant environmental and socio-economic benefits. Some of the studies include the extensive World Bank review of renewable energy technologies (Ahmed 1994), the UN-sponsored study on renewable energy (Johansson *et al.* 1993), the US government study on energy technologies for developing countries (OTA 1992), IIASA (International Institute for Applied Systems Analysis) studies on strategies for mitigating global warming (Nakicenovic *et al.* 1993), a series of reviews on renewable technologies in the journal *Energy Policy* (Jackson 1992), and a number of case studies from developing as well as industrialized countries (e.g. Elliott and Booth 1993; Moncada *et al.* 1993; Rajabapaiah *et al.* 1993; Ravindranath and Mukunda 1993). These studies are highlighted briefly in this chapter to present current thinking on the economics of bioenergy technologies, particularly in the light of environmental concerns at both local and global levels.

Economic viability is crucial for any large-scale dissemination or marketing of bioenergy options, for the following reasons:

1. Developing countries face severe scarcities of capital and the cost of capital is very high. For example, commercial bank interest rates in India for short-term loans are presently around 17.5–22 per cent and for long-term credit around 14–15 per cent.
2. Rational governments are expected to adopt a least-cost energy planning approach in selecting energy options.
3. The approach suggested for disseminating rural bioenergy centres includes private entrepreneurship as well as community partnership initiatives,

with some supportive funding from the government. For private enterprise or for the community to invest their resources there must be some financial profit motivation.

4. Bioenergy options face the competition of already established energy sources and technologies, particularly the large, centralized, power generation systems and vested industrial and government interests. These systems are also presently subsidized to varying degrees.

Before discussing the available evidence on the economics of bioenergy options, it is necessary to highlight those features of bioenergy options that make its comparison with conventional sources and technologies complex and difficult.

9.1 FEATURES OF BIOENERGY OPTIONS

Decentralized scale of operation

The bioenergy options considered in Chapter 6 were mainly biogas, liquid fuels, and producer-gas. A community biogas plant for cooking has to be at the village level as it is not feasible to transport dung (82 per cent water) over long distances and, also, the village community may not approve of it. The potential electricity generation capacity of village-community biogas systems is in the range 5–15 kW (Table 6.9). When the producer-gas option is considered, the emphasis is on kW-level decentralized systems (15–165 kW, Table 6.12). Medium-scale biomass electricity systems in the range 1–5 MW could also be considered as decentralized.

In comparison, the conventional power generation systems in India (coal-thermal and hydroelectric) are large systems in the capacity range of several hundred or a thousand megawatts. Direct comparisons are difficult since large systems invariably have some economies of scale. Some existing and proposed bioenergy systems outside India are in the range 5–50 MW in the USA, Europe, and Brazil.

Environmental benefits of bioenergy

Bioenergy options for heat or electricity generation are accompanied by significant environmental benefits: decreased air pollution; sustainable sources of biomass do not lead to any net carbon emission; if used to substitute fossil-fuel-based power, there will be a net carbon benefit by reducing carbon dioxide emissions; and if degraded lands are used for growing energy plantations, it is possible to conserve soil, rainwater, and biodiversity (Chapter 8). In comparison with bioenergy, both hydro- and coal-based thermal power systems have environmental problems such as greenhouse gas emissions, local air pollution, acid rain, waste disposal, and land submersion. Unfortunately, it is not yet possible to monetize on a comparable basis with conventional energies the significant environmental benefits of bioenergy. The positive environmental benefits of bioenergy systems and the environmental costs of conventional centralized systems are missed out in conventional financial analysis (Friedrich and Voss 1992).

Social and economic benefits

The social and economic benefits of bioenergy options such as local employment generation, development of skills, self-reliance, local control, and community participation in management of bioenergy systems cannot yet be monetized for conventional financial analysis. Again, there are social costs with conventional systems, such as the displacement of communities due to land submergence, which are often not included in conventional financial analysis.

Absence of a 'level playing field'

In conventional power generation and distribution systems there are direct, indirect, and hidden subsidies (Foley 1992*b*). In India, for example, LPG supplies (mainly to urban households) are subsidized by about 43 per cent (*Economic Times* 1994) and power supplies for agricultural pumping are fully subsidized since power tariffs for farmers do not exist or are only nominal (CMIE 1994). Electricity pricing in India is not based on long-run marginal costs, and pricing is more a political than a financial decision. Therefore, it is difficult to make a rational comparison of bioenergy options with the prices of grid electricity and fuels (kerosene and LPG) currently prevalent in countries such as India.

The difficulties are compounded as few detailed studies are available in developing countries on the financial aspects of conventional energy sources, such as coal-thermal or hydroelectric systems. One of the reasons being the unwillingness of state electricity utilities to publish the true financial status and the absence of systematic methodologies for energy pricing. Thus any financial analysis of the bioenergy options, and particularly comparisons with conventional energy options, must be viewed in this light. A number of case studies are included in this section to analyse the economics of different energy options:

(1) economic analysis of decentralized versus conventional centralized electricity generation systems in India, including decentralized bioenergy options;
(2) financial analysis of village-level rural bioenergy centres in India, including community biogas and producer-gas electricity options;
(3) evidence from international studies on financial analysis of bioenergy options and comparisons with other renewables and conventional systems;

(4) economics of advanced bioenergy systems; and
(5) case studies of demonstration or proposed advanced bioenergy systems in Brazil, Europe, and the USA.

The costs are presented in Indian rupees or $US1 = Rs31 or c., depending on the original study.

9.2 EVIDENCE FROM INDIA ON THE ECONOMICS OF BIOENERGY SYSTEMS

9.2.1 Centralized and decentralized electricity generation costs

1. Reddy *et al.* (1990) conducted a detailed financial analysis of decentralized power generation, conventional centralized power generation, and conservation options for India. The financial analysis was carried out by converting all prices (mainly from around 1990) to the 1986 level, and is based on discounted cash flow techniques. In the study the unit cost of energy (Rs/kWh) and the unit cost of power (Rs/kW) were estimated for a range of competing electricity generation as well as conservation options. The findings are reported here for some of the relevant options; for a detailed description of methodology and findings refer to Reddy *et al.* (1990). The cost of installed capacity (Rs/kW) and the cost of electricity (Rs/kWh) were calculated at a discount rate of 12 per cent (Figs 9.1. and 9.2). The technologies examined are those available during the period of analysis in India (around 1990).

Cost of power and electricity Among all the options, the investment required is least for the conservation option of improving the efficiency of irrigation pumps. This confirms that it is cheaper to conserve a kilowatt-hour than to generate a kilowatt-hour. Amongst all the power generation options, small hydroelectric is the least-cost option, followed by co-generation with sugarcane bagasse. It is interesting that the installed cost of power for other bioenergy options, namely biogas and producer-gas, are almost 10 per cent lower than the dominant coal and large hydroelectric options. Even when compared to natural gas-based power generation, the bioenergy options are competitive. The study also showed that at higher discount rates (20 and 25 per cent) the decentralized options, such as biogas and producer-gas electricity options, become even more attractive compared with the large centralized coal or hydroelectric power systems. This finding is very important in the light of the scarcity of capital and high interest rates in India.

The major conclusions of the study were that decentralized electricity options such as biogas and producer-gas are competitive in cost terms with large, centralized systems. The study further concluded that, based on the least-cost planning principle, the following order of preference is suggested

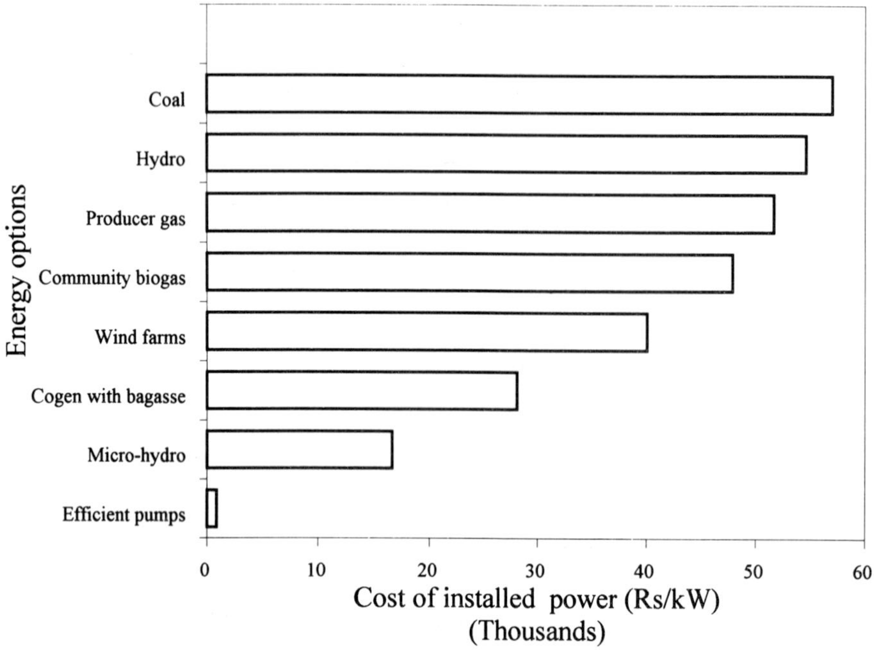

Fig. 9.1 Cost of power for different centralized and decentralized generation options in India (Reddy *et al.* 1990).

to meet the gap between demand and supply of electricity (not all options given in the study are included):

I conservation options, use all the limited potential;
II co-generation in sugar mills;
III decentralized renewable sources (such as micro-hydro and bioenergy options);
IV natural gas-based power system;
V hydroelectric followed by coal-based system;
VI nuclear option.

2. Sinha and Kandpal (1991) have attempted to identify the conditions under which different decentralized energy technologies become cost effective. The analysis is based on a comparison of the cost of delivered grid electricity to rural areas with the cost of delivering an equal amount of electricity using a decentralized technology. The cost of delivered electricity depends on:

(a) the investment required for extending a grid, determined by the distance of the village from the existing grid;
(b) the load factor;
(c) transmission and distribution losses; and
(d) the cost of power generation.

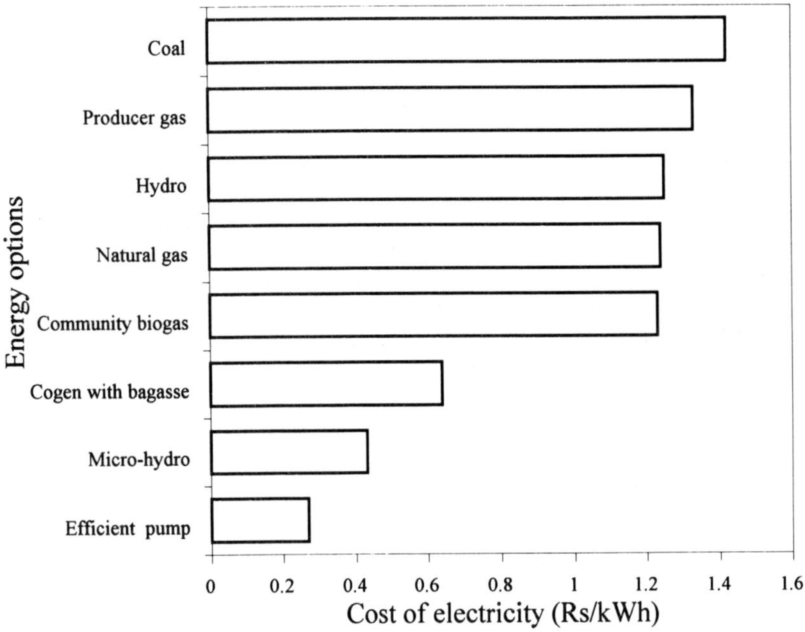

Fig. 9.2 Cost of electricity for different centralized and decentralized options in India (Reddy *et al.* 1990).

According to the study, for every kWh of electricity consumed in rural areas, 1.67 kWh needs to be generated at the busbar due to transmission and distribution losses. The cost of coal thermal power generation at the busbar was considered to be Rs0.35/kWh, while the cost of delivered electricity from the grid in rural areas in India is Rs0.58/kWh. The cost of delivered electricity in rural areas is higher because of the cost of the distribution network and its maintenance.

The load factor is crucial in rural electrification, and a study of four states showed an average load factor in the range 6–16 per cent (Ramesh *et al.* 1988). The cost per unit for various options with 10 and 25 kW installed capacity and varying load factors for a village at a distance of 5 km from a 33 kV grid, is given in Fig. 9.3. The cost of coal thermal power is compared with decentralized diesel and producer-gas options. For a typical distance of 5 km and a peak load of 10 kW, diesel power is cheaper than grid electricity for load factors less than 60 per cent. Biomass technologies become cheaper than grid electricity for load factors less than 30 per cent. The load factor is particularly low for agricultural pumping, at around 10 per cent (Ramesh *et al.* 1988). Thus for various rural applications under such low loads, bioenergy options are attractive.

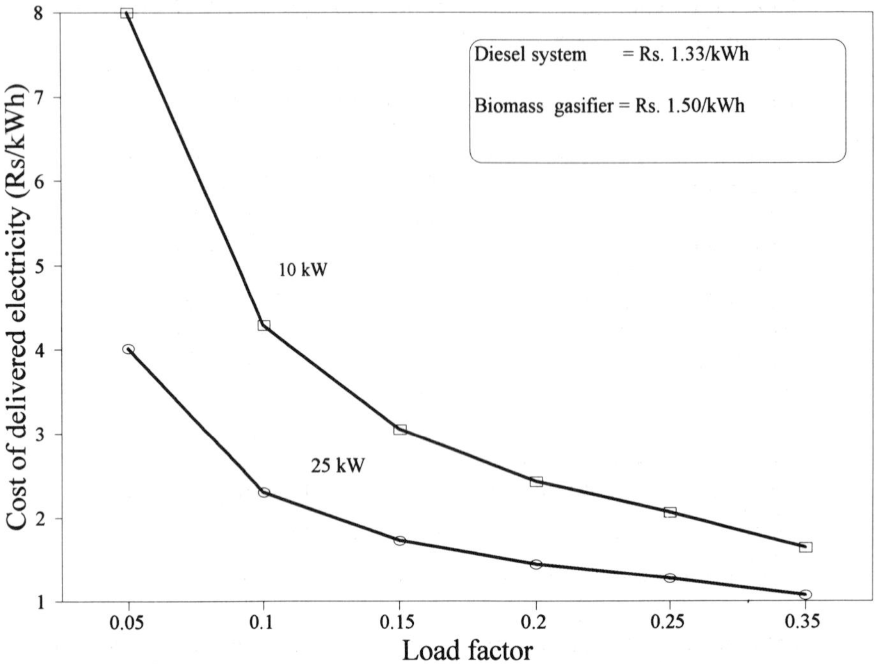

Fig. 9.3 Cost of electricity: centralized and decentralized options for an isolated village 5 km from a grid (Sinha and Kandpal 1991). The values for 10 kW and 25 kW are for centralized coal thermal power plant delivered electricity.

9.2.2 Biogas electricity: case study of Pura

A community biogas-based electricity system has been in operation in Pura village since 1987 for generating electricity to provide home lighting and pumping domestic water (Chapter 5). Its performance and economics have been monitored since then. An economic analysis of the Pura community biogas system based on the field operation is considered here (Rajabapaiah *et al.* 1993, 1994).

The biogas plant is a floating-drum KVIC design with a mild-steel drum gas-holder. The biogas fuels a diesel-engine generator system of 5 kW capacity. The biogas plant has the potential to generate 42 m^3/day of biogas. The capital cost of installed capacity, using 1992 prices, is Rs31 434 ($US 1014)/kW. Assuming a plant load factor of 63 per cent (15 h/day), the cost of electricity was estimated using the discounted cash flow technique. The capital costs are given in Table 9.1. To compare the electricity cost of the community biogas plant with that of centralized systems, the transmission and distribution losses involved in delivering electricity from a centralized system, as well as the auxiliary consumption of electricity, are included. The following conclusions were reached:

Table 9.1 Capital cost ($US) of a biogas-based electricity system in Pura village (India)

Item	Cost in 1989 ($US)[a]	Cost in 1992 ($US)[b]
Biogas plant	2554	2286 (70 866)[c]
Piping	166	139 (4309)
Sand filters	83	70 (2170)
7 hp diesel engine	754	633 (19 623)
5 kVA three-phase generator	1563	1314 (40 734)
Accessories, tools	415	350 (10 850)
Engine room	331	278 (8618)
Total	5866	5070 (15 7170)
Cost/kW of installed capacity	1173	1014

[a] Rajabapaiah *et al.* (1993), taking 1989 market prices and $US exchange rate (1 $US = Rs20).
[b] Rajabapaiah *et al.* (1994), taking 1992 market rates and 1993 $US exchange rate (1 $US = Rs31).
[c] Figures in parentheses are the values in Rs at the 1992 exchange rate.

1. As the capacity utilization increases (working hours/day), the cost of electricity declines from $0.25/kWh at 4.2 h/day to around $0.07/kWh at about 20 h/day.
2. At discount rates higher than 7.5 per cent, the unit cost of electricity from the biogas system is lower than that from a centralized power plant.
3. The cost advantage of the biogas-based electricity system increases with an increase in discount rates compared with the centralized generation system.

9.2.3 Producer-gas electricity: case study of Hosahalli

A comparative economic analysis of electricity generation between an integrated 'energy-forest wood gasifier' and an identical-capacity diesel-alone system was carried out using data from field operations (Ravindranath *et al.* 1990; Ravindranath and Mukunda 1993).

A 5 kW producer-gas duel-fuel system (conversion efficiency of about 18 per cent), using wood from a dedicated energy forest, has been in operation in Hosahalli village since 1988 (Chapter 5). Electricity is used for home lighting and pumping water for domestic consumption. The capital cost for the dual-fuel and diesel system is given in Table 9.2. The cost per kilowatt of installed capacity at 1993 prices for the small wood-gasifier system is Rs26 226 ($US846). The cost/kWh from the economic analysis based on life-cycle costing at a 12 per cent discount rate for both systems for different working hours per day is given in Fig. 9.4.

Table 9.2 Capital cost ($US) of a small producer-gas and diesel electricity system in Hosahalli village (India)

	Total capacity 3.5 kWe (1989 $US prices)[a]	Total capacity 3.5 kWe (1993 $US prices)[b]	Diesel system capacity 3.5 kWe (1989 $US prices)[a]
Gasifier	800	645	–
Building	250	352	352
Engine generator	1430	1451	1430
Accessories	300	161	300
Energy forest	300	352	–
Total	3080	2961	2082
Cost/kW	880	846	595
Cost/kW in Rs	17600	26226	11900

[a] 1989 prices and conversion rate (1 $US = Rs20) from Ravindranath *et al.* (1990).
[b] 1993 prices and conversion rate (1 $US = Rs31) from Ravindranath and Mukunda (1993).

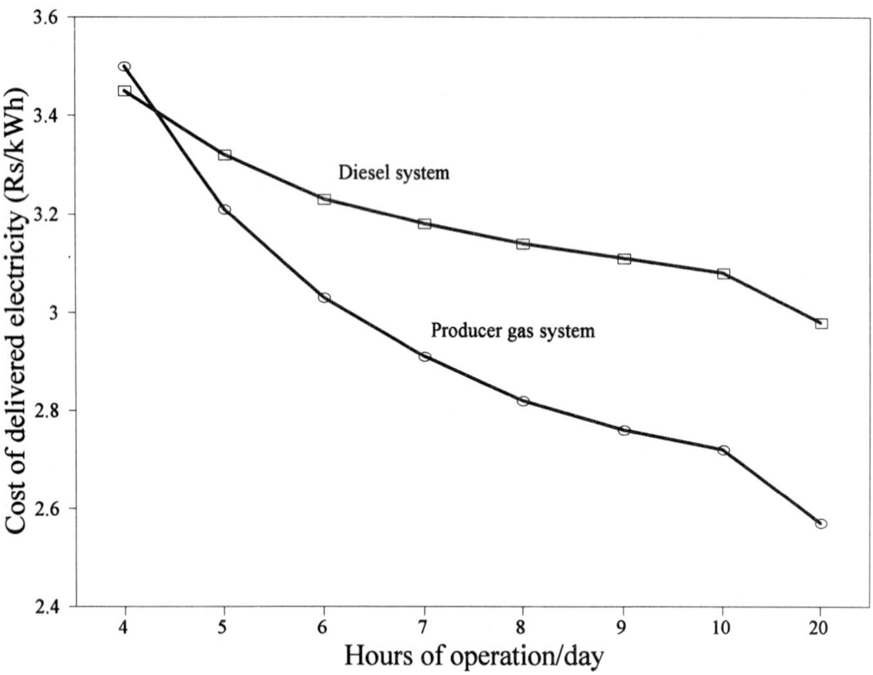

Fig. 9.4 Unit cost of electricity at different hours of capacity utilization for producer-gas electricity in Hosahalli village (Ravindranath *et al.* 1990).

1. The cost of electricity declines for both the diesel and producer-gas systems with an increase in working hours. For up to 5 hours' operation per day, the cost of producer-gas electricity is higher than that of the diesel system. But at a capacity utilization of over 5 hours per day, the producer-gas system becomes cheaper than the diesel system, given the relative costs of fuels.
2. The cost of electricity generation declined from $US0.15/kWh (Rs3/kWh) at 5 hours of operation to $US0.12 (Rs2.5/kWh) at 20 hours of operation per day.

Thus at low installed capacities the producer-gas system is cheaper than a diesel generation system of identical capacity at higher capacity utilization levels (see Chapter 6 for a discussion on capacity utilization).

9.2.4 Economics of 4.4 kW and 96 kW producer gas and diesel systems

Mukunda *et al.* (1993) have compared the economics of small and moderate-sized-dual-fuel producer-gas systems with diesel-alone systems of identical capacities for power generation. The open-top gasifier design considered in this study was developed at the ASTRA Centre in the Indian Institute of Science, Bangalore, and has a conversion efficiency of about 18 per cent. The rated capacities considered were 4.4 kW and 96 kW, which is a capacity range suited to the typical village situations considered in this study. Annual capacity utilizations of 2500 hours for the 4.4 kW system and 5000 hours for the 96 kW system were assumed. The life of the system was assumed to be 10 years, and that of the building to house the system as 40 years. The total life of the project was equated to the life of the building. The details of the financial analysis are given in Table 9.3, which shows the cost of electricity for the small as well as the moderate-sized kilowatt system, using the wood-gasifier dual-fuel mode as well as a diesel-alone mode.

The capital costs of diesel systems were lower than those of the dual-fuel systems by 18 per cent and 44 per cent, for the small and moderate-sized systems, respectively. However, the total life-cycle fuel cost for the gasifier system was less than 50 per cent of that of the diesel system, indicating the economic advantage of the dual-fuel system. The total cost of electricity for the wood-gasifier system was lower than the corresponding diesel system, at $0.15/kWh for the 3.5 kW load dual-fuel system and $0.07/kWh for the 80 kW load dual-fuel system.

The following conclusions were drawn from this analysis.

1. Wood-gasifier-based dual-fuel systems are 12 and 36 per cent more economical to operate than diesel-based generator systems for 3.5 kW and 80 kW load systems, respectively.
2. The economy of scale is obvious, and the cost of electricity declines by half when progressing from a 3.5 kW to an 80 kW load system.

Table 9.3 Financial analysis of diesèl and producer-gas electricity systems (1991 prices)[a]

Rated capacity kilowatts	4.4 kW dual-fuel (diesel alone)[b]	96 kW dual-fuel (diesel alone)[b]
Load in kW	3.5 (3.5)	80 (80)
Life of the plant in years	40 (40)	40 (40)
Capital cost per kW engine + gasifier	542.6 (500)	200 (187)
Present values of replacement of engine + gasifier	566.5 (522)	514.3 (232)
Capital cost/kW for cooling, cleaning, and control system	64 (0)	218 (0)
Present values of replacement of cooling, cleaning, and control system	67 (0)	228 (0)
Total capital cost/kW	1411 (1165)	1224 (688)
Life-cycle fuel cost/kW	1373 (3439)	3185 (6878)
Life-cycle operating and maintenance cost/kw	3037 (1879)	945 (744)
Operating cost/kWh	0.12 (0.15)	0.05 (0.07)
Total cost/kWh	0.15 (0.17)	0.07 (0.11)

Source: Mukunda *et al.* (1993).
[a] The costs are in 1991 $US at $1 = Rs20. At 6% discount rate.
[b] Figures in parentheses denote the engine running on diesel alone.

3. The pay-back period of both systems declines significantly when the hours of operation are increased. The pay-back period is sensitive to the cost of wood, although much less than to the hours of work per year.

The study concluded that the cost of electricity in the dual-fuel mode is lower than that in the diesel-only mode under the conditions used in the study. The cost of grid electricity from centralized sources during 1991 was $0.065 (Rs1.25)/kWh, which is subsidized; thus the cost of electricity using the 80 kW producer-gas system would be comparable with the grid electricity system.

9.2.5 Conclusions from Indian studies

The Indian case studies included different technologies, capacity levels and loads, rates of capacity utilization, different reference years, and varying capital and operational costs. In the Indian situation, comparison with diesel systems is justified as there are over 5.5 million diesel irrigation pumps besides a large number of residential, institutional, and industrial units using diesel for power generation. These latter units are used either as back-up systems or as the main source of electricity, due to power shortages, low reliability, and fluctuating voltages of grid power, in both rural as well as urban areas. The Indian case studies have the following features:

1. The Indian case studies are based on the current state of technology or off-the-shelf technology. No advanced or future technological developments were considered.
2. Mainly small-scale biogas or producer-gas systems (3.5–100 kW) were included.
3. The load factors were low and the transmission and distribution losses were a significant component of final costs.
4. Absence of a 'level playing field condition', since grid electricity and diesel are subsidized.
5. The producer-gas and biogas systems considered are operational, but the former is still being improved. With development it should be possible to cut costs further and increase conversion efficiencies.
6. The studies have not considered the environmental costs and benefits, foreign exchange premiums, social and economic benefits, and sustainability factors.

According to the learning curve concept, the bioenergy options are expected to become more attractive financially as more and more systems are marketed (Elliott and Booth, 1993). However, even with the limited examples from India, it is possible to draw the following conclusions:

1. Decentralized power generation systems, such as producer-gas, biogas, and micro-hydro, are cheaper to install and operate than conventional large coal-fired or hydroelectric systems for village electricity needs.
2. The bioenergy systems are either cheaper or comparable to diesel systems (at scales considered in the present study) at the higher rates of capacity utilization.
3. Economy of scale is obvious for the bioenergy systems and the levels of installed capacity projected for standard village sizes are in the feasibility range of 15–165 kW (Table 6.12).
4. Capacity utilization is critical for obtaining economies, and it is necessary to plan for at least 15 hours of operation per day. To deliver electricity to a range of services, as was discussed in Table 6.7, and with appropriate scheduling of activities, as presented in Chapter 6, it is possible to achieve over 15 hours of operation per day in rural bioenergy centres.
5. Decentralized options become attractive at higher discount rates, which are prevalent in India.

9.3 EVIDENCE FROM INTERNATIONAL STUDIES

A number of studies have analysed the economics of different bioenergy options in Brazil, USA, and Europe. Economic analyses has been carried out

Table 9.4 Cost of biomass energy systems for power generation from a World Bank review of renewable energy (Ahmed 1994)

Technology	Capacity	Capital cost ($/kW)	Fuel cost ($)	O & M cost ($/kWh)	Year	Cost of electricity in US cents/kWh (1990)	Method of estimation	Reference
Biomass gasifiers	37 MW	1600–1700	2.8/GJ	–	1990	6.0–7.0	Calculated cost	Elliott and Booth (1993)
Biomass gasifiers	37 MW	1200–1300	1–4/GJ	–	about 2000	4.0–6.0	Predicted cost	Elliott and Booth (1993)
Gas turbine	20 MW	1239	9.6/dry Mg (0.64/GJ)	–	1990	5	Cost estimated	Perlack *et al.* (1991)
Gasifiers	100–500 kW	429	65/dry t	0.01	1997	6.4	Projected cost	Grassi (1992)
Steam turbine	5–5000 kW	1900	2/GJ	–	1992	5.0–7.0	Operating exp.	OTA (1992)
Producer-gas IC engine	5–100 kW	420–680	2/GJ	–	1992	4.0–5.0	Operating exp.	OTA (1992)
Producer-gas IC engine	4.4 kW	1411	5/dry t	0.12	1991	15	Operating exp.	Mukunda *et al.* (1993)
Producer-gas IC engine	96 kW	1224	15/dry t	0.05	1991	7	Partly operating exp.	Mukunda *et al.* (1993)
Biogas IC engine	5 kW	1200	–	–	1992	11	Operating exp.	Rajabapaiaha *et al.* (1993)

O & M, operational and maintenance; exp., experience.

for different technologies, including the advanced biomass conversion systems. The relevant findings of these studies are used to understand the economics of bioenergy options. For details of methodologies and assumptions refer to the respective studies.

9.3.1 Electricity from biomass

Electricity generation is presently considered the most readily accessible route to the large-scale expansion of biomass for energy (Elliott and Booth 1993; Larson 1993; Patterson 1994). Biomass can be used for electricity generation through thermochemical processes or indirectly through the biochemical route. A summary of different studies showing the capital cost and total cost of electricity based on a review for the World Bank (Ahmed 1994) is given in Table 9.4. It includes both kilowatt- and megawatt-sized systems. The cost of electricity has been converted to 1990 price levels to assist comparison.

1. According to Elliott and Booth (1993), the capital cost of biomass electricity currently is in the range $1600–1700/kW, and is projected to decline by the year 2000. Perlack *et al.* (1991) also projected a similar value. According to OTA (1992) the capital cost for smaller systems (5–100 kW) is in the range of $420–680/kW, which is lower than the Indian examples (Tables 9.1–9.3).
2. All the studies estimated the cost of electricity at current prices to be 6–15 c./kWh, except OTA (1992) which projected a lower cost of 4–7 c./kWh. Elliott and Booth (1993) predicted that the cost of electricity will decline to 4–6 c./kWh by the year 2000.
3. The fuel cost of biomass, according to Elliott and Booth (1993) and OTA (1992), is expected to be in the range of $1–4/GJ by the year 2000. The current market price of woody feedstock in India is less than $1/GJ. Thus wood-based electricity systems will be very attractive in developing countries where labour costs are lower.

Reviewing the financial analysis of a large number of studies on biomass-based power generation systems, Ahmed (1994) concluded:

1. The cost of biomass electricity is declining and biomass gasifier/gas turbine technology appears to offer a cost-effective method of power generation in the future.
2. The costs are site-specific and vary with raw material costs, but they compare well with the costs of fossil-fired and even hydro-power generation in favourable situations. The costs in some situations are as low as 2–4 c./kWh.
3. Land requirements are a serious issue, especially in the light of the food-versus-fuel question. This issue is region- and site-specific and cannot be generalized.

4. As in the case of India or China, where degraded lands are likely to be used and where tree growing becomes part of land reclamation programmes (Perlack *et al.* 1991), the economic and food-versus-fuel issues take different dimensions (as discussed in Chapter 7).
5. Co-generation plants appear to be much more viable, especially if there are little or no fuel costs, e.g. sugarcane, and if surplus electricity can be sold to the grid system.

9.3.2 Advanced biomass and coal gasifier-based power generation systems

Major advances that have been made in coal gasification technology to match gas turbines to coal are also adaptable to biomass applications. A detailed analysis of advanced gasification systems has been made by Williams and Larson (1993) and this section is based on their study. A promising alternative to the steam-turbine cycle for biomass gasification is a set of biomass-integrated gasifier–gas turbine (BIG/GT) technologies that offer the potential for high thermodynamic efficiency and low unit capital costs. The GT systems are relatively insensitive to scale. The commercially available aero-derivative turbine cycle is the steam-injected gas turbine (STIG), and an advanced version of STIG is the intercooled steam-injected gas turbine (ISTIG). The biomass-gasifier–gas turbine plants are projected to have biomass-to-electricity efficiencies of 40–45 per cent and relatively low unit capital costs in the 5–100 MWe size (Larson 1993). The estimated capital, operation, and maintenance costs of BIG/STIG and BIG/ISTIG systems along with those of coal systems are given in Table 9.5.

The projected capital costs for the BIG/STIG, and particularly for the BIG/ISTIG, systems are competitive and even lower than the respective coal-fired systems. The total cost (including capital + operational + maintenance costs) for advanced biomass gasification systems are projected to be very attractive, at less than 3 c/kWh. With continued development, gas turbines fuelled by gasified biomass can be high-efficiency systems (in the 10–100 MWe range) and are likely to be able to compete economically in electricity production with large, modern coal or nuclear plants (Larson 1993). The BIG/GT systems (modular scales of 20–30 MW) have a number of features that make them attractive to developing countries, due to the low capital investment required and because significant parts of the systems could be manufactured locally (OTA 1992).

9.3.3 Economic performance of small gasifiers for power generation

Small gasifiers were disseminated in significant numbers in South-East Asia during the early 1980s (Chapter 5). The technical and economic performance

Table 9.5 Estimated capital ($US/kWh) and operation (USc./kWh) cost of advanced gasification technologies

Cost items	CIG/STIG[a]	BIG[b]/STIG	CIG/ISTIG[c]	BIG/ISTIG
Process capital cost	969	771	770	601
Total plant cost	1294	1031	1029	803
Total capital cost at 6 per cent discount rate	1410	1121	1121	873
Total cost at 12 per cent discount rate	1149	1152	1152	897
Busbar costs for power generation (US c./kWh)				
Fuel + labour + other fixed costs	0.86	0.62	0.73	0.52
Total variable O & M cost[d]	−0.16	0.10	−0.13	0.09
Capital cost at 6 per cent discount rate	2.36	2.04	1.92	1.64
Capital cost at 12 per cent discount rate	3.55	2.98	2.86	2.37

[a] CIG, coal-integrated gasifier; STIG, steam-injected gas turbine.
[b] BIG, biomass-integrated gasifier.
[c] ISTIG, intercooled steam-injected gas turbine.
[d] Negative values for total variable O & M costs are due to by-product credit for H_2SO_4.
Source: Williams and Larson (1993).

of this gasifier programme for power generation was monitored in a few countries, particularly in Indonesia and the Philippines by the Biomass Technology Group of The Netherlands for the World Bank and UNDP (BTG 1993). The investment and operational costs of small gasifiers in the range 10–40 kW are given in Table 9.6. The investment cost/kW for the imported gasifier was 2–4 times higher than the cost of locally produced gasifiers. The operational cost (fuel, labour, maintenance) for 6 out of the 10 gasifiers was less than 10 c./kWh. For two of the imported gasifiers it was 23 and 25 c./kWh, while for the locally manufactured gasifiers the operational cost was in the range 3–12 c./kWh.

An economic viability assessment of the gasifiers was made in the BTG study by comparing the breakeven diesel-fuel price (assumed to be $250/t). Diesel is taken as the next feasible substitute. The study concludes that at 1992 or 1993 prices of oil, biomass gasification is not an attractive option in the locations under study. Oil prices must rise by one-and-a-half to twofold for biomass gasification to become a financially attractive option. However, Indian studies (section 9.2) of gasifiers in the capacity range 5–80 kW showed that dual-fuel systems are more economical to operate than diesel-alone systems even using prevalent subsidized diesel prices in India.

Oil-importing developing countries such as India are always looking for

Table 9.6 Investment and operational cost of small-scale gasifiers from BTG (1993) (study for the World Bank/UNDP)

Site	Specific investment ($US/kW[a])	Specific investment ($US/kW (BGMP)[b])	Operational costs ($US/kWh)
Imported power gasifiers			
Sebubuk	2000	2300	0.07
Onesua	3600	4200	0.09
Mahe	850	850	0.25
Dogofiri	2600	2600	0.23
Local power gasifiers			
Balong	1150	1550	0.08
Majalengka	650	650	0.06
Lemoang	650	500	0.12
Nago	425	1400	0.04
Bolo	300	750	0.03
Itamarandiba	200	400	0.11

[a] Specific investment cost based on manufacturer maximal power output.
[b] Specific investment cost based on BTG-measured maximal power output.
BGMP, biomass gasifier monitoring project.

opportunities to reduce their dependence on imported oil. Furthermore, diesel prices are generally subsidized, thus it may be better to take actual delivered electricity costs from the existing grids for comparisons in order to draw conclusions.

9.4 PROPOSED ADVANCED BIOMASS GASIFIER POWER PROJECTS

Three specific project studies are available from Brazil, Italy, and the European Community, using advanced gasifiers.

Brazilian biomass power demonstration project

A demonstration project of a 30 MWe capacity, BIG/GT-based electricity generation system is underway in the north-eastern region of Brazil. A project appraisal report by Elliott and Booth (1993) is given in Table 9.7. The project is in the development and implementation phase, and operations should start by 1997. The projected capital cost is estimated to be $1300–1500/kW and the total cost of electricity, at a fuel cost of $2.6/GJ, is 5.1–5.6 c./kWh

Table 9.7 Cost and efficiency of Brazilian biomass power generation project (30 MW)

Cost details	Conventional technology—steam cycle	Emerging technology—BIG/GT cycle
Efficiency	20%	45%
Load factor	85%	85%
Capital cost	\$1800/kW	\$1300–1500
Cost of biomass	\$40/dry t or 2.6/GJ	\$40/dry t or 2.6/GJ
O & M cost	0.5 c./kWh	0.5 c./kWh
Fuel cost	3.6 c./kWh	1.6 c./kWh
Capital recovery	4.2 c./kWh	3.0–3.5 c./kWh
Total cost	8.3 c./kWh	5.1–5.6 c./kWh

Source: Elliott and Booth (1993).
BIG/GT, biomass-integrated gasifier–gas turbine.

for the BIG/GT cycles system. These costs compare favourably with the marginal costs projected for hydroelectric power projects in the Amazon region, and would involve lower capital investment per installed megawatt. The potential for biomass electricity in north-eastern Brazil is projected to be 41 TWh for sugarcane residues and up to 1400 TWh for wood from a dedicated tree plantation option, compared to the current generation level of 30 TWh annually (Carpentieri *et al.* 1993).

Producer-gas electricity option for Italy

A study prepared by MIT Energy Laboratory (USA) for ENEL (the electricity corporation of Italy) by Moncada *et al.* (1993) considered biomass-integrated gasification–turbine systems for Italy using already-available components. The capacity for the proposed system is 4.9 MWe. The study presented an economic analysis of the ENEL–CTRN, Pisa project aimed at demonstrating the feasibility of electricity production from biomass from integrated gasifier/gas–steam turbine cycle (B/IGCC) technology. It was a comprehensive analysis as it covered all aspects from growing the biomass to conversion into fuel and its utilization in electricity generation.

The biomass feedstocks considered included herbaceous crops and woody biomass plantations. Grassi and Bridgewater (1993) projected that the cost of woody biomass production will decline in Europe from the present level of \$US48/dry t to \$37/dry t in the future. The investment cost for this experimental plant is high at present (Table 9.8) but is projected to decline. Technical data for the ENEL plant are as follows:

(1) biomass productivity: 20–22 air-dry (10–15 per cent moisture) t/ha/yr;

Table 9.8 Cost of power generation for B/IGCC plants of capacity 4.9 MW (experimental plant) and 53 MW (commercial plant, after learning phase)

Component	4.9 MW experimental plant ($US/kW)	Component	53 MW commercial plant ($US/kW)
Dryer and hopper	72	Dryer and hopper	60
Gasifier: 3.3 t/h capacity	1608	Gasifier	434
Gas turbine: 3.91 MWe	426	Gas turbine (GE-6000)	273
Boiler + steam turbine: 1 MWe	278	Boiler + steam turbine	228
Electric generator	45	Electric generator	38
Civil works and auxiliaries	168	Engineering + civil works + auxiliary + other expenses	576
Total	2597	Total	1609

Source: Moncada *et al.* (1993).

(2) fuel size: biomass chips, diameter 15 mm and length 40 mm;
(3) plant size: 4.9 MWe (3.91 MWe from the gas cycle and 1.0 MWe from the steam cycle);
(4) capacity: 3.3 t of biomass/h;
(5) operation: 6000 h/yr;
(6) Life: 20 years (gasifier and power generation);
(7) Efficiency: net efficiency of the plants is 31 per cent;
(8) Reliability: 90%.

The cost of electricity from biomass is projected to be 15 c./kWh, compared to the European Union (EU) average of 11 c./kWh. The biomass option is just about competitive, but the adoption of B/STIG and B/ISTIG technology would reduce the cost further (Williams and Larson 1993). It should be noted that the cost figures are for the experimental plant. With learning-curve and technology developments in short rotation forestry and power generation technology, costs are likely to decline (Elliott and Booth 1993).

Electricity production from biomass in the EU

Grassi (1992) has considered various biomass-based options from the European Union perspective, and costings for some of the options are given in Table 9.9.

Direct combustion of biomass In comparison with the USA, where total biomass electric installed capacity using this technology is about 8000 MW,

Table 9.9 Projected electricity generation costs for biomass-based systems in the EU

Cost item	Advanced gasification	Ceramic gas turbine–aero-engine-derived gas turbine	Direct combustion of biomass in power station
Efficiency of the system	40%	48%	20%
Total investment ($US/kWe)	2400	1080	2400
Life of the system (years)	20	20	25
Operation (h/yr)	7000	4200	7000
Time for commercialization	3–5 years	–	–
Fuel cost (c./kWh)	3.1	5.0	6.2
Investment cost (c./kWh)	4.0	1.9	3.7
Operation cost (c./kWh)	1.1	1.2	1.1
Total cost (c./kWh)	8.2	8.1	11.0

Source: Grassi (1992).
Interest rate = 10%.
Cost of biomass = $US60/dry t.

the investment cost in Europe is relatively high, at $2400/kW, with the cost of production of electricity at 8 c./kWh. Grassi (1992) concluded that even though existing technology is already commercialized, the efficiency of conversion is low (20%) and the investment cost required is high, in addition to other problems, such as infrastructure limitations.

Decentralized electricity by direct combustion in turbines The first pilot project is being implemented by an ENEL–Daimler Benz–Ferruzzi grouping. This utilizes a ceramic turbine directly fired with powdered biomass fibre derived from C4-type crops, for example *Miscanthus*. The advantage of this highly efficient system (48 per cent) is for applications in the capacity of <500 kW. It can accept a range of biofuels, such as powdered biomass, slurries, up-graded bio-crude oil, and ethanol. The specific investment required is expected to be $1080/kWe and the total cost is projected to be 8.1 c./kWh. Grassi and Bridgewater (1993) predicted that with mass production the specific investment required could decline to a low of $240/kW installed, with an average cost of electricity production of 5.4 c./kWh. The technology is expected to be ready for commercialization in about 5 years.

Gasification using aero-engine-derived gas turbine–steam turbine combined cycles This system is expected to benefit from developments in the coal gasification technologies but with several advantages over coal, such as the absence of sulphur, low content of ash, and more reactive feedstock. These systems could be used for capacities of 1–30 MW with a system efficiency of 40 per cent. However, the costs are relatively high for systems with capacity below 10 MW. The investment cost is expected to be $2400/kW. At a biomass cost of $38/dry t, the operational cost (including fuel cost) is projected to be 5 c./kWh, which is comparable to electricity from coal at 4.6 c./kWh in the UK.

9.5 COSTS OF RENEWABLE ENERGY TECHNOLOGIES: IIASA ESTIMATES

IIASA

An IIASA study (Nakicenovic *et al.* 1993) has given comparative cost estimates of renewable energy technologies, including bioenergy technologies, at present and projected future levels (Table 9.10). The present costs are based on the best performing systems now in operation.

When renewable electricity generation options are compared, hydroelectricity is the least-cost option, while solar PV systems are six times more expensive. Wind energy systems are also competitive, at $0.07/kWh. When future costs are considered, solar PV and wind energy costs are likely to decline. Solar PV systems have been projected to be competitive with hydro and wind-based systems in the future. The cost of electricity generation using advanced biomass conversion systems is less uncertain and the projected price is competitive at $0.05/kWh. In contrast to biomass-based systems, hydro and

Table 9.10 Present and future costs of renewable energy technologies from IIASA study (costs for the year around 1992 for the best systems)

Energy technology	Present cost	Future cost
Electricity ($US/kWh)		
Hydro	0.05	0.05–0.10
Wind	0.07	0.03–0.06
Solar thermal (line-focus)	0.12	0.05–0.10
Solar thermal (point-focus)	0.25	0.04–0.08
Photovoltaic	0.30	0.04–0.08
Biomass/STIG	–	0.05
Tidal	–	0.08–0.10
Fuel ($US/GJ)		
Ethanol from sugar	8	6
Ethanol from wood	13	7
Methanol from wood	15	10
Biogas	–	1

Source: Nakicenovic *et al.* (1993).

wind-energy systems are very site- and season-specific and not all locations in the tropics are suitable.

Energy policy

Based on a number of studies on renewable energy, Jackson (1992) presented a set of economic costs of renewable energy options at current prices for the years 1990, 2000, and 2030 (Table 9.11). The cost of electricity from biomass sources was estimated to be 5–15 c./kWh in 1990 and is projected to decline to 5 and 4 c./kWh by 2000 and 2030, respectively. The cost of biomass electricity is projected to be lower than hydroelectricity and solar PV by the year 2000, along with feasibility in most locations. The costs of wind energy systems are also projected to be low but they are site- and season-specific.

9.6 ECONOMICS OF LIQUID FUELS FROM BIOMASS

Ethanol, methanol, and vegetable oils are liquid fuels obtained from biomass. With the exception of ethanol from sugarcane and maize, liquid-fuel programmes are not widely implemented commercially, mainly due to high costs. Brazil, the US, and Zimbabwe, for example, have considerable experience in the production of ethanol from biomass. Some of the costing information available is listed in Table 9.12, from the World Bank study by Ahmed (1994).

Ethanol from sugarcane

Brazil has the world's largest ethanol from sugarcane programme, with an annual production of about 12 billion litres for use in cars (Goldemberg *et al.* 1993). The cost of ethanol production in Brazil, based on data from 50 mills (Copersucar), was $11.5/GJ in 1990. The costs of ethanol have declined at about 4 per cent per annum since the 1970s, attributed to technology improvements and higher crop productivities (Ahmed 1994), and a continued

Table 9.11 Costs of electricity from renewables (cost in c./kWh—current prices)

	1990	2000	2030
Solar PV	30–100	10–15	4–6
Wind on shore	5–10	4–5	3–4
Hydroelectric	5–15	5–15	5–15
Biomass	5–15	5	4

Source: Jackson (1992).

Table 9.12 Cost of the liquid fuels ethanol and methanol from biomass

Raw material	Production cost as quoted ($US)	Production cost in 1990 $US/GJ	Year	Details	Reference
Ethanol					
Sugarcane	10.0–11.8/GJ (1990)	10.0–11.8	1992	Operating experience; average production cost in Brazil	OTA (1992)
Sugarcane	9.0/GJ (1990)	8.5	2000	Assumes BG/GT co-generation with surplus electricity and revenue credited against cost of ethanol	OTA (1992)
Cellulosic material	19–21/GJ	19–21	1992	Utilizing acid hydrolysis; technology commercially ready	OTA (1992)
Cellulosic material	15/GJ	15.0	2000	Technology could be developed	OTA (1992)
Sugarcane	0.229/1	11.5	1990	Anhydrous ethanol production; from 50 Copersucar mills (sample mean)	Goldemberg *et al.* (1993)
Corn	13/GJ	12.0	1992	Assuming the costs are in 1992 $; 12 per cent discount rate	Wyman *et al.* (1993)
Cellulosic material	12.6/GJ	13.3	1990	Estimate based on engineering study in 1990	Wyman *et al.* (1993)
Cellulosic material	7.5/GJ	7.9		Estimate; improved technology over the above	Wyman *et al.* (1993)
Methanol					
Cellulosic material	15.3/GJ	–	1989	790 tpd-biomass low-pressure gasifier	Wyman *et al.* (1993)
Cellulosic material	6.5/GJ	–	1989	5550 tpd low-pressure indirect gasifier with hot-gas conditioning	Wyman *et al.* (1993)

BG/GT, Biomass gasifier–gas turbine; tpd, tonnes per day.
Source: World Bank review (Ahmed 1994).

decrease in cost of production is feasible over the next several years (Goldemberg *et al.* 1993). Recent regulations provide opportunities for improving the economics of ethanol production through intensive use of the by-products—bagasse, stillage, and cane leaves—for electricity production. For example, for every tonne of sugarcane, 280 kg of bagasse is produced (50 per cent moisture). The cost of electricity through co-generation using bagasse as feedstock has been estimated to be 5 c./kWh. Adoption of advanced technologies such as bagasse gasification–gas turbines would lead to significant cost reductions. If the revenue from sale of surplus electricity is credited against the cost of ethanol, the projected cost of ethanol is $US8.5/GJ (OTA 1992).

Ethanol from lignocellulose (woody biomass)

Sugarcane requires good-quality land, large quantities of water, and intensive cultivation, all leading to high costs. For developing countries with high population densities, such as India, the sugarcane-ethanol (transportation fuel) option is not considered feasible and the lignocellulose option needs to be explored, since wood can be produced on degraded lands without irrigation. Using acid hydrolysis processes, ethanol production costs have been projected to be $19–21/GJ (OTA 1992). With improved biochemical technologies, the cost of ethanol could be reduced from $12.6/GJ to $7.5/GJ (Wyman *et al.* 1993).

Methanol from lignocellulose

The production of methanol from lignocellulose through the thermochemical process would typically convert 50–60 per cent of the energy content of biomass into methanol (OTA 1992). The cost of methanol production is estimated to range from $15.3/GJ, based on a low pressure oxygen gasifier, to $6.5/GJ, using low pressure indirect gasifiers with a hot-gas conditioning process at 12 per cent discount rate (Wyman *et al.* 1993). The cost of methanol from biomass falls between the cost of methanol from natural gas and from coal, at feedstock costs of $2/GJ for all fuels (Larson 1993). Methanol production from biomass is yet to be commercialized, and its adoption in many developing countries would await commercialization of the technology at feasible scales. Analysing the potential of liquid fuels from biomass, Wyman *et al.* (1993) concluded that technical opportunities exist which could eventually reduce the cost of ethanol and methanol to levels competitive with gasoline ($0.21/l) derived from oil at $25/barrel.

9.7 INTERNALIZATION OF ENVIRONMENTAL COSTS IN THE ECONOMIC ANALYSIS

Use of conventional sources of energy, particularly coal-fired thermal and large hydroelectricity, petroleum fuels, and the use of biomass in a non-renewable mode, is accompanied by a large number of environmental and social costs which were discussed in earlier chapters. These costs are not included in the economic evaluation or energy pricing. Renewable energy sources are, therefore, generally at an unjustified disadvantage with respect to their conventional competitors. Inclusion of external effects would improve the competitive position of renewable energy sources and thereby speed up their market penetration. If all costs are considered, renewables have considerably lower relative costs than market prices indicate (Hohmeyer 1992). The costs that should be included are: impact on human health (pollution), environmental damage (to flora and fauna), impact on the atmosphere and climate change, long-term costs of resource depletion (petroleum or coal), employment effects, subsidies for conventional sources (research and development, railways for transporting coal), and reallocation of human settlements. On the other hand, Foley (1992*a*) argued that renewable energy sources should be subjected to the same technical and financial analysis as the conventional options and the standards of analysis should not be lowered in order to justify the use of expensive or inappropriate electricity generating equipment simply because it does not rely on petroleum fuel.

Relative costs of electricity with internalization of some external costs

Based on a number of studies, Jackson (1992) made an attempt to partially internalize some of the environmental costs. The relative costs (roughly at current prices) for the important renewable energy options are given in Table 9.13. The values given are additional costs of generation from renewables compared to conventional electricity costs (at about 6 c./kWh). The negative values for renewables show that the options are cheaper, or have negative costs, compared to the conventional options. For example, biomass energy costs are currently in the range of −20 to + 1 c./kWh compared to conventional options, which cost around 6 c./kWh. Thus attempts to partially internalize some of the environmental costs show that renewable energy technologies are already competitive if environmental costs are attached to conventional fuels.

Table 9.13 Additional cost of electricity from renewables (with internalization of external costs) compared to conventional electricity costs in c./kWh (current prices)

Technology	1990	2000
Solar PV	+5 to +90	−15 to +6[a]
Wind on shore	−20 to +1	−21 to −4
Hydroelectric	−20 to +6	−20 to +6
Biomass	−20 to +1	−20 to +1

Source: Jackson (1992).
[a] Negative values indicate that if environmental costs and benefits are included, the cost of renewables will be negative.

Nevertheless, difficulties still continue to elude detailed attempts to 'internalize the externalities'. But with the information available it is possible to say that certain renewables which are not economically viable at present will become viable if external costs are included (Friedrich and Voss 1992). Based on a recent analysis, Hohmeyer (1992) also concluded, 'even without including all social costs and even with deliberate bias against renewable energy, the net social benefit (including some external costs) in monetary terms for wind and photovoltaics energy use is comparable with basic market prices of conventionally generated electricity'. However, the climate change debate has led to the concept of a greenhouse gas tax, particularly a so-called 'carbon tax'. In some countries, for example Denmark, Sweden, and the USA, there are incentives for the use of renewable energies to avoid emissions of pollutants, including CO_2 and sulphur. Negotiations are in progress in several countries for quantifying the damage to the environment, and to evolve greenhouse gas tax mechanisms. Some examples of existing incentives for renewable energy technologies, particularly bioenergy, are considered in the following section.

9.8 INCENTIVES TO USE RENEWABLE ENERGY

Sweden

A carbon tax equivalent to 25 ore/kg of CO_2 emission was introduced on the use of fossil fuels—oil, coal, natural gas, and LPG. However, use of fuels for electricity production were exempted. In addition to a carbon tax, the environmental cost of sulphur is also considered. The energy and environmental tax is 16 ore/kWh for coal and 15 ore/kWh for petrol ($US1 = Skr.5; NUTEK 1993).

USA

The implementation of PURPA (Public Utilities Regulatory Policy Act) in 1978 opened electricity markets to non-utility generation sources. Combined with renewable energy investment tax credits for 10 years, this has led to a capacity of about 8000 MWe of biomass electricity, 730 MWe of small hydroelectricity, and 1500 MWe of wind energy. Recently a programme of incentive payments of 1.5 c./kWh has been introduced for renewables such as biomass and wind.

Denmark

Denmark has embarked on a significant centralized biogas programme with a biogas production of 14 million m^3, corresponding to 0.5 PJ in 1992. A guaranteed favourable price exists for biogas energy. Fossil fuels are subjected to heavy taxes, while biogas energy is exempted. In addition, a state subsidy of Dkr.0.27/kWh ($US1 = Dkr.6.5) is granted for biogas-based electricity generation.

Brazil

In Brazil, the Proalcool (ethanol) programme received several incentives from its start in 1975, such as the granting of low-interest loans, the price of the ethanol being made attractive to consumers by selling it initially at the pump at 59 per cent of the price of gasoline, and guaranteeing purchase of an agreed quantity of ethanol by the government-controlled agency (Goldemberg *et al.* 1993). Such incentives led to a rapid growth of the ethanol industry from less than 1 billion litres in 1978 to about 12 billion litres/yr by 1989.

The lessons that can be derived for developing (and industrialized) countries which are subsidizing conventional energy sources is that they should provide appropriate incentives now for bioenergy options, if necessary, until the technology matures and scales of production become commercially viable. As will be shown later, financial incentives must be accompanied by relevant policy changes. According to the evidence presented in this chapter from several countries, bioenergy options are often already economically competitive or will soon be, and appropriate policy measures and the creation of a 'level playing field' alone may catalyse the spread of rural bioenergy centres and multi-megawatt capacity power generation using stand-alone or co-generation systems.

A start to internalizing some of the external environmental costs would be a consideration of sulphur and carbon emissions avoided by switching to biofuels from fossil fuels such as coal. Carbon and sulphur taxes to decrease emissions could be taken as 'income' for bioenergy systems.

Developing countries could promote bioenergy programmes, keeping in mind local benefits such as employment, reclamation of degraded lands, promotion of self-reliance, substitution of imported oil, and pollution abatement, and also claiming credit for avoiding greenhouse gas emissions and/or reducing net carbon emission. If such an approach is adopted, significant funding might be forthcoming in the future as climate change mitigation measures. In developing countries there is a need for large-scale energy generation to promote the quality of life as current energy use levels are so low.

9.9 IMPLICATIONS OF ECONOMIC ANALYSIS OF BIOENERGY SYSTEMS

In earlier sections evidence from India, Brazil, USA, and Europe was analysed and some of the implications are highlighted here.

1. Decentralized bioenergy options are competitive compared to centralized conventional options (Reddy *et al.* 1990; Sinha and Kandpal 1991), even at existing states of technology. With improvements in technology for efficient biomass production and conversion, the costs will decline even further.
2. Advanced biomass conversion systems will be commercially available in kilowatt as well as megawatt capacities, and will be viable even at the lower range capacities projected for decentralized systems, possibly by the year 2000. The advanced systems are to some extent scale-neutral (Williams and Larson 1993). High-efficiency systems (40–50 per cent), in the capacity range of <500 kW, are in the development stage and are likely to become available in about 5 years.
3. The megawatt-level systems planned for north-eastern Brazil and Italy are projected to be competitive, even using conventional economic analysis, and costs will decline further during the learning stages with further installations.
4. The two village-level bioenergy centres based on biogas and producer-gas systems in India have shown that they are financially comparable with conventional grid or diesel systems. The experience of these two systems shows that rural households are prepared to pay a higher price than the subsidized price of grid electricity, if reliable power is made available as a part of a package, along with their own participation in the management of the system.
5. In village situations in developing countries, power transmission costs and transmission and distribution losses are high, since power requirements are characterized by low and fluctuating loads. The cost of electricity is estimated conventionally for fossil and hydro systems without including such factors. Decentralized systems provide an opportunity to avoid significant costs and losses associated with grid-power transmission.
6. Energy pricing in developing countries does not reflect the economic, social, and environmental costs.
7. Many countries provide incentives for renewables, like the USA (incentive payment of 1.5 c./kWh), Britain (non-fossil-fuel obligation), and some countries have environmental taxes, such as Sweden and Denmark (a state subsidy of Dkr.0.27/kWh for biogas electricity, in addition to tax rebates for biogas electricity). This is a clear recognition of the tangible environmental benefits of renewable energies.
8. The economics of bioenergy systems depend on the costs of feedstocks, and are likely to improve due to the projected decline in wood production costs for Europe (Grassi and Bridgewater 1993), for the USA (Turnbull 1993), and for Brazil (Carpentieri *et al.* 1993; Elliott and Booth 1993).

In developing countries environmental awareness is growing and many strong environmental movements exist. Energy planners can no longer afford to ignore environmental problems. Environmental factors may cause delays or even cancellation of planned power projects in developing countries in the years to come (Myers *et al.* 1993). In India environmental consideration is already a component in energy planning, although not yet fully recognized (PC 1992). One or a combination of the following factors could catalyse the large-scale promotion of bioenergy:

1. The rate of technological advance measured in marginal cost of generation terms might be greater for biomass.
2. Environmental costs might be internalized or, put another way, the existing structure of energy taxation and subsidization may be changed to reflect social costs and benefits.
3. The relative structure of energy prices may change; for example a sharp escalation of oil prices, say by 2000 or later.

9.10 CURRENT AND FUTURE BIOMASS ENERGY PROGRAMMES

Bioenergy technologies are emerging technologies which have come a long way from open fires and wood stoves as a source of heat to modern sources such as electricity and liquid and gaseous fuels. In this chapter we have provided considerable evidence to show that bioenergy is a modernized fuel source with commercial potential and without the past stigma of being a primitive fuel. We present a number of significant efforts to expand and commercialize bioenergy in developing as well as industrialized countries.

India

The government has made a policy commitment in the official government plan document to promote renewable energy in the medium and long term (PC 1992). According to the latest reports, the target for biomass electricity generation has been fixed at 600 MW by the year 1997 (Krishnakumar 1993). The total potential for biomass electricity has been estimated to be in the range 17 000–60 000 MW (Cherail 1993), and the potential for family biogas plants has been fixed at 15 million, with the potential for community biogas plants being almost one for each of the 557 000 villages (MNES 1993*a*).

The allocation of funds for the bioenergy programme for the year 1992–93 was Rs790 million ($US25 million) for biomass conservation and conversion programmes. This accounts for 62 per cent of the total budget for all renewable energy technologies. About 2 million improved stoves and 160 000 biogas plants are planned annually. India has also planned a large afforestation programme (mainly for producing fuelwood) at the rate of 2 Mha annually with an expenditure of Rs17 billion or $US500 million (Rs8 billion from government + Rs8 billion from the private sector + Rs1 billion from external assistance). Recently, co-generation projects have been implemented in two sugar mills and at one rice-straw-based 10 MW power plant. There is also a plan to popularize wood gasifiers for village electrification and for pumping irrigation water on farmland. The ministry (MNES) has also invited proposals from industry for biomass electricity systems of installed capacity >5 MW for part funding (during 1994). Overall, the programme is severely constrained by funds since the total funding for all the renewable energy programmes for the year 1992–93 was only Rs1.28 billion ($US41 million), excluding afforestation.

Outside India

Many advanced biomass gasifier-based power and biogas programmes have been launched or are in the process of being launched:

(1) Sweden—in Varnamo, 6 MWe and 9 MWth plant (combined cycle);
(2) Finland—in Tampere, 6 MW;
(3) Brazil
    (a) in the north-east, 25–30 MW (BIG/GT)
    (b) already has a large ethanol programme of 12 billion litres/yr;
(4) USA
    (a) in Hawaii, 20 MWe by local authorities and US Department of Energy (BIG/GT)
    (b) in US biomass-based power generation systems of about 8000 MW capacity are already in operation;
(5) Italy—5 MWe B/IGCC plant planned by ENEL–CRTN, Pisa;
(6) Denmark—in addition to the existing 10 large centralized biogas plants, six or seven new large-scale biogas plants are planned. Many straw-fired heat and power plants exist, including co-firing with coal;
(7) Tanzania—in Dar es Salaam, an urban waste-based large programme of capacity 160 t of organic waste/day for electricity generation of 11.8 MWh/day; the project is funded by UNDP (GEF), and the Danish and the Tanzanian governments, and is in the process of implementation;
(8) China
    (a) China has the world's largest biogas programme for delivering cooking energy as well as electricity; the new focus seems to be on large- and medium-sized biogas plants and electricity generation, unlike in the past where the emphasis was largely on household plants
    (b) nearly 100 million improved stoves have also been built
    (c) China also has a significant biomass gasifier programme.

Biomass-based electricity, as well as liquid fuels, will be most relevant in the light of global efforts to find carbon-neutral energy sources. Bioenergy options are not only carbon neutral but also lead to a reduction in net CO_2

emissions when used to substitute for fossil fuels. Biomass also results in lower levels of sulphur emission and hence less acid rain. Furthermore, bioenergy systems are politically very attractive for developing countries due to significant local benefits such as employment, promotion of local control and self-reliance, land rehabilitation, and reduction in the national-level dependency on oil imports.

To conclude, it is worth quoting the Electric Power Research Institute (EPRI 1993) of the USA on the potential of biomass energy, 'Projections of emerging renewable energy markets in the first part of the 21st century consistently suggest that biomass energy systems will play a major role. The projections are based on the following facts: (1) vast availability of suitable land (tens of millions of acres) for growing energy crops, (2) environmental benefits (CO_2 neutral, low SO_2, and recyclable ash), and (3) favourable economics compared to coal-derived energy in the 1990s'. Patterson (1994) concluded that biomass electricity will have the biggest impact in tropical and sub-tropical regions in the coming decades.

10

Approaches to implementing bioenergy programmes

One of the important aspects to be considered in utilizing the bioenergy potential in India is the feature of dispersed rural human settlements encompassing 75 per cent of the country's population. Bioenergy options can provide a decentralized village-level, self-reliant energy system to meet village needs in a sustainable and environmentally sound way (Chapters 6 and 8). The next question we should ask is how can this be done. An approach to the implementation of rural bioenergy centre (RBC) schemes (which are at the heart of the proposed biomass-based energy programme) and of the large-scale bioenergy systems is considered in this chapter. Here we suggest an approach based largely on the Indian experience.

10.1 ECOSYSTEM APPROACH TO RURAL ENERGY

It is recognized that a village functions as an ecosystem or as an ecological unit with interactions and interlinkages among land, livestock, vegetation, water, and humans; energy is one of the important linking factors (Reddy 1981). Due to this interdependence, it is necessary to consider an ecosystem approach to village development (Agarwal and Narain 1989). A more detailed discussion of the ecosystem concept is beyond the scope of this book, but we stress the need to keep in mind the interlinkages among land, livestock, water, and human components while planning for rural energy. Some justification for this includes:

1. Land for energy plantations; while planning energy plantations, the grazing needs or grass production requirement of village livestock must be considered.
2. Fodder or grass availability determines the feeding levels of livestock and, in turn, cattle dung production and biogas production potential.
3. Availability of irrigation water (well, stream, farm-pond) determines the number of irrigation pumps and the electricity required for pumping.
4. The woody residues of some crops, such as mulberry, pigeon pea, corn, cotton, rice, and coconut, can be used as feedstocks in gasifiers.
5. The leaf litter from energy forests and some crop residues can be used as feedstock for biogas production, and the digested slurry is a valuable manure for crop production.

6. The land proposed for energy forests can become part of watershed development and enhance the water availability in water bodies for irrigation in the village ecosystem.

Rural energy and rural development

Any land-based activity, technology, or programme must aim at ecosystem-level planning. Biomass production (food, fodder, fuelwood) is dependent on the extent and quality of land and water availability. In India a beginning is being made on integrated land-use planning by adopting a watershed approach (PC 1992). Producing biomass for bioenergy options could become part of the land reclamation or watershed development programmes. The rural energy programmes, particularly bioenergy, must become a part of rural development programmes so that the village community can benefit from the interdependence of energy and other sectors of the village economy. By integrating bioenergy with other development programmes, such as crop irrigation and dairy farming, it is possible to enhance the success of such programmes. Such an approach will lead to increased incomes, which are crucial to make self-reliant systems financially sustainable.

10.2 RATIONALE FOR RURAL BIOENERGY CENTRES

The technical and operational feasibility of bioenergy options for meeting all the heating, lighting, and stationary shaft-power needs was shown in Chapter 6, with the beneficial impacts on local and global environments highlighted in Chapter 8. Thus one can start thinking realistically in terms of 'rural bioenergy centres (RBC)', a concept that is not new to India since it was proposed as far back as in 1980 in the form of rural energy centres (Reddy and Subramanian 1980), and the Ministry of Non-Conventional Energy (MNES 1993*a*) has a programme called 'Urjagram' (energy village). A beginning has been made with respect to establishing rural bioenergy centres in Pura and Hosahalli (Chapter 5). Presently, the scope of the Pura experiment is limited to biogas for lighting and pumping drinking-water, and at Hosahalli the experiment is limited to growing an energy forest, operating a wood gasifier, and providing electricity for lighting and pumping drinking-water.

The rural bioenergy centre concept would involve meeting the energy needs of heating, lighting, and stationary shaft-power-requiring services of a village, or a cluster of villages, through bioenergy options in a sustainable and participatory way. The services requiring energy in a typical village are listed in Table 6.7. The rural communities usually have a priority ranking of services requiring energy; for example, piped water supply for domestic use, cooking, lighting, irrigation water, and agro-processing.

The priority ranking may change from location to location, but many of the services would be common to most locations. The preferences of men may differ from those of women. Men in general are likely to give higher priority for irrigation water and a very low ranking to cooking. On the other hand, women may give higher ranking to a piped water supply and cooking energy than to irrigation-water pumping or agro-processing. To improve the quality of life, it is preferable to consider quality fuels for all the services.

The complementarity of biogas and producer-gas systems was discussed in Chapter 6. For example, wood from a dedicated energy plantation could be used for producer-gas electricity, and leaf biomass residues from the same plantation could be used as feedstock for biogas plants.

Thus it is desirable to adopt the concept of rural bioenergy centres to benefit from complementarity, to achieve economy of scale, to share facilities, and to reduce the fixed as well as operational costs. Experience from the Pura and Hosahalli village demonstrations has shown that it is not sufficient to provide drinking-water and lighting, it is equally important to provide higher-quality fuel for cooking and to provide reliable electricity for irrigation, agro-processing, and so on.

10.3 WILL ENERGY ALONE DELIVER DEVELOPMENT?

Would the generation and availability of electricity or biogas alone guarantee its use for cooking, lighting, or pumping water? The experience so far in India is that mere electrification of a village, that is, extending the transmission line to the middle of a village, has not ensured its use for providing services to the community. For example, at national level with 84 per cent of villages having an electricity supply, only 27 per cent of households use electricity for lighting.

There could be many reasons for limiting the use of energy services:

(1) the cost of obtaining the service is high—in India one needs to pay large deposits and fees to the utility and contractors to obtain electricity for homes; deposits are also required to obtain LPG cylinders;
(2) devices involve significant first costs; for example, a biogas plant requires an investment of Rs10 000–15 000;
(3) uncertain supplies and poor reliability of energy services, particularly electricity and kerosene for cooking in rural areas;
(4) non-availability of credit for obtaining energy services;
(5) complex procedures for obtaining electricity or LPG (cylinder and appliances) connections;
(6) lack of awareness regarding the services, their performance and potential benefits; for example, farmers in India are generally not aware of the potential of biogas and producer-gas options for pumping irrigation water.

Thus it is necessary to cut costs and improve access to modern fuels. Hibajene and Kaweme (1993) analysed the problems of rural electrification in Africa and suggested the following cost-reduction approaches to electrification of low-income households (they generally hold good for India also):

(1) roof-to-roof methods of connection, where electricity could be carried from house to house, rather than each house having an electricity connection from the main transmission pole;
(2) group schemes where all houses are connected simultaneously;
(3) clipping of cables within a house to the walls, rather than use of conduits;
(4) use of electricity dispensers which are charged periodically instead of an energy meter;
(5) involvement of private companies;
(6) reduce upfront costs and recover them in installments;
(7) provision of credit to group schemes.

A range of additional measures is required to enable rural users, particularly women (Bose 1993), to benefit from modern bioenergy services made available to them to promote development. Some examples of the measures required are as follows:

(1) ensuring guaranteed supplies to motivate households;
(2) simplifying the procedures involved;
(3) cost reductions for all services through development work on technologies and efficient delivery systems; and
(4) involvement of local communities, which would go a long way in promoting greater spread of bioenergy technologies.

10.4 DISSEMINATION STRATEGY FOR RURAL BIOENERGY CENTRES

The feasibility of RBCs has been demonstrated and now their economic, social, and environmental impacts merit wider recognition and dissemination. In India large-scale dissemination could theoretically mean up to half a million RBCs. We can consider a number of approaches for large-scale dissemination of RBCs:

(1) government programmes;
(2) private entrepreneurship-promoted programmes;
(3) community-partnership programmes.

Government

Implementation of an RBC plan requires constant interaction with local communities, for which government agencies are not the best suited, as has been seen in social forestry and community biogas programmes (Chapters 4 and 7). The RBC should not be a wholly government-managed programme for the following reasons:

1. Government officers will not have any stake in the long-term success or sustainability of the programme.
2. The cost of maintaining the government administrative structure would be very high.
3. Any government-administered programme is likely to have complex formalities and bureaucratic procedures, leading to delays in implementation and often loss of interest by NGOs and local communities.

The government has, however, a specific role to play in promoting bioenergy programmes via policy making, provision of funds, incentives, and so on.

Private entrepreneurship

This approach necessitates a profit motive for the entrepreneur. It is possible to visualize a situation where private entrepreneurs install and operate village-level biogas plants and producer-gas-based power generation systems and sell biogas for cooking and electricity to individual homes (home lighting), farms (agricultural pumping), and even to the local government (panchayat) for community services (such as pumping drinking-water or street lighting).

The private enterprise may require significant subsidies for services that are not commercial activities (such as cooking, street lighting, and pumping of drinking-water). Private enterprise requires the full backing and support of local communities. This necessitates suitable policy changes to permit such an arrangement.

A band of rural entrepreneurs who can undertake turnkey execution of the systems and maintenance contracts has to be developed.

Community partnership

The planning, installation, operation, and maintenance of RBCs requires technical skills. Local communities by themselves may not be able to undertake all the functions even if opportunities exist. Thus planning and installation may have to be left to entrepreneurs to undertake as turnkey operations. However, the utility will be owned by the village community. The local youth could be trained to operate and to undertake routine maintenance of the system, as has been demonstrated in Hosahalli and Pura villages in South India (Chapter 5). This approach may require some level of subsidy (particularly initial investment) and support from the government. Thus community partnership may be an ideal approach where local communities take decisions and enforce them, but all technical jobs are left to private enterprise on a turnkey basis and through a maintenance contract with entrepreneurs, which would ensure smooth servicing of the system.

It is necessary to have an open mind and create opportunities for both private enterprise and community partnership options and learn with experience.

10.5 DEMONSTRATION OF A RURAL BIOENERGY CENTRE

The first step before adopting any large-scale dissemination or commercialization of an RBC programme is to undertake demonstration projects, for the following reasons:

(1) to create awareness regarding the potential of RBCs to meet the energy needs of villages in diverse locations;
(2) to demonstrate the feasibility and economic viability of RBCs in different locations;
(3) to use the demonstration units to develop a methodology for large-scale expansion of RBCs in different socio-economic environments;
(4) to develop each RBC as a base for organizing training programmes for organizers, operators, and entrepreneurs undertaking implementation and management of RBCs in the administrative blocks; and
(5) to generate data on investment and operational costs.

Locations for demonstration

India has 557 000 villages and 3298 administrative blocks in 446 districts (CMIE 1992). It is not enough to have a few demonstration units for a large country like India. We consider it preferable to select two blocks per district and implement five demonstration units in a cluster for each of the 892 selected blocks. In fact, all developmental programmes are presently administered at the block level. A cluster approach, rather than having one unit per block or district, facilitates training, monitoring, and provision of support services, and also helps the learning process between units.

Phasing of demonstration units

India has significant experience in planning and implementing village-level programmes. Thus the demonstration phase need not be longer than, say, 5–7 years; first year for planning, planting the energy plantation, construction of the biogas plant, and setting up the infrastructure, and by the end of the second or third year it may be possible to harvest thin branches from the energy plantations and full-scale implementation could start. Monitoring and evaluation of the complete system could be during the fourth and fifth year.

In the demonstration phase the programme could be implemented in one block per district in the first and second years, and from the experience gained it could be extended in the third year to the second block in each district.

Training of personnel

Organizing training programmes at different levels, from planners to village operators to maintenance personnel, is essential, and these should be at the district or state levels. The first training could be at the state level for about three persons (including one person from an NGO) for each block before embarking on the planning of the RBCs. The training programme should include the following:

(1) methodology of conducting surveys, assessing the needs, organizing village institutions, implementation of projects;
(2) exposure to technologies and modes of implementation;
(3) field visit to sites of existing RBCs;
(4) methods of creating awareness in the community, explaining the programme and eliciting the community's participation;
(5) monitoring and evaluation of programmes.

Steps in planning and implementation of demonstration units

The demonstration programme should involve the following steps:

(1) forming a village bioenergy development council or society, with all the households as members and a small representative management committee for regular discussion;
(2) survey of land, livestock, population, water, and vegetation resources;
(3) survey of current sources of energy and the services derived from energies;
(4) discussions with the village community (women, farmers, entrepreneurs) in order to identify the energy services required by the local community and prioritizing the needs;
(5) defining the energy services and estimating the rates of requirements, seasonality, and location of delivery of such energy services;
(6) preparation of a bioenergy plan, defining the mix of technologies, in consultation with the village management committee, resources required, benefits, mode of sharing, roles and responsibilities of different actors, resource constraints, etc.;
(7) presentation of a plan to the village council for discussion in order to take final decisions;
(8) execution of the plan by professionals, turnkey jobs to entrepreneurs;
(9) organizing the sharing of the costs and benefits;
(10) monitoring the performance of the programme;
(11) organizing periodic meetings of the village council to report findings and to discuss actions required;
(12) modifying the programme operation based on the results from monitoring studies and suggestions of the village council;
(13) prepare feasibility reports and plans for expansion of the programme.

Monitoring and evaluation of the rural bioenergy centres

Monitoring and evaluation of the RBCs is very important. Most programmes lack this aspect, and even when there is an evaluation programme, the findings are rarely fed back to the field-level agencies, let alone to the village community. One of the main goals of the demonstration programmes is to learn from experience and to develop methodologies for large-scale dissemination of the programme. It is necessary to monitor the technical performance of the system, the financial aspects of the programme, energy plantation productivities, efficacy of mode of sharing of costs and benefits, and the functioning of village institutions.

Develop a methodology to establish rural bioenergy centres

Due to the variation in the socio-economic and resource status of different regions, it is unlikely to be feasible to adopt a uniform methodology and strategy for operationalizing the RBC plans for any country. For example, in some situations all woody biomass may come from the village commons and in others all of it may have to come from private lands; in some regions the irrigation potential is high and in the others it is low; in some regions traditional village social institutions exist for managing natural resources (such as water or forests) and they could be expected to manage the RBCs. There are region-specific variations and it is necessary to prepare appropriate strategies for different regions.

10.6 PHASING OF THE NATIONAL DISSEMINATION

The rate of dissemination of RBCs would depend on the annual rate of flow of funding and organizational capacity of the implementation agency. Every administrative block has, on average, about 169 villages. It would not be possible to implement a programme simultaneously in all the villages even if funding were available, due to the organizational limitations. The programme could be phased over a period of, say, 10–15 years. Currently in India the annual achievements under different programmes are: over 2 Mha of afforestation, over 2 million improved stoves, and 188 000 family biogas plants. India has also extended grid electricity to 21 000 villages/yr between 1980 and 1991 (TSL 1992). Most of these programmes are implemented through the relevant government departments. With adequate funding and the participation of local communities and private enterprise, it should be possible to implement the RBC programme in 10–15 years to cover all villages. India has adequate experience in large-scale dissemination of development programmes.

It would take about 2 years to fully operationalize an RBC in a village from the first contact with the community. The first component of the programme to be implemented is planting the energy forest. Given the fact that branches and thin shoots are preferred and not thick stems or trunks, it may be possible to harvest shoots in many locations by the end of second or third year. One should be careful not to make it into a target-chasing programme, since the quality of the programme will immediately suffer (Chapters 4 and 5). Only by ensuring a significant responsibility to the private entrepreneurs and the local community, and by arranging for a guarantee on the technical performance of the systems, can the quality and the performance of the programme be assured.

10.7 DIVERSITY OF SOURCES FOR FUNDING RURAL BIOENERGY CENTRES

The RBC programme at national level will not be feasible if it is wholly based on government subsidies or funding. If the government provides investment grants for power generation and drinking-water projects which benefit urban and industrial sectors, then there should be equal or higher levels of grants to RBCs, due to the significant benefits that accrue from them to local communities and the environment, particularly to rural women and the rural poor. A range of sources of funding is available:

(1) personal savings;
(2) loans to farmers or their families from banks, industry, or government;
(3) government grant or subsidy;
(4) private entrepreneur—commercial loans;
(5) international funding—loans, grants;
(6) state utilities—loans, grants;
(7) industry—loans to individual households or farmers.

The following broad guidelines could usually be adopted in decisions on funding:

1. Basic community services should be fully subsidized to promote equitable development, for example street lighting and drinking-water supply through community outlets.
2. Private household services leading to benefits at the individual family level should receive soft loans for installations (e.g. home lighting), households being expected to pay for the service at commercial rates.
3. For community installations, investment costs could be part grant and part loan from the government. Operational, maintenance, and replacement costs should be met by the beneficiaries.
4. Investment for commercial services (e.g. irrigation pumping, flour milling) should be treated as commercial loans and operated as a private enterprise, with economic pricing for the energy supplied.

Traditional systems such as wood stoves, kerosene lamps, and animal-drawn systems require a low initial investment, in contrast to modern bioenergy systems such as electrical connections, compact fluorescent lamps, and stoves for gaseous fuels. Rural households may not be in a position to make such large investments. Interest rates in the organized sector are presently about 21 per cent, whereas the rates in the informal sector are very high (upwards of 36 per cent) and currently credit is not easily available in rural areas. Arrangements should be made to provide credit at reasonable rates of interest to rural households and entrepreneurs. A list of infrastructure and investments required, the mode of funding, and possible sources of funding are given in Box 10.1. What is the rationale or motivation for different agencies to provide grants, subsidies, or credit?

Government agency Bioenergy options and energy-efficient devices provide higher quality of life through decreased human drudgery and increased productivity of resources (land, water, and human labour). Bioenergy options provide significant local and regional environmental benefits while providing direct benefits to rural women and the poor. These programmes are compatible with national goals of development. Full adoption of bioenergy as envisaged will lead to savings in kerosene (rural lighting and cooking) and diesel (in irrigation pumping) and thus to reduced imports of oil. Thus there is enough justification for the government to fund programmes as part of self-reliant and sustainable development efforts.

International agencies The motivation for international agencies to assist would be mainly for deriving global environmental benefits, such as carbon sequestration, reduction in carbon and sulphur emissions, and promotion of biodiversity.

Commercial banks Profit motivation.

State utilities Compared to hydro or coal-thermal electricity options, bioenergy options are environmentally friendly with little likelihood of opposition from local communities. They have none of the major environmental or social problems generally associated with coal and hydroelectric systems. Localized electricity generation reduces pressure on the grid, reduces transmission and distribution losses, and can increase revenue for the utility. Some bioenergy options already have lower costs.

10.8 INSTITUTIONAL ARRANGEMENTS FOR COST REDUCTIONS

A number of institutional measures are required to reduce establishment and operational costs. A strong community organization with adequate power to

Box 10.1 Potential sources of funding for investment capital for bioenergy systems

Service	Source of funding	Mode of funding
1. Community biogas plant for		
• cooking	Rural Development Department, village community	Part grant/part loan
• power generation	Electricity utility	Part grant/part loan
2. Piping biogas for cooking	Households Commercial banks	Credit to be recovered in instalments
3. Energy plantation for feedstock for electricity	International institutions Electricity utility	Farmland—interest subsidy, loan Common land—grant
4. Engine, genset, gasifier for community-level systems	Electricity utility Investment agencies	Part grant Part soft loan
5. Power transmission lines	State electricity utilities	Grant
6. Lighting connection for homes	Commercial banks Entrepreneurs	Loan to be recovered in instalments
7. Street lighting	Local panchayat (government)	Grant
8. Common taps	Local panchayat	Grant
9. Private taps	Household	Personal savings
10. Flour mill	Commercial banks	Loan to local entrepreneurs
11. Agro-processing	Commercial banks	Loan to local entrepreneurs
12. Irrigation pumping system; wells and piping	Commercial banks	Loans to farmers
13. Efficiency improvement measures for existing pumps	State electricity utilities	Part grant Part loan
14. Compact fluorescent lamps (CFL)	State electricity utilities Manufacturing companies	Part subsidy, part loan Interest subsidy
15. Large MW systems	Commercial banks International banks Government	Soft loan and incremental cost grant

take decisions and enforce them is necessary to plan rational measures to achieve cost reductions.

1. As seen earlier, a spread of energy-using activities over 15–18 hours would reduce the peak load and installed capacity. With adequate information, village committees could discuss rationally and agree to spread activities in order to reduce peak load and investment costs.
2. If there is a village institution, it could enforce measures to prevent overgrazing and regulate harvests from the energy plantations, thereby avoiding hiring of guards or putting up a fence.
3. The village community could decide to participate in energy plantation programmes by contributing labour, thereby reducing plantation establishment costs.
4. Village committees could decide about dung delivery to the biogas plant and collection of slurry at no cash cost.
5. Operation of community biogas plants and producer-gas systems requires trained persons. Local committees could allocate such work to village youths at agreed wage rates.
6. The cost of the transmission line and connections to houses could be reduced if the village houses agreed to take the cable from roof to roof, since the houses are usually adjacent to one another.

Several such measures involving many or all households or farmers would lead to significant cost reductions. A strong village-level institution is a precondition to achieving such cost reductions.

10.9 DISSEMINATION OF LARGE MEGAWATT-SCALE SYSTEMS

RBCs will be mostly at the level of a village, or at most a cluster of villages. However, the larger megawatt systems (1 MW, 5 MW, or 10 MW) would require a larger biomass catchment area, covering, for example, 32 villages per MW. This would require a catchment area of 8 km radius, assuming that 5 per cent of the geographical area would be dedicated to energy plantations (section 6.13.2). The potential number of megawatt systems in India at different capacities are: 1 MW × 16000 units, and 5 MW × 3200 units. The total installed capacity could be increased by increasing the efficiency of conversion, using modern systems, and by increasing the productivities of the energy plantations

These larger systems cannot be managed easily by local communities as in the case of RBCs. So there are two options, namely co-operative ventures or private industry.

Co-operative ventures

The farmers in a given catchment area could launch co-operative bioenergy systems, where each farmer who has land to dedicate to energy plantations would become a member of the utility. The whole system could be managed by professionals hired by the management committee of the utility co-operative. In India there is a long tradition of co-operative sugar mills, dairies, edible-oil mills, and cotton mills. Thus it is feasible to think of co-operative bioenergy utilities in the capacity range of 1–10 MW. Farmers who are members of a co-operative would dedicate land and supply biomass feedstock to the utility at a contracted price. The co-operative utility, with the help of professionals, would be able to provide technical assistance to farmers in order to obtain sustainable high productivities. Electricity could be sold to local areas and/or the grid.

Private-sector utilities

In India, until recently, most of the power generating utilities were under government control. With recent liberalization of the economy and the power sector, a number of private-sector-managed power generation systems are being started. It is feasible to attract private-sector investment for bioenergy utilities with appropriate policy initiatives and adequate financial incentives (see Chapter 12). The utility would survey and publicize the scheme whereby land-owning farmers in the catchment area could enter into agreements with the utility to dedicate land and to supply wood regularly to the utility at a contracted price. The utility would provide credit, technology, and other assistance to farmers for profitable forestry. The utility should generate and sell electricity at economic prices in order to make profits.

Commercial banks, governments, and multilateral agencies (such as the World Bank and development banks) should provide credit to both the co-operative as well as private-sector bioenergy utilities. These agencies could also provide the incremental costs until bioenergy systems achieve significant scales of manufacture and implementation. The megawatt systems will have to be wholly driven by the profit motive. However, financial incentives are required, due to the subsidies given to conventional systems and to the fact that bioenergy technologies provide significant environmental and socio-economic benefits. Many industrial and developing countries are already providing significant incentives to renewables (Chapter 9). To make the bioenergy system profitable in both the medium and long term, it will be necessary to adopt advanced biomass conversion technologies to achieve high conversion efficiencies.

10.10 FINANCIAL ARRANGEMENTS FOR RENEWABLE ENERGY IN INDIA

To promote institutional finance to renewable energy on attractive terms, there are two agencies in India; the Indian Renewable Energy Development Agency (IREDA) and the Programme for Acceleration of Commercialization of Energy Resources (PACER).

PACER

Set up in 1987, this agency aims at promoting the development of goal-oriented and market-responsive technological innovation in the Indian energy sector through financial assistance for a consortia of manufacturers, research institutions, and end users. PACER was set up as a part of the Industrial Credit and Investment Corporation of India, in collaboration with the US Agency for International Development. The priority areas of PACER include commercially attractive technologies to tap biomass and solar energy. By the beginning of January 1993, PACER had funded 16 consortium projects and eight research projects, with a total funding of Rs190 million. PACER assistance is limited to Rs3 million or 50 per cent of the estimated cost. Repayment to PACER is negotiated as a percentage of revenues accruing from project commercialization not exceeding 200 per cent of the PACER contribution.

PACER has funded bioenergy projects also. No information is available on the performance of the programmes funded or the approach adopted by PACER. It seems to be an innovative programme, especially for bioenergy projects which are at the final stages of research and development (R & D). Tie-up of the R & D institution, industry, and end users is an interesting concept. Here the participation of industry is crucial. However, the motivation for the industry is determined by a whole range of related policies and barriers, discussed in Chapter 11.

IREDA

This was set up in 1987 with the aim of extending financial assistance to promote, develop, and commercialize renewable energy technologies. IREDA has funded projects on biogas generation, biomass utilization for energy, and co-generation. It has received funding from the Indian government, the capital market, the Dutch government, and the World Bank. IREDA has sanctioned 327 projects worth Rs2.59 billion in terms of soft loans, and its contribution is normally 70–80 per cent of the project cost. The rate of interest is 9.8–10.5 per cent as against bank rates of 15–21 per cent. A moratorium period of 6 months to 3 years is permitted, depending on the technology. IREDA's recovery of loans is around 90 per cent and its success has been recognized by the World Bank, which is providing financial assistance (*India Today* 1994).

Both PACER and IREDA are expected to operate as commercial institutions. Both provide financial assistance to promote renewable energy through the participation of industry. Currently their operations are on a limited scale and they are expected to perform well despite all the given government controls over generation and distribution, and subsidized pricing of energy.

Direct government assistance for small programmes

The Ministry of Non-conventional Energy Sources (MNES) is providing grants directly or through state-level agencies for small-scale biomass gasification programmes, to cover initial investment costs to individuals, co-operative institutions, and state electrical utilities. For mechanical modes of applications, 80 per cent grants will be given by MNES for all end users. For electrical applications, 50 per cent grants for private users and 75 per cent grants for all community and public-sector units are available.

Assistance for megawatt-scale systems

MNES has ambitious plan for promotion of power generation using bioenergy technologies. It has devised a scheme to promote megawatt-scale bioenergy systems to establish their techno-economic feasibility. Some of the features are:

(1) minimum generation capacity for assistance should be 5 MW;
(2) MNES provides a non-recurring capital grant (or subsidy) of 30 per cent of project costs (up to Rs7 million/MW);
(3) the promoter's contribution must be at least 30 per cent;
(4) the local state electricity utilities must approve the project;
(5) availability of biomass has to be established;
(6) the ministry also provides small grants for preparing the project proposal, particularly the assessment of biomass availability.

This scheme was introduced in July 1994. We await the response of the industry.

Other incentives from government

(1) Income tax benefits;
(2) exemption from energy tax;
(3) 100 per cent depreciation of the plant during the first year of operations;
(4) low interest rates;
(5) excise duty exemption;
(6) exemption from central government sales tax and general sales tax.

Financial assistance (soft loans), interest rate subsidies, and tax concessions taken together have made little impact on the spread of renewable energy technologies, particularly bioenergy technologies, so far. But the situation may change in the years to come, once these incentives percolate to small entrepreneurs and NGOs. The barriers to bioenergy technologies are considered in Chapter 12.

10.11 CONCLUSIONS

An approach to the dissemination of bioenergy programmes was discussed by suggesting a rural bioenergy centre programme. A village ecosystem approach is suggested where rural energy planning becomes a part of the rural development programme. Along with the provision of energy, many policy changes and financing arrangements are necessary to ensure that rural communities derive the actual benefits, such as shifting to biogas for cooking, or electricity for lighting. The need for and the approach to a large demonstration programme for RBC, followed by large-scale dissemination programmes, was presented, highlighting the roles for private enterprise and local communities. Funding opportunities were highlighted by indicating roles for commercial banks, international agencies, and private enterprise. The significant financial incentives available in India for promoting bioenergy programmes were presented.

11

Implications for bioenergy in developing countries

In our analysis we have shown the need and the potential for bioenergy, and the possible economic, social, and environmental benefits for India. If bioenergy options are feasible for India, a country with a high population density and poor biomass resources (low forest and pasture area per capita), it should, at least theoretically, be even more feasible for many developing countries in sub-Saharan (SS) Africa, South-East Asia, and South America, with lower population densities and larger biomass resources (Table 1.2). Although it is tempting to undertake such an analysis at the macro level for other countries, one must be acutely aware of the limitations of such an exercise, especially if there is insufficient knowledge of the individual country. However, based on our analysis for India, some broad issues on rural energy, particularly the technical potential of bioenergy in some selected developing countries, are considered. No attempt is made to make any specific recommendation for any one country.

11.1 ENERGY CONSUMPTION PATTERNS IN SELECTED DEVELOPING COUNTRIES

Energy-use patterns in developing countries are characterized by low commercial energy use per capita and high dependence on biomass (see Table 1.2 for selected countries). Commercial energy use per capita varied for different developing countries; from 22 GJ/capita/yr in Brazil, to countries with moderate use levels such as India and Indonesia (9 GJ/capita/yr), and some countries in SS Africa (Tanzania and Zaire) and Sri Lanka with very low commercial energy-use levels (2–3 GJ/capita/yr), compared to 200 GJ/capita in OECD (Organization for Economic Co-operation and Development) countries. Developing countries would certainly like to increase their commercial energy consumption rapidly to improve the quality of life. This is obvious from the growth rate of commercial energy in some countries. At the same time, developing countries have a conflicting goal of reducing the dependence on imported oil, as most of them outside the Middle East (except Nigeria and Indonesia) depend largely on imported oil. Thus the challenge is to provide modern fuels equitably to all, particularly rural communities, with minimal dependence on external sources of energy.

Table 11.1 Bioenergy potential and land availability in selected developing countries

Countries	Population (million)	Cropland (Mha)	Forest and plantation (Mha)	Pasture (Mha)	Forest + pasture land (Mha)	Annual growth rate of electricity use per capita %	Annual growth rate of oil use per capita %	% of forest + pasture required for 0.5 t/capita biomass scenario[a]
Indonesia	178.2	21.9	118.2	11.8	130.0	5.2 (240)[b]	7.1 (174)[b]	8
Philippines	61.5	7.9	8.1	1.2	9.3	4.1 (422)	7.0 (185)	41
Sri Lanka	17.0	1.9	1.9	0.4	2.3	3.0 (183)	0.0 (73)	46
Ghana	14.9	2.7	10.2	5.0	15.2	3.4 (333)	1.5 (67)	6
Kenya	24.2	2.4	1.3	38.1	39.4	3.5 (134)	0.4 (91)	4
Senegal	70.0	2.3	7.8	3.1	10.9	0.7 (93)	7.8 (106)	4
Tanzania	24.5	3.3	33.7	35.0	68.7	0.3 (32)	0.3 (20)	2
Zaire	37.3	7.8	113.2	15.0	128.2	3.8 (170)	3.7 (10)	2
Zambia	8.1	5.2	32.3	30.0	62.3	0.0 (744)	0.0 (64)	1
Zimbabwe	9.8	2.8	9.0	4.8	13.8	3.5 (1075)	2.1 (58)	4
Brazil	150.4	59.9	568.0	184.2	752.2	3.1 (1643)	0.39 (388)	1

Source: population, cropland, pasture land from WRI (1992, 1994); forest and plantation area from FAO (1993*a*); energy data from UN (1992) and BP (1993).
[a] % of forest + pasture required = ((178.2 million (population) × 0.5 t wood/capita/yr)/(8 t/ha/yr of wood productivity))/130 Mha of forest + pasture land = 8.5% = 8%.
[b] Figures in brackets are per capita electricity (kWh/capita/yr) and oil (kg/capita/yr) use in 1990.

11.2 CURRENT OIL- AND ELECTRICITY-USE PATTERNS

In the developing countries, the population is growing at an annual rate of about 2–3 per cent, and it is particularly high in SS Africa and countries such as Bangladesh and the Philippines. It will be useful to compare the growth rate of energy with the population growth rate. The mean annual commercial energy growth rate between 1980 and 1992 was negative in countries such as Tanzania and Zambia, and less than the population growth rate in countries such as Sri Lanka, Senegal, and Ghana (WB 1994).

Oil

Using data from the UN (1992) and BP (1993) energy databases, the growth rates in oil use for the period between 1987 and 1990 and per capita consumption of oil for 1990 are given in Table 11.1. In some Asian countries oil consumption is growing at a high annual rate of about 7 per cent, but in Sri Lanka and SS Africa the growth rate is generally low. In some countries, such as Kenya, Tanzania, Zambia, Ghana, and Sri Lanka, the growth rate in oil use is lower than the population growth rate, which indicates that per capita availability is declining. All the countries in Table 11.1 (except Indonesia) are largely dependent on imports. The per capita consumption of oil is low in petroleum-importing countries such as India and also in a petroleum-exporting country like Indonesia (196 kg/capita/yr) as compared to OECD countries (1760 kg/capita/yr). Thus the current low consumption rates in developing countries cannot be reduced, and efficiency improvements alone may not lead to any significant improvements in energy services, particularly for rural populations which largely do not use them.

Electricity

Use levels are less than 200 kWh/capita/yr in many developing countries, such as Kenya, Senegal, Sri Lanka, Tanzania, and Zaire. In some countries like Brazil and Zimbabwe electricity use is higher than 1000 kWh/capita/yr. Such per capita national average electricity use levels usually do not indicate the extent of electricity use in rural areas. For example, in Zambia, although per capita electricity use is higher than in India and most developing countries, over 70 per cent of the electricity generated is used in energy-intensive copper production for export (Ranganathan 1992). In Zimbabwe the proportion of the rural population with electricity supply is only 0.25 per cent (even though the per capita annual use level is 1075 kWh). In Tanzania only 1 per cent of villages have electricity supplies. Even in Botswana, the richest country in SS Africa (per capita basis), only 3 per cent of rural households are connected to electricity (Ranganathan 1992).

The low levels of use and disparities between rural and urban areas are widespread in developing countries. The situation seems to be the same in Asia or Latin America, even though a relatively higher fraction of households have electricity supplies. The proportion of households not electrified in 1982 in Indonesia was 86 per cent; in the Philippines, 77 per cent; Peru, 87 per cent, Paraguay, 82 per cent, and Brazil, 33 per cent (WRI 1992). India's achievement in rural electrification is remarkable, with over 84 per cent of the half million villages being connected to the grid (in 1992). In the developing world many regions are virtually areas of darkness.

11.3 WHY DECENTRALIZED ELECTRICITY AND BIOENERGY?

Rural populations dominate in most of the developing countries. The overall population density in most of the developing countries is lower than in India (257 persons/km^2, Table 1.2) which indicates that the rural settlements are more dispersed than in India, for example in most of SS Africa, the density is less than 50 persons/km^2. The cost of electricity transmission and the transmission and distribution losses for rural areas were shown to be high for Indian villages in spite of relatively higher population densities. In countries with low population densities such as SS Africa, the cost of grid extension is very expensive and thus difficult to finance (Ranganathan 1992). The costs of electrification of villages (with a population in the range 1000–5000) in some countries, according to estimates by Lovejoy (1992), are: Bangladesh, \$70 000; Bolivia, \$105 000; Indonesia, \$125 000; Zambia, \$73 000. The cost of electrifying a village using grid power in most developing countries is very high compared to India (\$20 000 to 30 000). Thus decentralized electricity generation systems seem even more rational for such countries than for our analysis of India.

The next question is, what should be the source of energy for decentralized electricity systems? The first option that is usually discussed is diesel-based generation, which is widespread in many countries of SS Africa and South-East Asia. Increased use of diesel and kerosene would entail escalation in imports and foreign-exchange problems for all the oil-importing countries. The share of export earnings spent on energy imports can be very high in developing countries; for example, in SS Africa it was 74 per cent in Mozambique, 53 per cent in Tanzania, and 37 per cent in Kenya in 1987 (Davidson 1992). The use of diesel in remote areas also involves high transportation costs and contributes to large flows of money out of the village economy.

Renewable energies seem to be the clear choice to obviate some of these problems. Besides biomass, Africa has a large potential for micro-hydro alongside wind and solar options. Apart from cost considerations, the year-round requirement of electricity may impose limitations on intermittent renewables such as wind and micro-hydro since these sources, along with solar PV, would entail storage and/or integration of different electricity sources for continuous supply. In all likelihood, solar PV or wind-generation systems will be imported and installed by outside agencies, particularly from outside the country. Thus most of the wealth and employment generation will not be in rural or national locations.

Biomass as a source of energy does not have some of the limitations of wind, solar, and micro-hydro. For example, producer-gas systems can be operated at any time and biomass is amenable to producing both liquid and gaseous fuels, apart from electricity. Electricity systems could be set up in any location where biomass is available, at any scale all the way from 5 kW to 50 MW or more. Biomass production and processing also generate local wealth and employment.

11.4 BIOENERGY OPTIONS FOR DECENTRALIZED ELECTRICITY

The suitability of bioenergy options for decentralized electricity generation and the availability of technologies at various scales from kilowatt to megawatt sizes were shown in Chapters 5 and 6. In Africa both biogas and producer-gas options would be suitable. The human to cattle ratio in Africa as a whole is nearly the same as in India (3 : 1) but is lower in countries like Tanzania and Kenya (WRI 1994). Thus the biogas potential from dung in countries such as Kenya and Tanzania should theoretically be higher than in India. But in some locations of Africa the nomadic nature of cattle-rearing may not facilitate biogas production. The human to cattle ratio in other Asian countries is high: 12 in Indonesia and 14 in the Philippines. Thus in some Asian countries the biogas potential is likely to be lower than in India. Here only the producer-gas option is discussed, since data on the distribution of livestock populations in villages and dung yields are not available for other developing countries. The biogas option needs to be considered elsewhere with the help of detailed field studies, as is being increasingly explored in some SS African countries such as Tanzania.

Bioenergy options have to be analysed at the specific country level (and if possible at the country region level) as data are required on the size distribution of human settlements, services requiring energy, biomass feedstock, land availability, and so on. At this stage detailed analyses are not possible and at best only some indication of technical potentials can be established.

The current energy consumption in developing countries as a whole is estimated to be around 30 GJ/capita/yr if commercial energy and fuelwood are included (Table 1.2) and 35 GJ/capita/yr with all biomass, including crop residues and cattle dung (Woods and Hall 1994). However, the total energy use in most countries listed in Table 1.2 is in the range 12–25 GJ/capita/yr.

The share of fuelwood energy alone in total energy consumption varies from over 80 per cent in countries like Kenya, Tanzania, and Zaire to 30 and 25 per cent in India and Zimbabwe (Table 1.2). If crop residue and cattle dung are included, the mean biomass use could be taken as 10–15 GJ/capita/yr for developing countries. In SS Africa, biomass-use levels are higher than in India, since in Africa charcoal is a dominant fuel in urban areas. Charcoal kilns have an efficiency of around 20 per cent or less on a weight basis (Dutt and Ravindranath 1993; Hall and Mao 1994) and thus more firewood is required per capita in the charcoal mode compared to the direct fuelwood mode, and cooking practices could be different.

In this study, the potential of biomass to provide modern energies such as electricity and fuels to rural areas is explored at the current level of use of fuelwood (15 GJ/capita/yr). It is assumed that electricity will be used for lighting and shaft-power-requiring activities, while liquid and gaseous fuels would be used for cooking. It is difficult to estimate the per capita or per settlement requirement for electricity and liquid or gaseous fuels without the help of local studies. In rural India, for promoting the quality of life, the electricity requirement, using the norms given in Table 6.7, is estimated to be around 200 kWh/capita/yr (Table 6.11), assuming an efficient technology mix and 30 kg of kerosene (equivalent to 60 kg of methanol) being required for cooking. It may not be correct to assume such a figure for other countries as the activities requiring electricity, and cooking practices, may vary. On the other hand, one may reason that the units of electricity required for home lighting, street lighting, pumping water for domestic use, or milling of grains may not vary significantly between countries. But the electricity requirement for activities such as pumping irrigation water and agro-processing is certain to vary from location to location. The per capita electricity requirement of 200 kWh estimated for rural India includes irrigation pumping, which accounts for about 60 per cent of the total (Table 6.11). Thus if pumping irrigation water is reduced or excluded, the electricity requirement will be considerably less than 200 kWh/capita/yr. It is likely that the liquid fuel requirement may vary from region to region, depending on the cooking practices. To avoid these problems a simple approach of estimating the potential to produce modern energy using 1 t of woody biomass/capita/yr (equivalent to 15 GJ) is adopted.

11.5 BIOENERGY POTENTIAL OF 1 T OF WOODY BIOMASS (EQUIVALENT TO 15 GJ)

For convenience, of the 1 t of wood/capita/yr currently used, it is assumed that 0.5 t/capita/yr each will be used for electricity and liquid fuels generation. The quantity of electricity or liquid fuel generated will depend on the technology used and efficiency of conversion.

Electricity and methanol production potential

Using 18 and 36 per cent efficiencies of conversion for 0.5 t of wood, the potential for electricity generation using the producer-gas route is 384 kWh and 768 kWh/capita/yr, respectively. The methanol fuel production potential, using 0.5 t/capita/yr of lignocellulose feedstock via the thermochemical route, is 217 kg/capita/yr.

Producer-gas technology is available at various scales, but the technology for liquid fuel (methanol and ethanol) manufacture using lignocellulose at small and moderate scales is not yet commercially available, however, it could become available quite soon. Liquid fuel production from woody biomass is assumed as it is not considered feasible to dedicate good land for maize, sugarcane, or other crops for energy.

If the rural Indian liquid fuel (60 l/capita/yr) and electricity (about 200 kWh/capita/yr) requirement for ensuring a better quality of life is adopted, then 0.5 t of wood/capita/yr is adequate: with 0.25 t for electricity and 0.25 t for liquid fuel use. It should be noted that the current commercial energy-use levels given in Table 1.2 include urban and industrial sectors, which dominate commercial use—as shown, for example, in Zambia and Zimbabwe. Thus 192–384 kWh of electricity and 108 kg of methanol per capita from 0.5 tonne of wood would lead to a significant improvement in the quality of life in rural areas of SS Africa, Indonesia, and Brazil.

11.6 LAND AND WOOD RESOURCES

Can, for example, SS Africa, Indonesia, the Philippines, and Sri Lanka supply woody biomass sustainably? It is possible to argue that what is being contemplated is basically the use of currently used fuelwood and charcoal in the domestic sector for conversion to modern energy carriers. If the currently used fuelwood comes from sustainable removals, then there is no need for land to be dedicated for energy plantations. There is inadequate evidence as of now to show that rural fuelwood consumption is leading to deforestation. A study in Zambia concluded that the primary cause of deforestation is not tree cutting for fuelwood or charcoal (Macall 1987), although it is considered that fuelwood consumption is contributing to deforestation in highly populated areas of Africa (Davidson 1992). The inference could be that in large parts of SS Africa and Indonesia, for example, the current rural fuelwood harvests may be coming from sustainable harvests. Thus a shift to modern bioenergy will only lead to an improved quality of life without any non-sustainable pressure on woody biomass resources. Inevitably the situation may not be the same for all countries, and thus it is necessary to estimate the additional land required, assuming that 0.5 t of woody biomass/capita needs to be produced sustainably from dedicated lands for bioenergy.

The land required to produce 0.5 t wood/capita/yr, at a conservative productivity of 8 t/ha/yr, for the total rural population is expressed as a percentage of the existing forest and pasture land in order to consider land availability and the requirement for bioenergy. The area under degraded land is not available, and it is assumed that a part of forest and pasture land is degraded. Among Asian countries, only Indonesia has adequate land, requiring only 8 per cent of forest and pasture land. In the Philippines the potential is only moderate but cannot be ruled out, depending on local situations. When SS Africa, Brazil, or Indonesia are considered, land does not seem to be a constraint. In these countries even a scenario with 1 to 2 t wood/capita/yr can be considered in order to double or quadruple electricity and liquid fuel production. Here we again stress that only degraded forest and pasture land should be used, and the use of primary forests should be avoided, even though these countries have forest areas several times greater on per capita terms than India. Guidelines need to be developed to provide incentives for land to be selected for energy plantations so that only partially or fully degraded forest land or pasture land or highly degraded cropland would be used preferentially. In the case of pasture land, a silvipasture strategy could be adopted so that the current grass production is also assured. Using a correlation between rainfall and commercial plantation yields in Brazil, a study by Marrison and Larson (1994) projected that Africa could plausibly produce nearly 40 EJ/yr in plantations established on 165 Mha, which is equal to 10 per cent of the sum of pasture plus other land not required for agriculture, and excluding wilderness.

Technically, bioenergy has the potential to meet the rural energy needs of most of SS Africa and some countries in South-East Asia and Brazil. Even in countries such as the Philippines and Sri Lanka there could be many areas with low population density or with degraded land availability where bioenergy options are feasible. If wood productivity is increased to, say, 15–20 t/ha/yr or if forest and crop residues are used, the land requirement will decline. This would make bioenergy feasible in most developing countries. Patterson (1994) concluded that in the long run biomass electricity will have the biggest impact in tropical and sub-tropical regions, where high biomass productivity could be obtained and where the demand for electricity is growing at a fast rate. He also suggested that in some countries, it may be more acceptable to dedicate land for biomass for electricity than for hydroelectric power projects, where land would be permanently submerged and lost.

11.7 SOCIO-ECONOMIC FEASIBILITY

Technical feasibility is only one of several factors required to promote rational energy strategies. Technology or appropriate designs are already available or will soon become available in many developing as well as industrialized countries. As described in the case of India, bioenergy strategies promote many local social, economical, and environmental benefits, apart from global benefits. Undoubtedly there would be problems with respect to funding, and technology transfer and assimilation. Most countries need not develop their own designs and could benefit from research and development in such countries as China, Brazil, India, and South Africa, which have developed, or are experimenting with, bioenergy technologies, or have the potential for increased research and development.

The experience with bioenergy technologies, as well as with conventional technologies such as diesel-based generation systems in Africa, has shown that foreign assistance has failed to promote the assimilation of the technologies (Ranganathan 1992; BTG 1993). This need not necessarily be so if the outside technology assistance is based on proper evaluations in the field, involvement of local research institutions, training of all field and technical staff, and establishment of a local manufacturing base. In most developing countries, cars, trucks, radios, and televisions are operated and maintained in all regions, including remote villages. Thus, operation, maintenance, and management of the much simpler biogas- and producer-gas based bioenergy systems should be feasible. Overseas technicians obviously do not come to every village or town to repair the technologically complex cars, trucks, and television sets. They transfer to the host countries all the skills, and often even the manufacturing base for spares if the scale of operation is economically viable. It appears to us that the arguments about the absence of skills or the problems in transfer of skills are largely unfounded, and what is required is only a systematic effort to transfer the technology and skills.

Funding is certainly a problem for rural electrification in general, and particularly for bioenergy technologies. But the reduced costs due to savings in the expensive transmission network must be included in the calculations. Thus governments should take initiatives to evaluate the existing bioenergy and other renewable energy options for the range of economic, social, and environmental benefits.

Developing countries themselves should evaluate and take decisions to allocate adequate funding to promote the availability of quality fuels in rural areas. Africa, theoretically, provides an excellent opportunity for bioenergy options, since virtually all of rural SS Africa is still outside the grid network, unlike in India where most villages are already connected to the grid. Infrastructural problems in Africa may still hinder the widespread adoption of modern bioenergy systems. Economic and environmental benefits should drive bioenergy and not subsidies, although subsidies may be required initially until the infrastructure and a significant scale of operation are established.

11.8 ISLAND COUNTRIES AND BIOENERGY

Amongst developing countries there are island countries such as Mauritius, Maldives, and Fiji, and some large countries such as Indonesia and the Philippines which have a large number of inhabited islands. Indonesia and the Philippines have 6000 and 2773 inhabited islands, respectively. Even smaller countries, such as Fiji, have 110 such islands. Meeting the energy needs of small islands with modern conventional fuels involves transporting petroleum fuels for long distances over water. Transmitting electricity may not be feasible at all, and thus diesel generators are used. Many of these islands could be made self-reliant with respect to electricity by shifting to biomass systems. Bioenergy options could also be a starting point for development in such islands, with improvement of local skills, local employment, and wealth generation. Detailed studies of small islands should be promoted to establish current energy-use patterns, energy sources and supplies, costs of transmitting diesel and electricity, and to compare different options, including modern bioenergy options for improving the quality of life.

12

Bioenergy: barriers, policy options, and guidelines

Some of the modern bioenergy technologies have been in the portfolio of energy planners for over a decade. They have received substantial government support in countries such as Austria, Brazil, China, France, India, Sweden, USA, and Zimbabwe, to name a few. There are some successful examples from developing countries, such as the ethanol programmes in Brazil and Zimbabwe, the biogas programme in China (and to some extent in India), bagasse electricity in Mauritius, and the afforestation programme in India. However, apart from the ethanol programmes, they have yet to make significant impacts on the energy supplies in these and other developing countries. Interestingly, bioenergy technologies have made a significant entry into the energy supplies of a number of industrialized countries, such as the USA (electricity), Denmark (biogas), Sweden (combined heat and power), and Austria (heat and power and liquid fuels).

India has a good technical base for the bioenergy technologies which have been indigenously developed and for which significant demonstration programmes have been implemented, with examples of both success and failure. However, in spite of significant experience, bioenergy technology opportunities have remained on the fringe of the Indian energy planners' agenda. The energy planners have neither given up the bioenergy programmes nor have they been allocating adequate funding in the planning process (PC 1992). The present study of India has shown that it is possible to meet the energy needs of the rural population of a large and populated country using biomass feedstocks. The study of other developing countries, such as Kenya, Tanzania, and Indonesia, have also shown that by dedicating only 0.5–1 t of woody biomass/capita/yr for bioenergy it is possible to meet the energy demands of the rural population, and that the potential even exists for meeting a part of non-rural demands.

According to the Renewables Intensive Global Energy Scenario (RIGES) in the studies of Johansson *et al.* (1993) and also the studies of the World Energy Council (WEC 1993), bioenergy options have a crucial role to play in the energy future, since they also offer significant environmental and socio-economic benefits to developing countries. The big question is why the renewable energies, particularly bioenergy technologies, are not spreading through market mechanisms or through the government utilities, even though they are shown to be becoming financially attractive (WB 1992; Ahmed

1994). It is necessary to understand the barriers to the large-scale spread of bioenergy in order to realize its full potential.

Studies have attempted to understand the barriers to implementing energy efficiency technologies (Reddy 1991; Azimi and Lohani 1992; ADB 1993) as well as the barriers to the spread of renewable energy technologies (Hall 1991; Reddy *et al.* 1991; Foley 1992*b*; Williams 1994). An attempt is made here to understand the constraints and to suggest policy options for decentralized rural bioenergy centres as well as large-scale bioenergy systems for electricity and liquid fuels. It is also necessary to learn from the experience in industrialized countries since renewable energies, particularly bioenergy systems, may become increasingly important in an environmentally conscious world where sustainability and equity may dictate energy policies in the future.

12.1 BARRIERS TO BIOENERGY

Biomass energy still faces many barriers, economic, institutional, and technical (Hall 1991). One could also add attitudinal barriers and national and international vested interests. Wood has the reputation of being a low-quality fuel and historically has been granted poor status by governments (Murray and de Montalembert 1992; Williams 1994). Biomass energy is thus often considered problematic because it has many facets, and interacts with so many different areas of interest, such as land use, forestry, agriculture, domestic animals, and social factors. It may be worthwhile discussing all the different categories of barriers, although it may look like a shopping list. All barriers are not likely to exist at all locations, or for any one technology option. There could be one or two dominant problems operating in a given situation; for example, the government policy of not permitting private utilities to sell electricity at economic prices is a dominant factor in some situations, such as the large-scale rice-husk-based power generation system in the Punjab case study (Raju Associates 1992; Chapter 5) and the lack of training and education in the operation and maintenance of biogas plants as in the Marathwada example (Dyalchand and Murthy 1988; Chapter 4). The extensive listing of potential barriers would be of help to policy makers as a checklist to be kept in mind for specific locations and to help prepare mitigation options. A detailed listing is given in Box 12.1 and some critical barriers from the list are discussed below.

Attitudinal barriers

One of the first hurdles to be crossed is the lack of understanding or negative attitude of planners, policy makers, and industry that bioenergy is a technology fit only for a few demonstration projects, after which it can be forgotten. Of

Box 12.1 Potential barriers to bioenergy projects

Attitudinal barriers or ignorance

1. Standpoint among energy planners, policy makers, and entrepreneurs that bioenergy is a primitive technology
2. Commercialization of modern bioenergy options are not feasible and fit only for a few demonstration programmes
3. High investment cost
4. Food versus fuel conflict and non-availability of land
5. Rural bioenergy centres not feasible as village communities are divided, unequal, and lack skills
6. Developing countries lack the capacity to adopt and manage modern bioenergy technologies, let alone develop them indigenously

Technical barriers

1. Technology still needs development for cost reduction and reliability
2. Multifeed systems not yet available
3. Need to reduce 'botheration factor' in preparing feedstock
4. Energy plantations: absence of a package of practices and quality seed material or clones for high yields

Financial barriers

1. Insignificant funding to bioenergy options compared to conventional systems for R & D as well as dissemination
2. Scarcity of capital and high cost of capital
3. Large international funding only for conventional energy options and to some extent to solar PV and wind energy

Institutional barriers

1. Uncertain land tenure for farmers and local communities to undertake long-term commitment for supply of woody biomass
2. Absence of institutional arrangements to empower local communities to manage RBCs and energy plantations
3. Irrational energy pricing, no consideration of LRMC principle
4. High subsidies to electricity and petroleum products (LPG, kerosene, and diesel)
5. Restrictions on sale of electricity by private utilities at economic prices
6. None or inadequate arrangements to inform entrepreneurs and consumers fully on the potential benefits of bioenergy systems

LRMC, long-run marginal costing.
RBC, rural bioenergy centre.

equal concern are the feelings that bioenergy technologies are high cost and that there is a conflict between biomass fuel and food production. These attitudes are usually due to the absence of detailed analyses, significant and visible field demonstration projects, the lack of information to decision makers, entrepreneurs, and consumers, and the failures of initial attempts. Thus the first step should be to implement a significant number of demonstration projects and to provide evidence and information to the policy makers and entrepreneurs in order to dispel apprehensions.

Technical barriers

It is evident that in the past technologies that were not fully developed and/or which were not suitable were shipped to developing countries and thus encountered serious technical problems (Foley 1992*b*; BTG 1993). In India, small producer-gas or fixed-dome biogas designs are still in the development stage. High-efficiency, multifeed, long-performing, and user-friendly producer-gas and biogas designs are required. In the case of liquid fuels, small- and moderate-scale systems are not yet available. High-efficiency large-scale gasifier systems are also not yet commercially available and advanced gasifiers/turbine systems are still undergoing demonstration in Sweden and Finland. Importing technology is not a solution but, in fact, can be a barrier to indigenous development in countries such as Brazil, China, and India.

Energy pricing and subsidies

In most developing countries electricity as well as liquid and gaseous fuels receive substantial subsidies (Foley 1992*a*; Leach 1992; Ranganathan 1993). In India electricity is subsidized not only for rural but also for urban consumers, and it is virtually free for agricultural pumping (CMIE 1994). In such a situation it is impossible for any renewable energy options to compete and be viable. Presently, most energy pricing in developing countries is not based on long-run marginal costing principles and is largely a political decision, although this situation is slowly changing. The economic comparisons of decentralized electricity generation must always be made with the marginal cost of delivering a unit of centrally generated electricity to an irrigation pump or a village house. No industrialist or entrepreneur would invest money in a bioenergy project if the output has to compete with highly subsidized energy supplies. The existing subsidies provided for renewable energy technologies are woefully inadequate to compete with highly subsidized electricity supplies for agricultural pumping or subsidized kerosene and LPG for domestic use. Environmental benefits (both local and global) and social benefits (local employment and self-reliance) of bioenergy are not given any consideration in current comparisons between conventional and bioenergy options.

Institutional barriers

These barriers are critical and have been highlighted in a number of chapters. The main institutional barriers are government subsidies to conventional sources of energy, restrictions on the sale of electricity at economic prices, issues related to land tenure, and state control of energy utilities and other related agencies. The absence of strong institutional arrangements to empower local communities, or of policies permitting arrangements between entrepreneurs and farmers for long-term supplies of biomass, could prevent village communities and farmers from taking an active interest in rural bioenergy centres. Provisions of subsidies alone, for example, to bioenergy systems will not be enough if other policies remain unchanged, such as the ban on selling electricity or the high subsidies for grid electricity. Appropriate institutional arrangements and policies by themselves could, however, contribute significantly to the spread of many renewable energy technologies (Jackson 1992).

Financial barriers

In most developing countries capital is scarce and expensive and thus entrepreneurs are reluctant to invest in uncertain projects in the face of unfriendly institutional and pricing policies. There is also a problem for rural households deriving benefits from biogas for cooking, since at present fuelwood is generally gathered at no cash cost and a change to biogas would involve expenditure. Similarly, the first cost of obtaining electricity for domestic use is high and is a barrier, as has been shown in both India and Zambia. It is essential to reduce biogas system costs through the community biogas option and involve local communities in their operation and management, as well as to reduce the first cost of electrical connections for homes through alternative techniques.

International assistance

Multilateral funding agencies such as the World Bank and other development banks continue to fund conventional sources such as coal-based power generation and large hydroelectric schemes in developing countries. Often renewable energies, such as solar PV and wind energy programmes, are promoted to some extent in developing countries irrespective of their economics, flow of benefits, or competitiveness with bioenergy options.

Bioenergy technologies are thus caught in the chicken and egg dilemma—as long as they are not produced on a large scale they are unlikely to be economically attractive, and as long as they are not economically attractive they are unlikely to be produced on a large scale. However, economic analysis of small and large bioenergy projects in Chapter 9 showed that bioenergy options in many situations are already attractive and rational energy pricing and suitable energy policies would be adequate to stimulate their spread. The barriers specific to a rural bioenergy centre option and large biomass electricity and liquid fuel options are summarized here.

Rural bioenergy centre (RBC) option

(1) Absence of strong institutional arrangements to empower local communities to adopt an RBC option;
(2) high subsidies to conventional electricity and petroleum fuels in rural areas and inadequate funding and credit for bioenergy programmes, and the high cost of credit;
(3) absence of clear policies to make villages self-reliant in energy even though feasible options exist and have been demonstrated.

Large-scale bioenergy programmes

(1) Non-economic pricing of energy and subsidies for conventional systems;
(2) no assured market for electricity, biogas, or liquid fuels;
(3) a technology, although available, not yet commercially available off-the-shelf in most countries.

Barriers for individuals, communities, and industry

The major actors involved are raw material producers, bioenergy producers, consumers, equipment manufacturers, and managers (utility owners, entrepreneurs, or village communities). The specific barriers faced by each are listed in Box 12.2. Land tenurial uncertainties, absence of an assured market, and low productivities are likely to affect biomass producers for both village-level and large bioenergy systems. Absence of an assured market, or even lack of freedom to sell energy, and existing subsidies for conventional sources demotivate investors in bioenergy. Low market volumes, apathy of state utilities, and competition from large conventional energy-equipment manufacturers locally as well globally are likely to affect the manufacturers of bioenergy equipment.

12.2 SOCIAL FACTORS: BARRIERS OR CATALYSTS TO BIOENERGY?

Community action is an important precondition to the success of bioenergy-based development systems, both at the village level as well as with large systems. Even if energy planners and the government are committed to renewable energies such as bioenergy (PC 1992), the questions that may be raised are whether or not decentralized bioenergy systems (such as RBC) are feasible given the inequities within villages, lack of skilled manpower to

Box 12.2 Potential barriers faced by various actors implementing bioenergy projects

Raw material producers

1. Land tenurial uncertainties
2. Absence of package of practices and quality seed material or clones for high and sustainable woody biomass productivity
3. Lack of information to farmers on potential benefits of using degraded lands for energy plantations
4. Restrictions on harvesting and transportation
5. Supplying dung to a centralized plant and collecting sludge afterwards

Biomass electricity, biogas, and liquid fuel producers

1. Absence of an assured market
2. Absence of economic pricing
3. Restrictions on the transmission and sale of energy
4. Scarcity of capital and high cost of capital
5. High subsidies to electricity and petroleum fuels
6. No access to technology, servicing, repair, or guarantee

Equipment manufacturers

1. Low market volumes
2. Disinterest of state utilities
3. Inadequate credit arrangements

Managers of bioenergy systems

1. No clear government policy to promote RBC or megawatt-size systems
2. No institutional arrangements for empowering local communities to manage rural bioenergy centres

Consumers

1. High first cost
2. Non-availability of credit
3. Absence of easy access, lack of information on bioenergy options

install and operate systems, the absence of strong community organizations to manage the systems, and the investment requirements. This is the gist of the argument often used by planners and administrators to dismiss such decentralized bioenergy systems. However, our experience indicates that such apprehensions are unfounded. In the absence of any government legislation or well-established community institution in Pura and Methan, the village communities have co-operated to contribute dung and operate biogas plants (Venkataramana 1992; Rajabapaiah *et al.* 1993); protect an energy forest, operate a gasifier, pay for services, and manage the system in Hosahalli (Srinivas *et al.* 1992); and to contribute free labour for preparing *Lantana* feedstock for gasification and pay for electricity in Himalayan villages (Joshi *et al.* 1993). This has been possible due to the participation of local communities in the decision-making process and management of the systems. Genuine participation and commitment of the community can be ensured only if some of the following aspects are considered in the programme implementation:

1. The programme must meet the felt needs of the community and not the perceived needs of project officers or planners. Vivid examples are Pura and Hosahalli, where in both the villages the first priority of the village community was shortage of potable drinking-water. Thus the community commitment was limited in Pura when the first phase of the project attempted to provide biogas only for cooking, and similarly in Hosahalli when, in the first phase, electricity was provided only for lighting. However, after the introduction of the community drinking-water service, community commitment was forthcoming in both the villages. Not that cooking energy was not a problem in Pura or that kerosene lighting was not a problem in Hosahalli, they were simply not the main immediate concerns at that time.

2. Involvement of the community at all stages. The community must be involved right from the beginning and at all stages thereafter. An example is that of social forestry (Agarwal and Narain 1989), where the Forest Departments tried to involve the panchayats (village-level governments) who were asked to transfer the land required to the Department who, in turn, raised the plantations on their own, without involving the panchayats or the communities themselves, and then after 3 years, tried to transfer the tree plantations back to the panchayats. Not surprisingly, most panchayats refused to accept the plantations for fear that they would not be able to protect and manage the plantations until maturity, since they feared that their communities would destroy the plantations. In one case this did indeed happen within days of transfer to a panchayat. The Forest Departments protected the plantations at very high cost and with a heavy hand without much local co-operation. The reasons for these problems were simply that the local communities were not sure of the flow of benefits and the land tenure. According to another national-level study of social forestry (PC 1987), in many villages the panchayats or village communities refused to hand over land to the Forest Department for fear that they would lose their grazing land and access to biomass from the land. Such problems might not have arisen if the local communities were involved in all decisions right from the beginning.

3. Flow of benefit to all. No programme that requires community action can succeed if only a few people in the village benefit. The essence of the experience from Pura and Hosahalli is that the whole community (every household) has nearly equal access to benefits such as domestic lighting and drinking-water from street taps. In Hosahalli, all houses, irrespective of their size or economic status, were given two light points, and every household

had to collect water from street taps. Certainly not all services are required by all the households or at the same rate. For example, large houses and small huts cannot be expected to have identical requirements for lighting; initially it worked in Hosahalli (for 3 years) and at later stages some large houses demanded more light points, which was rational. But what is fundamental is that certain critical energy services must be accessible to all to ensure a given quality of life and beyond that there are bound to be differences in the intensity of energy requirement. Thus for sustainable success, the involvement of the entire community is necessary and, whatever bioenergy systems are selected, access must be provided to some crucial services, such as drinking-water, home lighting, flour milling, and street lighting, to all in the community.

4. Flow of information. A village community or 'gramasabha' (village council) cannot be expected to take a rational decision on any issue without adequate information on the options available, their impacts, funding availability, and its roles and responsibilities. Again we can look back on the 'social forestry' programme. According to an Orissa State government evaluation, the local communities in 82 per cent of the villages did not know how the produce from social forestry plantations would be used or distributed and, more importantly, most of the communities did not expect any share from the final output (Government of Orissa 1987). In Maharashtra State neither the panchayat leaders nor the Forest Department explained clearly to the villagers that the social forestry plantations were a community resource, thus the communities believed that the trees on the common lands were government trees (Agarwal and Narain 1989). In reality all the social forestry plantations on village commons are meant exclusively for the benefit of local communities. Discussion with the village communities and provision of information on the purpose of forestry and the mode of distribution of benefits would have gone a considerable way in alleviating the fears of the community and prevented the complete failure of such projects in many situations.

From these examples it is obvious that appropriate guidelines to involve local communities would not only help smooth the implementation of bioenergy programmes, but would also help in promoting the equitable distribution of benefits and possibly reduce investment and operational costs. This would lead to the rational scheduling of activities requiring electricity, in order to reduce the installed capacity at village level, the protection of energy forest (as in Hosahalli), and the voluntary contribution of labour for preparation of feedstock (as in the Himalayan villages).

12.3 POLICY OPTIONS FOR BIOENERGY

The costs of all commercial forms of renewable energy have declined significantly over the past two decades and the commercial development of renewables may thus be justified even on non-environmental grounds (Jackson 1992; WB 1992). A recent review of renewable energy concluded that given adequate support, renewables could meet much of the world's growing energy needs (Johansson *et al.* 1993). However, not all appropriate or rational bioenergy policy options are easily implementable, so it is desirable to separate out the strategies based on the levels of feasibility and impact.

12.3.1 High-feasibility and high-impact strategies

Such strategies are very important since they should form the first set of strategies to be implemented by countries wanting to promote bioenergy programmes.

Preparation of status reports of suitable bioenergy technologies

Each country should prepare status reports of bioenergy technologies available in their country including those that are currently available and those that will soon become available in other countries for import. National governments could prepare such studies with or without outside help, on the lines of country studies presently being funded for assessing the inventories of greenhouse gas emissions and formation of mitigation strategies in the post-UNCED (Rio Conference 1992) period.

Rational methodology for energy pricing

Countries should develop rational methods of energy pricing. In developing countries the marginal costs of electricity generation are always higher than the average costs, since energy demands are rapidly growing and require large additions to generation capacity (Azimi and Lohani 1992). Energy pricing must be based on long-run marginal costing (LRMC) principles. Given the high cost of transmission (electricity or petroleum fuels) and high transmission and distribution losses of electricity, it is necessary to consider the total cost of delivering a unit of energy at the end-use point. The time has now arrived to include at least some of the environmental and social costs in energy-pricing methodologies. Developing countries, as well as industrialized countries, cannot afford to ignore environmental costs, particularly to the local environment at present and to the global environment in the long run. A rational methodology for energy pricing, which can initiate a national debate encompassing both short-term and long-term impacts of subsidies, is necessary.

In India many state governments have begun to rethink about the very high subsidies on power supplies for irrigation. This would have a considerable impact on promoting rational energy pricing, which is the first step in promoting renewable energy technologies. Price subsidies to infrastructure almost always benefit the non-poor disproportionately (WB 1994). Studies to understand the short- and long-term impacts of the existing subsidies are required: those who benefit and lose out from subsidies (as it is ultimately the rural poor who suffer in the long run; Ranganathan 1993), the impact on the financial status of utilities, and impacts on efficiency of use of energy and on the environment. There is also the consequence of wastage of irrigation water as farmers do not pay for electricity.

Funding for research and development

The level of funding for bioenergy development is inadequate (as in India) or absent (as in many smaller countries). Funding should be enhanced for clearly defined areas. Larger and richer developing countries, such as Brazil, China, India, Indonesia, and South Africa, can afford to dedicate the funding required to develop bioenergy technologies. This may lead to the development in a few years of competitive technologies that are compatible with the situations in developing countries. Smaller or poorer developing countries could benefit from such a development.

Land tenure policies

National governments could pass legislation to give freedom to the managements of rural bioenergy centres (basically village communities) and large bioenergy systems so that they can negotiate with farmers or the village community and enter into long-term contracts for supply of biomass for electricity generation or liquid fuel production.

Credit for appliances, energy systems, and electrical connections

Lack of financing for the first cost of modern fuel devices and the high cost of finance restricts the poor from benefiting from high-quality energy sources (Leach 1992; Ranganathan 1993). It is also essential to promote finance for bioenergy producers and consumers. Many countries, including India, provide some interest subsidies and grants for producers but there is a further need to provide credit to consumers for obtaining electrical connections or buying energy-efficient devices.

Large demonstration programme

Such a programme is essential to generate information, create awareness in planners and policy makers, change attitudes, develop methods for large-scale dissemination, and to evaluate the performance of technologies in the field. The demonstration units should be widespread and they should not be isolated attempts, since they need to be highly visible and to make an impact. Policy makers are unlikely to believe the feasibility of projects based on an isolated success story. Funding may not be a problem in many developing countries, but the multilateral agencies, UN, and other donors should contribute to practical bioenergy demonstration programmes in developing countries on a substantial scale. Dissemination programmes could be concentrated in villages and locations not connected to the electricity grid in the first phase, followed by other locations. Planning is necessary for rural bioenergy centres, a few megawatt-scale systems, and a few lignocellulose-based methanol systems is necessary.

Involvement of entrepreneurs in the implementation of bioenergy programmes

In most developing countries, governments have so far played a dominant role in the dissemination of alternative energy programmes, and their performance is often poor (Chapters 4 and 5). The Sirsi case study in South India (Chapter 4) showed how involvement of private entrepreneurs in the dissemination of biogas programmes led to higher performance and fewer failures. Governments should involve private entrepreneurs on a turnkey basis for the implementation of bioenergy programmes and the maintenance of major equipment under contractual conditions.

Removal of restrictions on fuelwood production and transportation

Restrictions on production, harvesting, transportation, and processing of fuelwood should be lifted. This would enable farmers and village communities to undertake fuelwood production with long-term commitments to supply to utilities.

Removal of restrictions on the sale of electricity, biogas, and liquid biofuels

The example of the rice-husk-based electricity system in Punjab, India (Chapter 5) showed that restriction on the sale of electricity at an economic price was the major constraint. Governments could lift the restrictions on the sale of energy at economic prices by private enterprise or village rural bioenergy centres. This should not pose any problems as most developing countries face shortages of electricity and also import petroleum fuels. With many developing countries adopting economic liberalization policies, it should be easier now than in the past.

Ensuring a minimum internal rate of return to investors

In the face of financial problems and severe shortages of power, and with the liberalization of economies, countries such as India are offering guaranteed minimum internal rates of return to private investors, particularly to multinational power corporations. Under the existing state controls and subsidies, no private entrepreneur would invest. However, there is now an opportunity to provide financial incentives to entrepreneurs interested in bioenergy projects. In fact, like many European countries and the USA, there could be additional renewable energy incentives to those given to conventional energy projects in order to allow for environmental and social benefits.

12.3.2 Low-feasibility and high-impact strategies

Several rational policies could have large impacts but are difficult to implement. A good example is the subsidized or free electricity for irrigation pumping in India, where energy experts and planners recommend rational pricing mechanisms. Even politicians may agree in private, but in public they advocate highly subsidized electricity and other inputs to farmers, and subsidized kerosene and LPG to urban consumers. In fact, political parties compete with one another to provide such subsidies to farmers to win their support (*The Economist* 1994). Slowly the situation is changing, due to the realization that subsidizing rural electricity for uses which are productive only disadvantages the users themselves in the long run, by encouraging utilities to reduce the quality and quantity of the power supply. Subsidies should be avoided for productive uses since electricity pays for itself (Ranganathan 1993). The present situation may not last much longer as State electrical utilities in India are in a serious financial situation and may be forced to adopt more rational policies.

Examples of similar high-impact policies which have a low feasibility of being implemented are listed here.

Adopting long-run marginal costing (LRMC) for electricity

The LRMC principle, although essential, will take time to find acceptance in developing countries. As long as urban consumers are not charged at LRMC, it is unfair and difficult to charge LRMC for rural users (Ranganathan 1993).

Removal of subsidies

As discussed earlier, it is difficult to remove subsidies entirely, due to political reasons, but some categories of users could be made to pay the LRMC price; for example, urban users of electricity, LPG, and kerosene, and large farmers using electricity for irrigation.

Empowering local institutions

There is little disagreement on this at the political level but it has not occurred to any significant extent, even though politicians extol the concept of decentralization of power. The major losers of genuine decentralization would be staff at middle and lower levels in the government and utilities. Even if an Act is passed in parliament, there is little likelihood that it would become fully operational at the lower administration levels, at least in the short term.

Reallocation of funding from conventional energy sources to bioenergy

The funding allocated to renewable energy is less than 2 per cent of that allocated to conventional energy in India (PC 1992). It is very difficult to convince the sceptical conventional energy planners to reallocate a large amount of funding to renewable energy. Bureaucrats usually find it much more convenient and beneficial (to themselves) to fund, say, one 1000 MW power plant than 2000 × 0.5 MW power plants.

12.3.3 High-feasibility and low-impact policies

These are policies or programmes that are likely to find favour with conventional energy planners and administrators but which are more or less a continuation of the *status quo*.

Incentives to isolated bioenergy programmes

Governments are ready to accept this as a type of 'charity' since the funding required is small and it will also satisfy their conscience. However, such programmes are not likely to make a significant impact. The experience so far in India and other countries has shown that such half-hearted attempts will not promote alternative energy options but instead may harm them, as such programmes often fail, thereby creating unease among potential users.

Government-administered implementation of bioenergy programmes

National and provincial governments usually prefer to retain full control over the implementation of programmes. Thus implementation of any alternative energy programme through government operations is more likely to find favour with the bureaucracy than any through private enterprise or local community. This stems from the bureaucratic mistrust of local communities and desire to retain control over funds.

Import of designs, hardware, and expertise

Governments of developing countries usually readily accept any proposal to import designs and hardware, particularly if it comes as a grant, aid, or soft loan. Thus bioenergy programmes involving technology imports have higher chances of government acceptance and implementation than indigenously developed technologies. Indigenous manufacturers cannot provide grants, unlike their counterparts in industrialized countries where governments promote private manufacturers, keeping long-term markets in mind. This may change soon as countries such as Brazil, China, and India have adequate technical capabilities, and in fact are leaders in various bioenergy technologies.

The categorization of options along a hierarchy of feasibility and impacts is easier said than done. What is not feasible today may become feasible sooner or later if, for example, large international funding becomes available for bioenergy as a climate-change mitigation measure, and if governments of developing countries welcome such opportunities. A few hundred successful demonstration projects in a country may suddenly lead to a rush toward bioenergy options. The ranking may also change with technological developments, changes in government, and progress in the global debate on sustainable development. If advanced, high-efficiency biomass gasifier systems for electricity generation at kilowatt-scales (<500 kW) become commercially available with high efficiencies, land will not be a constraint, even in countries with high population densities and low plant productivities. Thus the situation is dynamic, influenced by technological developments and liberalization of developing-country economies and global environmental concerns.

12.4 WHAT IS FEASIBLE IN THE NEAR TERM?

No matter how simple or rational a policy may be, it may not be implemented. For example, in India the replacement of high-friction foot valves and suction lines in irrigation pumps by low-friction types is known to lead to a saving of about 30 per cent of electricity used in irrigation-water pumping (Reddy *et al.* 1991). It has also been shown to be highly cost effective (ADB 1993). Each pump uses an average of about 5362 kWh of electricity per year and there are over 9 million electric pumps. It would make good economic sense even if the State electricity utilities replaced all the high-friction foot valves and suction lines free of charge. However, neither the farmers nor the electrical utilities have adopted any serious programme of large-scale replacement.

Thus it is insufficient to show how bioenergy is a rational energy option from the point of promoting development using the economic, equity, and sustainability yardsticks. It must be acknowledged that there are many barriers to implementation, as discussed earlier. Each country should bear these factors in mind when preparing feasible short- and medium-term action plans. The feasibility criteria should consider the possibility of policy changes and the funding required for a given option.

We attempt to give an example of a set of actions which may be feasible in the short term:

1. Research and development should be promoted to develop efficient, user-friendly, and low-cost bioenergy systems.
2. Develop and debate a rational methodology for energy pricing.
3. Launch a large demonstration programme for promoting rural bioenergy centres (RBCs).
4. Plan RBCs, starting with unelectrified villages.
5. Provide significant financial incentives for bioenergy options.
6. Plan for a few large-capacity electricity and liquid fuel projects.

In a period of 5 years or so, with the establishment of a large number of RBC demonstration units and a few megawatt-scale systems, the correct atmosphere to launch a national-level programme may prevail. What can be implemented successfully and what problems have to be solved would become apparent. With many local benefits accruing from successful bioenergy programmes, politicians may realize the potential economic and political dividends from policies promoting bioenergy options, in the short term.

12.5 ROLES OF DIFFERENT 'ACTORS'

The planning, funding, implementation, and operation of bioenergy systems involve the active participation of a number of 'actors' or institutions, from the village poor to possibly international funding institutions. Their roles are listed in Box 12.3. The dominant actors are likely to be the governments themselves, acting as inhibitors (as they are mainly doing now) or promoters of bioenergy programmes. Examples of active promotion are many: the PURPA act in the USA leading to over 8000 MW of biomass electricity generation capacity, various incentives in Brazil leading eventually to the production of 12 billion litres of bioethanol annually, government support in Zimbabwe for a 40 million litre/year bioethanol plant, promotion of efficient stoves in Kenya resulting in over half a million being produced in the past 6–7 years, and the subsidies to family biogas programmes leading to 1.8 million biogas plants in India and several million in China. The role of governments should be that of promoter and/or facilitator. The governmental

Box 12.3 Roles of different actors in promoting bioenergy options in developing countries

Governments
1. Formulate policies to promote bioenergy and to create a 'level playing field' with other energy sources
2. Provide financial incentives to bioenergy utilities, both decentralized and megawatt systems
3. Ensure a market for bioenergy; both biomass electricity and liquid fuels
4. Institutional and tenurial arrangements to empower farmers and communities to manage bioenergy systems
5. Promote rational energy pricing
6. Promote R & D and technology transfer

Commercial banks and funding agencies
1. Provide credit to entrepreneurs as well as to households and farmers for end-use devices
2. Provide financial management services to entrepreneurs and rural bioenergy centres

3. Arrange for education and motivation of private entrepreneurs on the commercial opportunities of bioenergy, possibly with funding or incentives from governments and donor agencies

International agencies

1. Develop methodologies for assessment of bioenergy and other renewable energy and conventional energy options incorporating externalities, socio-economic and environmental costs and benefits
2. Provide grants for significant demonstration programmes in developing countries and assist technology transfer
3. Provide incremental costs to bioenergy options due to potential global benefits (after the demonstration phase)

Private entrepreneurs

1. Use all the government concessions, incentives, and credit from commercial banks to set up rural bioenergy centres as well as megawatt systems
2. Enter into long-term contracts with farmers and village communities for the supply of wood and animal dung
3. Establish equipment manufacturing units and service units and undertake turnkey contracts to set up and maintain rural bioenergy centres

Village communities

1. Form village institutions to manage energy plantations and rural bioenergy centres
2. Arrange for village youth to be trained in the operation and essential maintenance of machinery
3. Enter into long-term contracts with farmers to supply biomass
4. Contribute to cost reduction measures through labour contribution, scheduling of activities for equitable load, etc.

Non-governmental organizations

1. Assist village communities to form effective village institutions to manage energy centres
2. Implement and participate in demonstration programmes
3. Communicate benefits to local communities and enable them to participate in rural energy programmes

R & D institutions

1. Work towards specific goal-oriented programmes
2. Develop multi-feed biogas and producer-gas systems
3. Work on cost reductions
4. Increase life of systems and make them user friendly
5. Assist technology transfer, demonstration, and evaluation
6. Assist in training, operation, repair, and maintenance personnel

R & D, research and development.

role is crucial to safeguard the interests of the poor and the environment (WB 1994). Governments could provide financial incentives such as tax benefits to renewable energy systems, impose taxes on conventional sources of energy, provide interest subsidies, frame policies to remove restrictions on bioenergy (like the sale of electricity), make institutional arrangements (empowering rural communities), and finally promote research and development. What governments should not do is implement and manage the programmes themselves.

Multilateral financial institutions which currently fund conventional energy programmes (World Bank, development banks, etc.) should evaluate bioenergy programmes and fund them for their benefit to local and global environments, besides the other benefits already discussed. Ahmed (1994) has made an economic assessment of bioenergy options for the World Bank, and has shown that bioenergy systems are already competitive with conventional systems in many situations. The World Bank also recognizes that renewable energy programmes must be promoted commercially (WB 1992). It is time that such intentions are transformed into actions, particularly in the light of global concerns about the environment and sustainable development.

The role of private enterprise, farmers, and village communities is very clear since they should plan, implement, manage, and benefit from bioenergy programmes. Entrepreneurs could undertake turnkey installation and maintenance of bioenergy systems, as in the case of family biogas plants in the Sirsi area (Chapter 4) or the community biogas plant for cooking in Methan village (Chapter 5). The role of village communities is well demonstrated by the Pura biogas and Hosahalli producer-gas electricity systems (Chapter 5), where they manage the systems with technical and financial assistance from outside agencies. In industrialized countries, private enterprise is at the forefront of bioenergy programmes; for example, farmers managing large biogas systems in Denmark and district heating in Austria, large engineering companies in Sweden and Finland, utility companies and oil multinationals in Brazil, France, UK, and USA.

The industrialized countries have a three-part role to play in promoting renewable energy:

(1) as a source of some advanced energy technologies;
(2) as a source of technical assistance and innovative financing for some countries; and
(3) as an example by adopting these technologies in their own countries as part of sustainable development programmes and as a demonstration of their commitment (WRI 1994).

12.6 WHO GAINS AND WHO LOSES?

In any programme of shifting from a conventional approach and technology to alternatives, there will be some losers and some gainers. For example, transfer of decision-making power (on species to be planted and mode of distribution of products) from the Forest Department to village communities will lead to removal of authority and control over financial resources and products from the Forest Department. Thus foresters may be seen as the losers and are likely to scuttle all serious attempts towards community management of forest resources. Similarly, a large-scale shift to rural bioenergy centres to meet all the electricity requirements of rural areas will lead to loss of control over electricity generation and distribution by State electricity utilities and also by large equipment manufacturers and contractors involved in civil constructions, transportation, and so on. It is necessary to recognize those standing to lose or gain from any large-scale shift to bioenergy options, in order to understand the vested interests involved and to plan to overcome the related problems.

Rural bioenergy centres

If rural bioenergy centres are implemented, as suggested in Chapter 6, through the full participation of local communities, with a potential to generate about 100 TWh of electricity and to provide biogas for cooking, some of the gainers and losers are as follows.

Gainers

1. Village communities, since they gain control over the resources, products, and sharing mechanisms.
2. Village women and the poor, as they obtain energy services such as piped water, improved lighting, cooking gas, etc.
3. Village communities gain in the short as well as long term through employment and land reclamation.
4. Rural poor, through income generation in growing and harvesting energy forests, transportation of feedstock, and operation of biogas and producer-gas systems.
5. Farmers, who use their degraded land for producing biomass feedstocks.
6. Equipment manufacturers of biogas and gasifier systems and contractors.

Losers

1. Forest Department staff; loss of control over land, wood resources, and funds.
2. Local electrical utility authorities and staff; loss of distribution controls.
3. Equipment manufacturers of conventional energy systems.
4. Local politicians who used their power to favour selected producers, consumers, and dealers.

Large bioelectricity systems

Surprisingly, implementation of large-scale producer-gas systems would result in nearly identical losers and gainers as with rural bioenergy centre programmes:

Gainers

1. Farmers, who could use degraded land for commercial production of wood for electricity systems.
2. Industries, who manufacture new hardware.
3. Entrepreneurs of the systems.

Losers

1. Forest Department, electrical utilities, bureaucrats, and possibly even politicians.
2. Equipment manufacturers and dealers in large conventional energy systems.

Currently there are large national and multinational power generation equipment manufacturing corporations who have strong commercial interests in the multibillion dollar power generation equipment industry, spanning all the continents. They would be concerned about competition and/or could participate in a new phase of decentralized electricity generation projects. Similarly, the large-scale spread of bio-ethanol or bio-methanol programmes would affect the petroleum-producing, importing, and processing industries. It may be necessary to encourage different national and multinational corporations in the manufacturing and distribution of bioenergy equipment. Otherwise it may be difficult for small-scale bioenergy equipment manufacturers to counter the influence of large power generation and transmission equipment manufacturing corporations on multilateral funding agencies, such as the World Bank and the Asian Development Bank, as well as the governments of donor countries. One possible option is that these large corporations could undertake the manufacture of bioenergy equipment themselves. An analogy could be found in the manufacturers of chlorofluorocarbons (CFCs), where, with the signing of the Ozone Treaty, the same manufacturers became manufacturers of the substitute, thereby having a vested interest in the spread of ozone-friendly technology.

12.7 INTERNATIONAL ASSISTANCE AND COLLABORATION

International assistance and collaboration are important for technology transfer and investment capital, since developing countries have differing technical and financial capabilities for research and development as well as for the dissemination of bioenergy technologies. Bioenergy technologies present an interesting contrast since, unlike in other fields (e.g. solar PV or wind energy, where the USA and Western Europe dominate), industrialized countries do not have all the expertise, as many developing countries are leaders in some aspects of bioenergy technologies. China and India lead in small-scale biogas systems, whereas Denmark has excelled in large-scale biogas plants for heat and power. In the case of producer-gas, India has achieved significant expertise in small-scale systems, whereas with medium-scale systems, Finland and Sweden have expertise and experience. The US has vast experience with constructing hundreds of steam-generating biomass power plants. In the case of liquid fuels (ethanol) Brazil, USA, and Zimbabwe are leaders. Therefore, expertise and leadership are geographically dispersed and distributed between developing and industrialized countries, so that technology transfer could be a two-way affair between these two groups.

However, the experience with small-scale producer-gas technology transfer in the 1980s has shown that simply transferring the technology to developing countries without adequate involvement of local institutions and training has failed (BTG 1993). The ideal option is to develop the technologies in as many countries as is practical. However, not all countries have adequate technical and financial capability to undertake the necessary research and development. Thus all major or large developing countries ought to develop technologies on their own and share it with neighbouring countries, or those with similar socio-economic and resource situations. Technology transfer from one country to another should be accompanied by adequate involvement of the host country's technical institutions and entrepreneurs.

Flow of financial assistance will be largely from industrialized (OECD) to developing countries, and in exceptional cases from advanced or large developing countries such as Brazil, China, or India to other less developed countries. Since bioenergy technologies provide significant global benefits in terms of reducing carbon emissions and in promotion of *in situ* biodiversity, industrialized countries should promote and fund bioenergy projects in developing countries. An attraction for industrialized countries is that the cost of obtaining global benefits per dollar invested could be lower than in their own countries. For example, the establishment cost of a forestry option for carbon sequestration is lower in developing countries as compared to that in North American or European countries. A study integrating over a period of 100 years showed that the cost of sequestering carbon in European Community countries is around $US21.4/t of carbon, whereas it would be around $US2.4/t of carbon sequestered in tropical Latin America (Nakicenovic *et al.* 1993). It was shown in Chapter 8 that the bioenergy approach provides continuous carbon benefit (emission reduction) when sustainable wood production is used for biomass electricity or liquid fuels to substitute for fossil fuels, which contrasts with the carbon sequestration approach. Bioenergy projects could become a significant component of 'joint implementation programmes' in the climate-change mitigation strategies.

The OECD countries could provide funds at three levels:

(1) for research and development on bioenergy technologies in their countries as well as in selected developing countries; even countries such as China and India could undertake collaborative work;

(2) initially they could fund demonstration and field trial programmes through grants; and
(3) later, assuming successful field trials, they could fund the incremental cost of shifting from conventional to bioenergy programmes.

There could be questions of conflicts of global, national, or local interests. Planners and bureaucrats in developing countries may prefer a few large power plants rather than thousands of rural bioenergy centres or hundreds of megawatt-scale systems. Influential agencies such as the World Bank and other development banks could play a crucial role in promoting bioenergy options based on strict environment impact assessments and social cost-benefit analyses. Technologies to be transferred should be mature and efficient with potential for local replicability, should generate economic benefit to the host country, and must be cost-effective. At present bioenergy technologies have certain disadvantages in common with any emerging technology, and having to compete with large corporations promoting conventional energy technologies, they will need significant incentives during the learning-curve phase.

12.8 GUIDELINES FOR BIOENERGY PROGRAMMES

Experience of the development and spread of bioenergy technologies over the years has highlighted the need for some broad guidelines or a checklist that planners and managers of RBCs or large bioenergy systems must be aware of and use whenever possible. If the planners and managers of many of the failed programmes had some clear guidelines from the start, many of the failures could perhaps have been avoided.

It is neither feasible nor desirable to give a uniform and universal set of guidelines for a given mix of technologies or programmes for different countries, and probably even within a country. At best what can be attempted is to list a range of guidelines based on experience from a large number of programmes, focusing on local, national, and global concerns. Planners would naturally consider only those relevant to the local political, social, and environmental situations, and, with experience, hopefully add new guidelines. Some guidelines already exist, such as the detailed set that Shell and WWF (1993) have developed on tree plantation programmes, and more general ones published by the National Audubon Society (Beyea *et al*. 1992; OTA 1992). Early recognition and discussion of the guidelines is necessary among the various actors involved, particularly with local managers and local communities. Obviously, the listing of all possible guidelines appears too idealistic, as in reality compromises and trade-offs are necessary since it is unlikely that all the guidelines would be strictly followed in any location. However, flouting critical guidelines would probably lead to failures and disasters. Thus the local decision makers may have to define a minimum set of critical guidelines along with a set of desirable guidelines. The guidelines are grouped into environmental, social and institutional, technological, and economic ones, and are simply listed here, since they have been already discussed at various points in earlier sections and chapters.

12.8.1 Social and institutional guidelines

1. *Strong village institutions*

   Village-level institutions, along with adequate power to take decisions and enforce them, are a necessary precondition for all bioenergy programmes requiring land and the participation of local communities.

2. *Decision-making process for RBCs*

   All major decisions on sharing costs, benefits and defining responsibilities for different persons and agencies, and enforcing discipline should be taken by discussion and consensus through a village council.

3. *Management committee*

   The village council should appoint a small management committee for taking day-to-day decisions and supervising the operation of the system.

4. *Participation at all stages*

   The village management committee must be involved at all stages, starting from planning, and all major decisions should be taken at the village council level.

5. *Equitable sharing of as many benefits as possible*

   Certain services, such as drinking-water, domestic lighting, and cooking energy, must be made accessible to all at affordable prices. Street taps must be freely accessible to all. Every house must get electricity, although the number of lighting points may vary depending on size. In the case of irrigation water, it is desirable to ensure equitable distribution of power and water from common sources among farmers and, of course, farmers must have access to power at commercial rates for private irrigation systems.

6. *Land tenure*

   Local communities must have full long-term tenure to all products from the common land and water. It must be legally ensured and proper agreements must exist.

7. *Commercial turnkey installation and maintenance*

   The village council and the funding agency could jointly decide to hand over the installation and major maintenance of the system to a commercial organization, although the day-to-day operation and minor maintenance should be by trained local community members or a local entrepreneur.

12.8.2 Economic guidelines

1. *Pricing of energy*

   The pricing must be based on long-run marginal costing. The costing procedure could consider government subsidies (capital or operational), incentives, avoided costs, and, if possible, environmental costs and benefits.

2. *Free services*

   Basic services such as drinking-water through community taps, and lighting for village schools, hospitals, temples and streets should be free in all villages for the community, but the cost of such services could be recovered from local or provincial governments.

3. *Commercial pricing for services*

   All services other than the few mentioned above must be charged at an economic price in order to make profits.

4. *Local labour employment*

   Local labour should be used for as many operations as possible, except for specialized technical jobs, in the initial stages.

5. *Realistic projection of prices for inputs and outputs*

   It is necessary to ensure realistic projections of the prices of all inputs and outputs, the supply of raw materials, and the demand for energy from the planning stage onwards.

6. *Subsidize incremental cost*

   Until the modern bioenergies are fully developed, and until level playing conditions are created, the incremental cost involved in establishing bioenergy systems could be subsidized.

12.8.3 Technical guidelines

1. *Indigenous technology*

   Large developing countries should develop the technologies themselves. In cases where necessary, this could be done initially in collaboration with research institutions in more advanced developing countries and, only in specific situations, with institutions in industrialized countries.

2. *Local manufacturing base*

   After prototype testing, all the components and spares should be manufactured locally in developing countries, depending on the scale of operation.

3. *Spares and support services*

   All the spares and repair services should be accessible locally to the bioenergy units.

4. *Training local people in technology*

   Local entrepreneurs, operators, and servicing personnel must be thoroughly trained. Only major repairs and replacement should be left to centralized servicing agencies.

5. *Manufacturer/entrepreneur guarantee*

   The manufacturers of gasifiers, biogas drums, turbines, stoves, and electrical equipment, etc. must provide guarantees for their equipment and design performance. In the case of installed or *in situ* constructed systems, such as biogas plants or improved stoves, the contractor should provide a guarantee of performance of the system for a specific period.

6. *Field testing for suitability and reliability*

   All technologies must be adequately field tested for significant periods for performance, reliability, and user-friendliness, before any large-scale dissemination or commercialization is attempted.

7. *Systematic survey and planning*

   A systematic survey of the availability of resources (such as land, biomass, livestock), productivities (biomass or cattle dung), demand for energy, manpower available, technical services available, irrigation potential, agro-processing potential, etc. is a prerequisite before launching any programme.

12.8.4 Environmental guidelines

Most of the environmental guidelines pertain to biomass production and disposal of wastes resulting from bioenergy conversion.

1. *Exclusion of primary forests*

   In all countries primary forests should not be used as a source of biomass feedstock for electricity or liquid fuels and not even as a sustainable source of woody biomass. Also, they should not be displaced by artificial plantations.

2. *Biomass from degraded lands*

   Developing countries have degraded forest, common, and private lands which should be used for biomass production for energy in a sustainable way. A multi-objective forestry programme may be preferable to one designated only for biomass energy. This could comprise land reclamation, employment generation, biodiversity improvement, fruit production, and so on.

3. *Long-term sustainability criteria*

   Details of the silvicultural and management practices for long-term sustainability were discussed earlier (Chapter 7). Some practices are:

   (a) inclusion of multiple tree species and multiple clones for a given species;
   (b) inclusion of nitrogen-fixing tree species, native tree species, shallow- and deep-rooted species;
   (c) all unused twigs, leaves, and reproductive parts should be left on the site;
   (d) biogas sludge and the ash from producer-gas operations should be returned back to the plantation site, and there should be minimum physical disturbance to soil.

4. *Measures for biodiversity*

   (a) use multiple species, or multiple clones of a given species;
   (b) leave a fraction of land area in each landscape for succession of indigenous vegetation through natural regeneration;
   (c) leave a few gaps or islands of natural vegetation in a plot;
   (d) use optimum spacing and density of trees;
   (e) practise spot-weeding and fertilizer application.

5. *Environmental impact assessment*

   Conduct environmental impact assessments for the plantation component as well as for the whole bioenergy programme, to ensure that all costs, benefits, and impacts of various practices are evaluated.

13
Conclusions and implications

The population of the world has nearly crossed the 6 billion mark and its growth in the developing world continues to be about 2–3 per cent annually. Correspondingly, the total primary energy consumption in developing countries increased by 33 per cent between 1981 and 1991 (at 3712 Mtoe in 1991; OECD 1993). Consumption of coal increased by 56 per cent between 1982 and 1992 in developing countries (BP 1993). The fuelwood demand in developing countries is increasing and currently it is around 1.3×10^9 m^3 (FAO 1993*b*) in tropical countries.

The tropical deforestation rate during the period 1981–90 was higher than in the previous decade—15 Mha annually during the 1980s (FAO 1993*a*). There are threats of declines in biodiversity and of climate change. Land degradation is threatening biomass production potentials. At the same time, a change in the awareness and attitudes of many countries towards the issues of environmental degradation and sustainable development has occurred over the past decade, leading to the UNCED meeting in Rio de Janeiro in 1992. This led to Conventions on Climate Change, Biodiversity and Desertification, and most countries have now declared their intention of pursuing a path of sustainable development. Energy is a critical factor in promoting development. There are basic questions as to sustainability of the current rates of growth of commercial sources of energy based on fossil fuels. Furthermore, the current energy-use patterns, which are focused on commercial energy, have led to inequities within and between rural and urban areas in developing countries. We should also mention the large inequities between developing and industrialized countries in commercial energy use.

Energy has been a major contributor to environmental degradation: local air pollution (indoor from cooking stoves and regional from coal power stations), land submergence (for hydroelectricity), global warming (fossil fuel emissions), deforestation or forest degradation (for charcoal and fuelwood), acid rain (fossil fuel emissions), and so on. There has been a search for alternative sources of energy which are more equitably distributed between and within countries than fossil fuels, and which are environmentally more benign.

Renewable energy

In the past decade the potential role of renewables in energy production, resulting from technological developments and significant cost reductions, has been recognized by governments, donor agencies, and researchers such as the Planning Commission of India (PC 1992), the World Bank (WB 1992; Ahmed 1994), World Energy Council (WEC 1993), UN-sponsored studies (Johansson *et al.* 1993), and a series of studies in the *Energy Policy* journal dedicated to renewable energy (Jackson 1992).

Many renewables will probably have a place in the development plans of different countries. The benefits of renewables do not warrant repetition here. The advantages of bioenergy have been highlighted for developing countries in earlier chapters: local benefits (employment, self-reliance, local control, land reclamation) and regional or global benefits (reduction in air pollution, net reduction in carbon emissions, increased biodiversity in degraded lands, reduced pressure on forests). Bioenergy options are universally applicable, as they can be established wherever biomass can be grown, with little or no seasonality or storage constraints (unlike most other renewable forms of energy). Bioenergy is also a versatile energy source with potential for conversion to liquid and gaseous fuels, and to electricity. The generation of electricity is a promising option for rapidly expanding the use of biomass for energy at scales from a few kilowatts to a few tens of megawatts, being suitable particularly for rural locations and decentralized applications. Bioenergy also provides economic advantages for rural applications, such as reduction of costs and losses inherent in transmission and distribution of electricity to dispersed low-load situations.

Of course, no renewable energy option is without its own problems. There are fears over the demand for biomass feedstock leading to devegetation, and of monoculture energy plantations leading to loss of biodiversity. There are also doubts about the costs and reliability of the technologies and the skills required to operate and manage bioenergy systems, particularly in rural areas. Any analysis of the potential of and the barriers to bioenergy must include practical guidelines to develop policy options for its promotion. Bioenergy options must also be subjected to comparative analysis of social, economic, and environmental costs and benefits in parallel with other renewable and conventional energy options.

Environmental benefits alone will not sell a technology, particularly in developing countries, but there is no denying the fact that environmental degradation will leave the poor worse off than the rich. Thus even developing countries cannot afford to ignore the environmental aspects of energy production and use. Since conventional energy pricing does not include social and environmental costs and benefits or the features of rural energy requirements, where the majority of people in developing countries live, renewables start off at a disadvantage. Good and acceptable methodologies for quantifying and incorporating social and environmental factors are not yet available for energy analysis. In the case of bioenergy technologies they are location-specific and are linked to local land, socio-economic, and resource situations. This requires detailed analysis at the regional level within a country, or at least at each country level with the highlighting of regional differences. The present study is one such attempt to explore the potential, possibilities, and problems of bioenergy options for India.

Indian case study

We have analysed the current patterns of energy use which show the still predominant role of biomass energy, in spite of the growing supplies of commercial energy, particularly in rural areas. The non-sustainable features of a commercial energy-driven strategy, the continued inequity between rural and urban areas, and the potential damaging impacts to the environment were highlighted for the Indian situation. The focus of the present analysis is rural India, where 75 per cent of the 844 million population live, largely not benefiting from modern commercial energies or from energy planning. The more recent goals of development planning have clearly indicated the trends towards a greater emphasis on renewable energy and decentralized systems, and the promotion of employment and community management. The energy strategy for India should not necessarily be solely to increase the kilograms of oil or kilowatt-hours of electricity use per capita, but to deliver the energy services required by a housewife, farmer, transporter, or an entrepreneur to promote development.

A tentative set of services requiring quality energy in any typical village situation was defined to include cooking, domestic lighting, community services, agricultural pumping, and all other shaft-power requiring activities (such as agro-processing). The main goal is to provide quality modern fuels, such as gaseous or liquid fuels, for cooking in all homes; electricity for lighting for all homes; community services in all villages (pumping drinking-water, street lighting, and flour milling), and electricity for all shaft-power requiring activities (pumping irrigation water and agro-processing) in all villages. The analysis of bioenergy options showed that it is possible to meet the energy requirements of all services that are necessary to promote development and quality of life in rural areas. Technologies for decentralized electricity generation through producer-gas systems, and biogas from cattle dung (and plant leaf litter) for cooking, are already available off-the-shelf for different scales of operation, although significant scope exists for improvements in efficiency of conversion, use of multiple feedstock, and user friendliness. However, liquid fuel technology using lignocellulose feedstock, at the scales feasible for India, and high-efficiency megawatt-scale producer-gas systems are not yet available. At present, research and development efforts are under way, and efficient and viable bioenergy technologies will soon become available at competitive prices. A concept of a rural bioenergy centre is suggested, where all the village energy requirements are considered in an integrated way and bioenergy options are matched to the needs. The bioenergy programmes must become an integral part of the overall rural development programmes through genuine empowerment and participation of local communities.

Land and sustainable biomass supply

The need for dedicated energy plantations for generating about 122 Mt of woody biomass annually for decentralized electricity was stressed. It is not desirable to use the existing natural forests (even for supposedly sustainable modes of removal) given the low area under forest (64 Mha, or <0.1 ha of forest/capita) and the potential threat to biodiversity in the process of extraction. Currently 181 Mt of woody biomass is used nationwide annually in the rural domestic sector for cooking. With the provision of biogas and liquid fuels, the fuelwood requirement for cooking would become redundant. One could argue that this biomass, which comes largely from sustainable sources, could be used as feedstock.

An analysis of land availability shows that the area under crops has stabilized since 1970 in India, even though the population has been growing at over 2 per cent annually (Chapter 7). Crop productivity, although increasing, is still lower than in China and South Korea. There is large potential to double or even treble food production without any increases in the area under cropping, by increasing cropping intensity and crop productivity. A land-use strategy to produce food, fodder, timber, and fuel (for bioenergy) shows that land is not a constraint in India to meet all these biomass needs of the growing population. A part of farmers' croplands and much of village commons (pasture land), and forest lands (less than <10 per cent tree crown) are degraded and not suitable for profitable cropping and thus the non-crop area has remained nearly constant over the past 20 years. Such lands have not been converted to croplands as one would expect with rising populations. Even at a low energy conversion efficiency of 18 per cent, only 15–30 Mha (productivity 8–4 t/ha/yr, respectively) are adequate for meeting biomass needs of energy for 98 per cent of rural settlements. Energy plantations with appropriate species mixes and density could also provide other products, such as grass for livestock, leaves for biogas, and other valuable tree products (fruits, seeds, etc.). The land required would decrease with increases in efficiency of energy conversion as well as plantation productivity, the potential exists for both. The leaf litter by-product from village energy plantations is suggested as a feedstock for biogas generation, along with cattle dung for cooking. The main premise is that bioenergy could provide the necessary incentives to reclaim currently degraded lands, which have little possibility of being rehabilitated via conventional afforestation or fodder development programmes. Thus bioenergy programmes could become a part of a rural development programme package involving reclamation of degraded lands and watersheds, local employment generation (production and processing of wood feedstock, operation of biogas and producer-gas systems), promotion of self-reliance, and improving the quality of life, especially for women and the rural poor.

High biomass productivity is necessary for profitability to growers and for reduced land requirements, but such yields must be sustainable. To some extent the profitability of biomass production for energy in degraded lands could be increased by including high-value tree species yielding fruit, flower, and industrial raw materials as by-products. Currently, forestry or plantation research in India has no significant results to show in practice for high biomass yield or its sustainability. A productivity of 8 t/ha/yr or more is feasible with appropriate seed quality, species mix, silvicultural practices, and soil and water conservation measures. Sustainable biomass production at moderate productivities seems to be feasible, and a set of guidelines that would contribute to sustainability are suggested.

Large bioenergy systems

The potential for megawatt-scale biomass electricity generation and liquid fuel systems cannot be ignored since they could be of significant economic benefit to farmers who own degraded lands that could be dedicated to biomass production. Furthermore, these systems could lead to reduced oil imports and replacement of coal in power generation. We calculate that not all existing degraded lands are required for village-level decentralized producer-gas electricity systems or timber production. Surplus degraded land would be available. Similar to sugar co-operatives, where growers in a given area set up a sugar mill to process their cane, farmers owning degraded lands could set up megawatt-scale electricity generation units and supply feedstock. A potential of 16 000–32 000 MW of installed capacity is shown to be feasible by dedicating only 5 per cent of India's geographical area, depending on the efficiency of conversion. For a 1 MW system, biomass feedstock needs to come from a radius of only 8 km, at a productivity of 8 t/ha/yr by dedicating only 5 per cent of the geographical area for energy plantations, assuming only 18 per cent conversion efficiency. The motivation for the farmers would be the same as in the case of a sugar mill, i.e. profits and employment but from largely unused lands. The large potential for co-generation in sugar mills has been recognized by the government to be at least 3500 MW; this option needs to be pursued on a priority basis since bagasse is already available in a centralized location and is a by-product that is presently used inefficiently.

Institutional and financial policy options

We recognize that technical potential is only one of, or rather the first of, many other conditions necessary for the promotion of any new options. The significant environmental and social benefits of bioenergy alone are unlikely to lead to any large-scale shifts to bioenergy. Even provision of financial subsidies alone will not promote the bioenergy options, in the absence of a whole range of policy measures. For example, no amount of subsidy on investment cost alone for a megawatt-sized electricity plant will help if there are restrictions on the sale of electricity to buyers at economic prices, or if it has to face state-owned utilities providing free or highly subsidized electricity. The barriers and the problems in large-scale commercialization or dissemination of bioenergy options are presented in Chapter 12, along with a set of policy measures, the roles of the different actors involved, and a set of guidelines. However, one must be realistic as energy policies are more often based on short-term political considerations rather than rational financial or environmental criteria. In developing countries, energy procurement, production, distribution, and pricing are largely controlled by State utilities, and high subsidies on electricity and petroleum fuels are prevalent. There is a growing realization that such price subsidies to infrastructure almost always benefit the non-poor disproportionately (WB 1994). A bioenergy entrepreneur, farmer, or village community will face many obstacles in the production and transportation of biomass, and in pricing inputs as well as outputs such as electricity and biogas. The removal of existing restrictions on energy production and sale, the empowerment of rural communities (to which the government is committed on paper), along with the promotion of private initiatives and appropriate financial incentives, may act as catalysts for decentralized bioenergy production.

The mere generation of electricity, biogas, or liquid fuels may not ensure its use or the flow of benefits to rural households, particularly the poor. Subsidizing access to infrastructure or services is often more beneficial to the poor than price subsidies (WB 1994). There is a need to develop technical as well as financial instruments to make them accessible to the poor, for example changing the mode of electrical wiring, doing away with energy meters, adoption of roof-to-roof connections, provision of soft credit for electrical connections, and subsidies on first cost.

Implications for developing countries

Bioenergy analysis for India has some implications for other developing countries, since the majority of them are expected to provide even more fertile ground for bioenergy options than India. Such developing countries are characterized by:

(1) lower population densities;
(2) higher biomass resource potential (large areas under forest and pasture land per capita);
(3) geographically dispersed human settlements;
(4) very low spread of grid electricity in rural areas; and
(5) larger dependence on imported oil and outflow of export earnings leading to foreign-exchange crises.

In SS Africa, conventional energy or development plans based on imported oil and centralized grid electricity seem to provide little hope for most rural populations. The suitability of decentralized energy production and the high potential of bioenergy to provide quality fuels is shown for developing countries. Land does not seem to be a constraint for sustainable biomass feedstock production. The bioenergy potential could also be attractive for countries with large numbers of inhabited islands, such as Indonesia, the Philippines, and Fiji.

Not all countries of SS Africa or South and South-East Asia have the technical capability or need to develop all the bioenergy technologies. Larger and relatively technology-rich developing countries such as Brazil, India, China, and South Africa could take a lead in technology development. Other smaller or less resource-rich developing countries could evaluate the technologies and adopt the appropriate ones. International assistance from industrialized countries could be more effective than it has been so far, even though many in developing countries are still reticent about the technical implications. In many regions international assistance has been deemed to have failed to promote rational energy policies or rural electrification which benefits the poor. Today many industrialized donor countries are more favourable to renewables than ever before, and even the World Bank is making a concerted effort to be environmentally acceptable.

Bioenergy has clear advantages over other renewables and in order to convince the energy planners and entrepreneurs in developing countries its trump cards must be emphasized: being suitable to very many locations, promotion of self-reliance, possible local manufacturing base for technology, and also that it could become a part of an integrated rural development package leading to local benefits such as land reclamation and employment generation. Such attributes are not necessarily applicable to solar PV, wind, or other high technologies. The environmental advantages of bioenergy technologies should be clearly favourable to overseas assistance planners. Bioenergy technologies must achieve a significant scale of operation to be economically viable and visible. It is here that agencies such as the World Bank and other development banks should provide investment capital and incremental costs to bioenergy, to achieve significant, viable, and visible scales of operation. The time is right and ripe for large-scale implementation of bioenergy programmes on a much wider scale than in the past decade or so. Sufficient funding and policy support is necessary so that bioenergy can play a key role in energy provision in the next century and contribute to sustainable and equitable development.

References

ABE (1985). *Towards a perspective on energy demand and supply in India in 2004/05*. Advisory Board on Energy, Government of India, New Delhi.

ADB (1991). *Environmental consideration in energy development*. Asian Development Bank, Manila.

ADB (1993). *Energy end use – an environmentally sound development pathway*. Asian Development Bank, Manila.

Agarwal, A. and Narain, S. (1989). *Towards green villages*. Centre for Science and Environment, New Delhi.

Ahmed, K. (1994). *Renewable energy technologies: a review of the status and costs of selected technologies*, World Bank, Technical Paper, No. 240. World Bank, Washington.

Ahuja, D. R. (1990). Research needs for improving biofuel burning cookstove technologies. *Natural Resources Forum*, **14**, 125–34.

Alam, M., Dunkerley, J., Gopi, K. N., Ramsay, W., and Davis, E. (1985). *Fuelwood in urban market*. Concept Publishing Company, New Delhi, India.

Arnold, J. E. M. and Stewart, W. C. (1991). *Common property resource management in India*. Oxford Forestry Institute, Oxford.

Azimi, A. M. and Lohani, B. N. (1992). Barriers to energy end use efficiency. *Energy Policy*, **20,** 533–45.

Baguant, J., Manrakhan, D., and Panray, R. B. (1991). *Exploring the possibility of using low grade ethanol as a kerosene substitute for cooking*. SAREC/IDRC Workshop on energy policy and planning resources programme, Eastern/Southern Africa, NIR, Botswana.

Barnard, G. W. (1990). Use of agricultural residues. In *Bioenergy and the environment* (ed. J. Pasztor and L. A. Kristoferson), pp. 85–112. Westview Press, Boulder, Colorado.

Bekkering, T. D. (1992). Using tropical forests to fix atmospheric carbon; the potential in theory and practice *Ambio,* **21,** 414–19.

Bernardo, F. P. and Kilayko, G. U. (1990). Promoting rural energy technology: the case study of gasifiers in the Philippines. *World Development,* **18,** 565–74.

Betters, D. R., Wright, L. L., and Couto, L. (1991). Short rotation woody crop plantations in Brazil and United States. *Biomass and Bioenergy,* **1**, 305–16.

Beyea, J., Cook, J., Hall, D. O., Socolow, R. H., and Williams, R. H. (1992). *Toward ecological guidelines for large-scale biomass energy development*. National Audubon Society, New York.

Bhat, D. M. and Ravindranath, N. H. (1994). *Evaluation of social forestry programme in Sirsi*. CES, Indian Institute of Science, Bangalore.

Bhat, D. M., Prasad, S. N., Hegde, M., and Saldhana, C. J. (1984). *Plant diversity studies in Uttara Kannada district*, CES Technical Report No. 9. Centre for Ecological Sciences, Indian Institute of Science, Bangalore.

Bhat, P. R. and Ravindranath, N. H. (1994). *Assessment of biogas programme in Sirsi region of Western Ghats*. CES, Indian Institute of Science, Bangalore.

Bhide, S. (1992). Wood demand in Haryana's household sector. In *The price of forests* (ed. A. Agarwal). Centre for Science and Environment, New Delhi.

BMTC (1992). *Regional workshop on measures to strengthen house construction activity*, Background Paper. BMTC, Madras.

Bose, R. K., Puri, C., and Joshi, V. (1991). Energy profiles of three unelectrified villages in Eastern Uttar Pradesh of India. *Biomass and Bioenergy,* **1,** 99–109.

Bose, S. (1993). *Money, energy and welfare*. Oxford University Press, New Delhi.

Bowonder, B., Rao, N. P., Dasgupta, B., and Prasad, S. S. R. (1985). Energy use in eight rural communities in India. *World Development,* **13,** 1263–86.

Bowonder, B., Prasad, S. S. R., and Unni, N. V. M. (1988). Dynamics of fuelwood prices in India. *World Development*, **16,** 1218.

BP (1993). *B.P. Statistical review of world energy*. British Petroleum Co., London.

BTG (1993). *UNDP/WB small-scale biomass gasifier monitoring report*. Biomass Technology Group, University of Twente, The Netherlands.

Campinhos, E. and Ikemori, Y. K. (1989). Selection and management of the basic population of *Eucalyptus grandis* and *E. urophylla* established at Aracruz for the long term breeding programme, *Proceedings of IUFRO Conference, Pattaya, Thailand.*

Cannell, M. G. R. (1982). *World forest biomass and primary production data*. Academic Press, London.

Carpentieri, A. E., Larson, E. D., and Woods, J. (1993). Future biomass based electricity supply in Brazil. *Biomass and Bioenergy*, **4,** 149–76.

CEA (1991). *Projections for electrical energy consumption up to 2006–7*. Central Electrical Authority, Ministry of Power, New Delhi.

Central Forestry Commission (1981). *India's Forests 1980*. Department of Agriculture and Cooperation, Ministry of Agriculture, New Delhi.

Chambers, R., Saxena, N. C., and Shaw, T. (1989). *To the hands of the poor – water and trees*. Oxford and IBH Publishing, New Delhi.

Chanakya, H. N., Borgaonkar, S., Meena, G, and Jagadish, K. S. (1993). Solid phase fermentation of untreated leaf biomass to biogas. *Biomass and Bioenergy*, **5,** 378.

Chaturvedi, A. N. (1990). *Trees and fuelwood from non-forest land*. FAO, Bangkok.

Chaturvedi, P. (1993). *Bioenergy production and utilization in India, expert consultation on biofuels for sustainable development; their potential as suitable to fossil fuels and CO_2 emission reduction*. FAO, Rome.

Cherail, K. (1993). Spurt forecast in non-conventional energy. *Down to Earth*, 15 May, 5–7.

CMIE (1992). *Basic statistics relating to Indian economy*. Centre for Monitoring Indian Economy, Bombay.

CMIE (1994). *Current energy scene in India*. Centre for Monitoring Indian Economy, Bombay.

Comarmond, G. (1992). *Cogeneration – the Mauritian experience*. AFREPREN, Gaborone.

Davidson, O. R. (1992). Energy issues in Sub-Saharan Africa – future directions. *Annual Review of Energy and the Environment*, **17,** 359–403.

Daxiong, Q., Shuhua, G., Baofen, L., and Gehua, W. (1990). Diffusion and innovation in the Chinese biogas program. *World Development*, **18,** 555–63.

DEA (1991). *The Danish programme on centralised biomass plants*. Danish Energy Agency, Copenhagen, Denmark.

Debell, D. S., Whitesell, C. D., and Suhubert, T. H. (1989). Using N_2-fixing *Albizia* to increase growth of *Eucalyptus* plantations in Hawaii. *Forest Science*, **35,** 64–75.

Deccan Herald (1994). Grain need by 2000 may be 180 Mt. 13 August, Bangalore.

DES (1985). *A case study report on marketing farm forestry produce of* Eucalyptus and Casuarina *in Bangalore and Kolar district*. Directorate of Economics *and* Statistics, Government of Karnataka, Bangalore.

Dixon, R. K., Brown, S., Houghton, R. A., Solomon, A. M., Trexler, M. C., and Wisniewski, J. (1994). Carbon pools and flux of global forest ecosystems. *Science*, **263,** 185–90.

DNES (1992). *Annual Report 1991–92*. Department of Non-Conventional Energy Sources, New Delhi.

Dutt, G. S. and Ravindranath, N. H. (1993). Bioenergy: direct applications in cooking. In *Renewable energy* (ed. T. B. Johansson, H. Kelly, A. K. N. Reddy, and R. H. Williams). Island Press, Washington, pp. 653–697.

Dyalchand, A. and Murthy, N. (1988). District level management systems for biogas programme. *Economic and Political Weekly,* 28 May, M-80–4.

Economic Times (1993). Balance of payments: Key indicator, 24 February, Bangalore.

Economic Times (1994). Government hikes LPG prices, import duty slashed, 12 January, Bangalore.

Elliott, P. (1994). Brazilian biomass power project. *Energy for Sustainable Development*, **1,** 41–3.

Elliott, P. and Booth, R. (1993). *Brazilian biomass power demonstration project*. Shell Centre, London.

EPA (1990). *Analysis of the economic and environmental effects of ethanol as an automotive fuel*. Special Report, Office of Mobile Sources, Environmental Protection Agency, USA.

EPRI (1993) *Host utility announcement*, HU-102164, Electric Power Research Institute, Palo Alto, California.

Evans, J. (1990). Long term productivity of forest plantations – status in 1990. *Proceedings 14 FRO World Congress, Montreal, Canada*, **2,** 165–80.

Evans, J. (1992). *Plantation forestry in the tropics*. Clarendon Press, Oxford.

Evans, J. and Hibberd, B. (1993). *Tree plantation review; Operations-Study-9*. Shell International Petroleum Company, Surrey.

FAO (1993*a*). *Forest resources assessment 1990; tropical countries*. FAO, Rome.

FAO (1993*b*). *Forestry statistics today and tomorrow*. FAO, Rome.

FAO (1993*c*). *Status and development issues of the brick industry in Asia, regional wood energy development programme in Asia*. FAO, Bangkok.

FAO (1993*d*). *Mixed and pure forest plantations in the tropics and subtropics*, Forestry Paper 103. FAO, Rome.

Foley, G. (1985). Wood-fuel and conventional fuel demands in the developing world. *Ambio*, **14,** 253–8.

Foley, G. (1992*a*). Rural electrification in the developing world. *Energy Policy*, **20,** 145–52.

Foley, G. (1992*b*). Renewable energy in third world development assistance. *Energy Policy*, **20,** 355–64.

Friedrich, R. and Voss, A. (1992). External costs of electricity generation. *Energy Policy*, **21,** 114–22.

FSI (1988). *State of forest report 1987*. Forest Survey of India, Ministry of Environment and Forests, Dehra Dun, India.

FSI (1990). *State of forest report 1989*. Forest Survey of India, Ministry of Environment and Forests, Dehra Dun, India.

FSI (1992). *State of forest report 1991*. Forest Survey of India, Ministry of Environment and Forests, Dehra Dun, India.

FSI (1994). *State of forest report 1993*. Forest Survey of India, Ministry of Environment and Forests, Dehra Dun, India.

Gadgil, M. and Guha, R. (1992). *This fissured land – An ecological history of India*. Oxford University Press, New Delhi.

Gadgil, M. and Sinha, M. (1986). The biomass budget of Karnataka. In *State of environment report* (ed. C. J. Saldanha), pp. 19–30. Center for Taxonomic Studies, Bangalore.

Garg, V. K. (1992). Litter production and its nutrient concentration in some fuelwood trees grown on sodic soil. *Biomass and Bioenergy*, **5,** 323–8.

Goldemberg, J. and Macedo, I. C. (1994). Brazilian alcohol program: an overview. *Energy for Sustainable Development*, **1,** 17–22.

Goldemberg, J., Johansson, T. B., Reddy, A. K. N., and Williams, R. H. (1988). *Energy for a sustainable world*. Wiley Eastern Limited, New Delhi.

Goldemberg, J., Monaco, L. C., and Macedo, I. C. (1993). The Brazilian fuel-alcohol program. In *Renewable energy* (ed. T. B. Johansson, H. Kelly, A. K. N. Reddy, and R. H. Williams), pp. 841–863. Island Press, Washington.

Government of Orissa (1987). *Evaluation of Orissa social forestry project phase I*. Bhubaneswar, India.

Grassi, G. (1992). R & D Biomass programme of the CEC. In *Proceedings of the 2nd World Renewable Energy Congress, 13–18 September 1992, Renewable Energy Technology and the Environment*, pp. 1238–1245. Reading, UK.

Grassi, G. and Bridgewater, A. V. (1993). The opportunities for electricity production from biomass by advanced thermal conversion technologies. *Biomass and Bioenergy*, **5,** 339–45.

Gupta, S. and Ravindranath, N. H. (1994). *Financial analysis of cooking energy options for India*. CES, Indian Institute of Science, Bangalore.

Gurumurthy, K., Raturi, D. P., and Bhandari, (1984). Biomass production in energy plantations of *Prosopis juliflora*. *Indian Forester*, September, 879–92.

Hall, D. O. (1991). Biomass energy. *Energy Policy*, **19,** 711–37.

Hall, D. O. (1993). Biomass energy. Chapter 4, In *Tree plantation review* (ed. Shell/WWF). Shell International Petroleum Company Limited and WWF, Surrey.

Hall, D. O. and Mao, Y. S. (1994). *Biomass energy and coal in Africa*. Zed Books Ltd, London.

Hall, D. O., Mynick, H. E., and Williams, R. H. (1990). *Carbon sequestration vs. fossil fuel substitution*, CEES Report No. 255. Princeton University, Princeton.

Hall, D. O., Mynick, H. E., and Williams, R. H. (1991). Cooling the greenhouse with bioenergy. *Nature*, **353,** 11–12.

Hall, D. O., Woods, J., and Scurlock, J. M. O. (1992). Biomass production and data. In *Photosynthesis and production in a changing environment: a field and laboratory manual* (ed. D. O. Hall *et al.*), Appendix C. Chapman & Hall, London.

Hall, D. O., Rosillo-Calle, F., Williams, R. H., and Woods, J. (1993). Biomass for energy: supply prospects. In *Renewable energy* (ed. T. B. Johansson, H. Kelly, A. K. N. Reddy, and R. H. Williams), pp. 593–651. Island Press, Washington.

Hibajene, S. H. and Kaweme, S. (1993). *Electrification of low income households*. SEI and Ministry of Energy and Water Development, Zambia.

Hohmeyer, O. (1992). Renewables and full cost of energy. *Energy Policy*, **20**, 365–76.

Houghton, R. A. (1991). Tropical deforestation and atmospheric carbon dioxide. *Climatic Change*, **19**, 99–118.

IIPO (1990). *Report of the survival rate of trees*. Indian Institute of Public Opinion, New Delhi.

Indian Express (1993). Subsidy on petroleum fuels, 30 July, Bangalore.

India Today (1994). Renewable energy for sustainable development, 31 August, 85–101.

IPCC (1990). *Climate change, IPCC impact assessment*. Intergovernmental Panel for Climate Change, Australian Government Publishing Service, Canberra.

IPCC (1991). *Climate change: the IPCC response strategies*. Intergovernmental Panel for Climate Change, Island Press, Washington DC.

IPCC (1992). *Climate change: 1992 supplementary report to IPCC scientific assessment*. Intergovernmental Panel for Climate Change, Cambridge University Press.

Jackson, T. (1992). Renewable energy; summary paper for renewable series. *Energy Policy*, **20**, 861–83.

Jagadish, K. S. and Ravindranath, N. H. (1990). Evaluation of the ASTRA stove programme. In *Bringing stoves to the people – an assessment of impact* (ed. S. Joseph, K. Krishna Prasad, and H. B. Van der Zann), pp. 33–44. ACTs Press Publication, Nairobi.

Jain, B. C. (1989). Assessment of current status and the potential for commercial exploitation of biomass gasification in India. *Biomass*, **18**, 205–19.

Jain, R. K. (1991). Fuelwood characteristics from Central India. *Biomass and Bioenergy*, **1**, 181–3.

Jodha, N. S. (1990). *Rural common property resources: contributions and crisis*. Society for Promotion of Wasteland Development, New Delhi.

Johansson, T. B., Kelly, H., Reddy, A. K. N., and Williams, R. H. (1993). *Renewable energy*. Island Press, Washington.

Joshi, A. P., Badola, R. C., and Singh, D. P. (1993). Biomass gasification for villages: towards second phase. In *Recent advances in biomass gasification and combustion* (ed. P. J. Paul and H. S. Mukunda), pp. 274–9. Interline Publishers, Bangalore.

Joshi, V., Sadafulla, P., and Pal, R. C. (1989). *Evaluation of improved stoves in Tamil Nadu, Rajasthan and West Bengal*. Tata Energy Research Institute (TERI), New Delhi.

Joshi, V., Raman, P., Mande, S. P., and Kishore, V. V. N. (1992*a*). *Techno-economic viability of the mobile unit for repair and maintenance of biogas plants*. TERI, New Delhi.

Joshi, V., Sinha, C. S., Karuppasawmy, M., Srivastava, K. K., and Singh, P. B. (1992*b*). *Rural energy data base*. TERI, New Delhi.

Kadekodi, G. and Ravindranath, N. H. (1994). *Economic and policy analysis of forestry as a carbon sequestration option for India*. Centre for Ecological Sciences, Indian Institute of Science, Bangalore.

Kapoor, R. P. (1992). Investment in afforestation on trends and prospects. In *The price of forests* (ed. A. Agarwal), pp. 173–7. Centre for Science and Environment, New Delhi.

Krishnakumar, S. (1993). India to optimize bioenergy use. In *Managing biodiversity for energy* (ed. K. Chaturvedi), pp. 11–14. Indian Association for the Advancement of Science, New Delhi.

Kulakarni, A., Sant, G., and Krishnayya, J. G. (1994). Urbanization in search of energy in three Indian cities. *Energy*, **19**, 540–60.

Kumar, A. and Joshi, M. C. (1972). The effects of grazing on the structure and productivity of the vegetation near Pilani, Rajasthan, India. *Ecology*, **60**, 665–79.

Kutty, K. (1992). Wood resources in homesteads of Kerala. In *The price of forests* (ed. A. Agarwal), pp. 107–10. Centre for Science and Environment, New Delhi.

Larson, E. D. (1993). Technology for electricity and fuels from biomass. *Annual Review of Energy and Environment*, **18**, 567–630.

Leach, G. (1987). *Household energy in South India*. Elsevier Applied Science, Essex.

Leach, G. (1992). The energy transition. *Energy Policy*, **20**, 116–23.

Leach, G. (1993). *Pakistan household energy strategy study (HESS)*. Prepared for the Government of Pakistan under UNDP Pak/88/036 by Energy Sector Management Assistance Programme in Association with the Energy Wing.
Leach, G. and Mearns, R. (1988). *Beyond the woodfuel crisis*. Earthscan Publications, London.
Ledig, F. T. and Kitzmiller, J. H. (1992). Genetic strategies for afforestation in the face of global climate change. *Forest Ecology Management,* **50,** 153–69.
Lewis, N. B. (1967). Regeneration of man-made forests. *Proceedings of the World Symposium on man-made forests and their industrial importance,* pp. 321–43. FAO, Rome.
Lokras, S. S., Babu, D. S. S., Bhogle, S., Jagadish, K. S., and Kumar, R. (1983). Development of an improved three pan stove. *Proceedings of the ASTRA Seminar*. Indian Institute of Science, Bangalore.
Lovejoy, D. (1992). Electrification of rural areas by solar PV. *Natural Resource Forum,* **16,** 102–10.
Lundgren, B. (1980). *Plantation forestry in tropical countries – physical, biological potentials, risks*, Rural Development Studies, No. 8. Swedish University of Agricultural Sciences, International Rural Development Center, Uppsala, Sweden.
Macall, M. K. (1987). *Zambia wood energy study*. University of Twentee, The Netherlands.
Macauley, M. (1989). Fuelwood use in urban areas; a case study in Raipur, India. *Energy Journal*, **10,** 157–80.
Marland, G. and Marland, S. (1992). Should we store carbon in trees? *Water, Air and Soil Pollution*, **64,** 181–5.
Marrison, C. I. and Larson, E. D. (1994). An estimate of the biomass energy production potential in Africa. Centre for Energy and Environmental Studies, Princeton University (Draft), USA.
Melillo, J. M., McGuire, A. D., Kicklighter, D. W., Moore, B., Vorosmarty, C. J., and Schloss, A. L. (1993). Global climate change and terrestrial net primary production. *Nature*, **363,** 234–9.
Ministry of Agriculture (1992). *Land use statistics*. Government of India, New Delhi.
Ministry of Commerce (1992). *Monthly statistics of the foreign trade of India, I and II*. Government of India, New Delhi.
Ministry of the Environment (1986). *Report of the inter-ministerial group on wood substitution*. Government of India, New Delhi.
Ministry of the Environment (1990). *Developing India's wastelands*. Government of India, New Delhi.
Mitchell, J. F. B., Manabe, S., Meleshko, V., and Tokioka, T. (1990). Equilibrium climate change – and its implications for the future. In *Climate Change: the IPCC Assessment* (ed. J. T. Houghton, G. J. Jenkins and J. J. Ephraums), pp. 131–72. Cambridge University Press, Cambridge.
Mitra, A. P. (1992). *Greenhouse gas emissions in India*, Scientific report No.4. Council of Scientific and Industrial Research and Ministry of Environment and Forests, New Delhi.
MNES (1993*a*). *Annual report 1992–93*. Ministry of Non-Conventional Energy Sources, New Delhi.
MNES (1993*b*). *Urja Bharathi 3*. Ministry of Non-Conventional Energy Sources, New Delhi, India.
MNES (1993*c*). *Report of task force formulation of national programme on biomass based co-generation in India*. Ministry of Non-Conventional Energy Sources, New Delhi.

Mohan, M. (1984). *Forest economies and valuation*. Medhawi Publishers, Dehra Dun, India.

Moncada, P., Schenler, W. W., and Connors, S. R. (1993). *Bio-electricity options*, MIT-Energy Laboratory Report MIT-EL-93-004 WP. MIT, Mass., USA.

Mooney, H. A., Drake, B. G., Luxmoore, R. J., Oechel, W. C., and Pitelka, L. F. (1991). Predicting ecosystem responses to elevated CO_2 concentrations. *Bioscience*, **41,** 96–104.

Mrindoko, B. J. (1993). Making of a biogas project in Tanzania – Takagas Project. Paper presented at *International Conference on Medium and Large Scale Biogas Plants in Developing Countries*. DANIDA, Taastrup, Denmark.

Mukunda, H. S., Dasappa, S., and Shrinivasa, U. (1993). Open top wood gasifier. In *Renewable energy* (ed. T. B. Johansson, H. Kelly, A. K. N. Reddy, and R. H. Williams) pp. 699–728. Island Press, Washington.

Murry, C. H. and de Montalembert, M. (1992). Wood still a neglected energy source. *Energy Policy*, **20,** 516–21.

Muthurasu, D., Gnanavel, R., Karuppaswamy, M., Kumar, S., and Haridasan, T. M. (1993). Biomass gasifier in rice mill: prospects and economics in Madhurai district. In *Recent advances in biomass gasification and combustion* (ed. P. J. Paul and H. S. Mukunda), pp. 291–300. Interline Publishing, Bangalore.

Myers, S., Goldman, N., Martin, N., and Friedman, R. (1993). Prospects for power sector in nine developing countries. *Energy Policy*, **21,** 1123–32.

Nakicenovic, N. *et al.* (1993). Long term strategy for mitigating global warming. *Energy*, **18,** 401–609.

NCA (1976). *Report of the National Commission on Agriculture*. Ministry of Agriculture and Irrigation, New Delhi.

NCAER (1981). *Report of rural energy consumption in Northern India*. National Council for Applied Economic Research, New Delhi.

NCAER (1985). *Domestic fuel survey with special reference to Kerosene*. National Council for Applied Economic Research, New Delhi.

NCAER (1989). *Evaluation survey study of household biogas plants set up during 1984–85 and 1985–86*. National Council for Applied Economic Research, New Delhi.

NCAER (1992). *Evaluation survey of household biogas plants set up during seventh five year plan*. National Council for Applied Economic Research, New Delhi.

NIOH (1980). *Comparative epidemiological studies of the effect of air pollutants*. National Institute of Occupation Health, Ahmedabad.

NIRD (1991). *Rural development statistics 1990*. National Institute of Rural Development, Hyderabad.

Nishanka, S. K. and Misra, M. K. (1990). Ecological study of an Indian village ecosystem – energetics. *Biomass*, **23,** 319–26.

Nobel, P. S. (1991). *Physico-chemical and environmental plant physiology*. Academic Press, London.

NSS (1992). National Sample Survey. *Sarvekshana*, **3**.

NUTEK (1993). *Energy in Sweden 1993*. NUTEK, Stockholm, Sweden.

NWDB (1986). *Report of the committee on fuelwood*. National Wasteland Development Board, Ministry of Environment, New Delhi.

NWDB (1991). *Guidelines for microlevel planning*. National Wasteland Development Board, Government of India, New Delhi.

OECD (1993). *Energy statistics and balances of non-OECD countries 1990–91*. Organization for Economic Co-operation and Development, Paris.

OTA (1992). *Fueling development-energy technologies for developing countries*. Office of Technology Assessment, Congress of US, Washington.

Panda, A., Dabas, M., Varalaxmi, V., Shah, V. N., and Gupta, J. R. (1992). *Grass yield under community participation in Haryana*. TERI, New Delhi.

Pandya, S. M. and Sidha, V. K. (1987). Seasonal variation in plant biomass and primary production in desert vegetation at Khavda of Kutch district (Gujarat), India. *Tropical Ecology*, **28,** 57–68.

Parikh, J. K., Reddy, D. S., and Banerjee, R. (1994). *Planning for demand side management in the electricity sector*. Tata–McGraw–Hill, New Delhi.

Patterson, W. (1994). *Power from plants*. Earthscan, London.

Paul, P. J. and Mukunda, H. S. (1993). *Recent advances in biomass gasification and combustion*. Interline Publishing, Bangalore.

PC (1985). *Seventh five year plan 1985–1990*. Planning Commission, Government of India, New Delhi.

PC (1987). *Social forestry programme: evaluation report*. Programme Evaluation Organization, New Delhi.

PC (1989). *Agro-climatic regional planning: an overview*. Government of India, New Delhi.

PC (1992). *Report of the Eighth Five Year Plan*. Government of India, New Delhi.

Perlack, R. D., Ranney, J. W., and Russell, M. (1991). *Biomass Energy Development in Yunnan Province, China (preliminary evaluation)*, Oak Ridge National Laboratory Report, ORNL/TM – 1179. ORNL, Oak Ridge, Tennessee.

Poffenberger, M. and Singh, S. (1992). Forest management partnerships: regenerating India's forests. *Unasylva*, **43**, 46.

Puri, C. (1988). *Analysis of energy consumption and production pattern in Dhanwas: a survey report*. TERI, New Delhi.

Rajabapaiah, P., Jayakumar, S., and Reddy, A. K. N. (1993). Biogas electricity; the Pura village case study. In *Renewable energy* (ed. T. B. Johnansson, H. Kelly, A. K. N. Reddy, and R. H. Williams), pp. 787–816. Island Press, Washington.

Rajabapaiah, P., Somashekar, H. I., and Reddy, A. K. N. (1994). Biogas and rural energy centers. Paper presented at seminar on watershed development and biomass energy, Indian Institute of Science, Bangalore.

Rajagopalan, M. and Demaine, H. (1994). Issues in energy subsidies for irrigation pumping. *Energy Policy*, **22,** 89–95.

Raju Associates (1992). *Rice husk based cogeneration project of an integrated rice processing plant in Punjab in India – a case study*. Prepared for BUN and WINROCK International, New Delhi.

Ramakrishna, J. (1991). *The Indian national improved stove programme, country study II, improved biomass cookstove project*. East–West Centre, Hawaii, USA.

Ramakrishna, J., Durgaprasad, M. B., and Smith, K. R. (1989). Cooking in India: the impact of improved stoves on indoor air quality. *Environment International*, **15,** 341–52.

Ramesh, S., Sabharwal, S. C., Bhagat, G., and Sinha, C. S. (1988). *Study of decentralized energy options for the rural sector of India*. Report prepared for Rural Electrification Corporation, TERI, New Delhi.

Ranganathan, V. (1992). *Rural electrification in Africa*. AFREPEN, Zed Books Ltd, London.

Ranganathan, V. (1993). Rural electrification revisited. *Energy Policy*, **21,** 142–51.

Ranganathan, V., Rao, S. S., and Prabhu, G. S. (1993). *Demand and supply of fuelwood in Karnataka*. Indian Institute of Management, Bangalore.

Ranney, J. W. (1992). Environmental considerations in biomass resource management for a greenhouse-constrained society. In *Technologies for a greenhouse-constrained society* (ed. M. A. Kuliahsa, A. Zuckher and K. J. Ballew), pp. 519–42. Lewis Publ., Boca Raton, US.

Ravindranath, N. H. (1993). Biomass gasification: environmentally sound technology for decentralized power generation; a case study from India. *Biomass and Bioenergy*, **4**, 49–60.

Ravindranath, N. H. and Chanakya, H. N. (1986). Biomass based energy system for a South Indian village. *Biomass*, **9**, 215–33.

Ravindranath, N. H. and Gadgil, M. (1991). *Natural resource management and sustainable agricultural development*. CES and ASTRA, Indian Institute of Science, Bangalore.

Ravindranath, N. H. and Mukunda, H. S. (1993). *Rural energy center based on energy forest-wood gasifier system at Hosahalli village*. ASTRA, Indian Institute of Science, Bangalore.

Ravindranath, N. H. and Ramakrishna, J. (1994). *Energy options for cooking in India*, Center for Ecological Sciences, and ASTRA, Indian Institute of Science, Bangalore. (Submitted to *Energy Policy*.)

Ravindranath, N. H. and Somashekar. H. I. (1991). Biomass production, utilization and conservation in a semiarid area. In *Karnataka state of environment report 5* (ed. C. J. Saldhana), pp. 21–32. Center for Taxonomic Studies, Bangalore.

Ravindranath, N. H., Nagaraju, S. M., Channeswarappa, A., Balakrishna, M., Balachandran, B., and Reddy, A. K. N. (1981). Indian village agricultural ecosystem – case study of Ungra village part 1: main observations. *Biomass*, **1**, 61–76.

Ravindranath, N. H., Babu, D. S. S., and Shailaja, R. (1989*a*). Conservation of biomass; potential of a fuel efficient wood stove. *Energy Management*, **13**, 14–20.

Ravindranath, N. H., Babu, D. S. S., Shailaja, R., and Jagadish, K. S. (1989*b*). *Cooking practices using ASTRA stove in different food zones of Karnataka* ASTRA, Indian Institute of Science, Bangalore.

Ravindranath, N. H., Shailaja, R., and Revankar, A. (1989*c*). Dissemination and evaluation of fuel efficient and smokeless ASTRA stove in Karnataka. *Energy Environment Monitor*, **5**, 48–60.

Ravindranath, N. H., Duttaprasad, H. L., and Somashekar, H. I. (1990). A rural energy system based on energy forest and wood gasifier. *Current Science*, **59**, 557–60.

Ravindranath, N. H., Nayak, M. M., Hiriyur, R. S., and Dinesh, C. R. (1991). The status and use of tree biomass in a semi arid village ecosystem. *Biomass and Bioenergy*, **1**, 9–16.

Ravindranath, N. H., Chanakya, H. N., and Rajabapaiah, P. (1992*a*). *Biogas energy options for India: constraints and strategies*. ASTRA, Indian Institute of Science, Bangalore.

Ravindranath, N. H., Somashekar, B. S., and Gadgil, M. (1992*b*). *Carbon emission and sequestration in forests: case studies for India and China*, Lawerence Berkley Laboratory, LBL:32759 UC-402. LBL Berkley, USA.

Ravindranath, N. H., Somashekar, H. I., Shailaja, R., Parthasarathy, C. K., and Jagadish, K. S. (1992*c*). Study of tree plantation resources in a semi-arid region of Karnataka. *Energy Environment Monitor*, **8**, 71–7.

Ravindranath, N. H., Nambi, V. A., and Gadgil, M. (1994*a*). *Natural regeneration; a forestry option for degraded land of India*. CES, Indian Institute of Science, Bangalore.

Ravindranath, N. H., Nambi, V. A., Indrani, S., and Vinod, V. R. (1994*b*). Prosopis

juliflora *and household energy*. CES Technical Report, Indian Institute of Science, Bangalore, India.
Reddy, A. K. N. (1981). An Indian village agricultural ecosystem – case study of Ungra village Part II. *Biomass*, **1,** 77–88.
Reddy, A. K. N. (1982). Rural energy consumption patterns – a field study. *Biomass*, **2,** 255–80.
Reddy, A. K. N. (1991). Barriers to improvements in energy efficiency. *Energy Policy*, **19,** 953–61.
Reddy, A. K. N. and Ravindranath, N. H. (1987). Biomass and village energy and rural development. In *Biomass: regenerable energy* (ed. D. O. Hall and R. P. Overend), pp. 387–412. John Wiley and Sons, Chichester.
Reddy, A. K. N. and Reddy, B. S. (1983). Energy in a stratified society: case study of firewood in Bangalore. *Economic and Political Weekly*, **41,** 1757.
Reddy, A. K. N. and Reddy, B. S. (1994). Substitution of energy sources for cooking in Bangalore. *Energy*, **19,** 561–72.
Reddy, A. K. N. and Subramanian, D. K. (1980). Design of rural energy centres. *Proceedings of the Indian Academy of Sciences*, **C II,** 395–416.
Reddy, A. K. N., Sumithra, G. D., Balachandra, P., and D'sa, A. (1990). Comparative costs of electricity conservation; centralised and decentralized electricity generation. *Economic and Political Weekly*, **3,** 1201–16.
Reddy, A. K. N., Sumithra, G. D., Balachandra, P., and D'sa, A. (1991). A development focussed end use oriented electricity scenario for Karnataka. *Economic and Political Weekly*, **14,** 891–910, 983–1001.
Rosillo-Calle, F. (1990). Liquid fuels. In *Bioenergy and environment*, (ed. J. Pasztor and L. A. Kristoferson), pp. 113–62. Westview Press, USA.
Sampson, R. N. (1993). Forest management and biomass in the USA. *Water, Air and Soil Pollution*, **70,** 519–32.
Saxena, N. C. (1991). Farm forestry in India; a literature review. In *Proceedings of the international workshop on socio-economic aspects of tree growing*. Institute of Rural Management, Anand, India.
Saxena, N. C. (1992). Farm forestry and land-use in India; some policy issues. *Ambio*, **24,** 420–5.
Seebauer, M. (1992). *Review of social forestry programmes in India*. GWB Gesselschaft Für Walderhaltung und Waldbewirtschaftung GMBH, Michelstadt, Germany.
Shailaja, R. and Ravindranath, N. H. (1990). Women and rural environment. In *State of environment report IV* (ed. C. J. Saldanha), pp. 7–16. Center for Taxonomic Studies, Bangalore.
Shailaja, R., Ravindranath, N. H., Somashekar, H. I., and Jagadish, K. S. (1994). Biomass generation in mixed tree plantations. *Energy and Sustainable Development*, **1**, 51–5.
Shangbin, G. (1993). Development of biomass and biogas programmes in China. Paper presented at *International Conference on Medium and Large Scale Biogas Plants in Developing Countries*. DANIDA, Taastrup, Denmark.
Sharma, A., Prasad, R., Venkataramana, P., and Joshi, V. (1991). *Study of energy use and environmental effects in the Garwhal region of the Central Himalaya and an action plan for mitigation*. Tata Energy Research Institute (TERI), New Delhi.
Sharma, R. B., Chandra, M., and Singh, K. P. (1990). Variations in forest biomass and productivity. Chapter 5 in *Environmental degradation at Obra-Renukoot-Sing-*

rauli Area and its impact on nature and derived ecosystems Final Report. (ed. Singh, J. S., *et al.*). MAB Project, Banaras Hindu University, Varanasi.

Shastry, C. M. and Ravindranath, N. H. (1994). *Evaluation of improved ASTRA stove in Sirsi*. CES, Indian Institute of Science, Bangalore.

Shell and WWF (1993). Plantation guidelines. In *Tree plantation review*. Shell International Petroleum Company and WWF, Surrey.

Siddayao, C. M. (1992). Energy investments and environmental implications. *Energy Policy*, **20,** 223–32.

Singh, B. (1988). Biomass potentials of some firewood shrubs of North India. *Biomass*, **16,** 189–203.

Singh, D. V. (1993). Energy-use pattern of rural households in Himalayas: case of Himachal Pradesh. *Energy Environment Monitor*, **9,** 11–14.

Singh, O., Sharma, D. C., and Rawat, J. K. (1992). Production and decomposition of leaf litter in sal, teak, eucalyptus and poplar forest in Uttar Pradesh. *Indian Forester*, **119,** 112–27.

Sinha, C. S. and Kandpal, T. C. (1991). Decentralized versus grid electricity for rural India: the economic factors. *Energy Policy*, **19,** 441–8.

Smith, K. R. (1991). Biomass cookstoves in global perspectives: energy, health and global warming. In *Indoor air pollution from biomass fuel*, working papers from a WHO consultation. WHO, Geneva.

Smith, K. R. (1993). Fuel combustion, air pollution exposure and health: the situation in developing countries. *Annual Review of Energy and Environment*, **18,** 529–56.

Smith, K. R., Apte, M. G., Yuquing, M., Wongsekiarttirat, W., and Kulakarni, A. (1994). Air pollution and the energy ladder in Asian cities. *Energy*, **19,** 587–600.

SPWD (1984). *Developing India's wasted lands*. Society for the Promotion of Wasteland Development, New Delhi.

Srinivas, S. N., Ravindranath, N. H., Dasappa, S., Shrinivasa, U., and Mukunda, H. S. (1992). Wood gasifier based rural power generation system: a case study. *Pacific and Asian Journal of Energy*, **2,** 81–91.

Subramanya, N. S., Rao, K. V. P., and Ramprasad, M. S. (1993). Development strategies for effective dissemination of gasifiers, sterling engines. In *Recent advances in biomass gasification and combustion* (ed. P. J. Paul and H. S. Mukunda). Interline Publishing, Bangalore, India.

TERI (1992*a*). *Fuelish trends and wise choices: options for the future*. Tata Energy Research Institute, New Delhi.

TERI (1992*b*). *TERI energy data directory and yearbook 1990–91*. Tata Energy Research Institute, New Delhi.

TERI (1993). *Rays of hope – renewable energy for sustainable development*. Tata Energy Research Institute, New Delhi.

The Economist. (1994). Where India's reforms get stuck, 22 January, London.

The Observer (1994). India will need 220 Mt of food grain by 2000: report, 5 August, New Delhi.

Tiwari, A. K. and Singh, J. S. (1987). Analysis of forest land use and vegetation in a part of Central Himalaya using aerial photographs. *Environmental Conservation*, **144**, 233–44.

Tiwari, D. N. (1994). Forestry research for sustainable development. *Indian Forester*, **120**, 3–11.

TSL (1992). *Statistical outline of India*. Tata Services Limited, Bombay.

Turnbull, J. H. (1993). Use of biomass in electric power generation: the California experience. *Biomass and Bioenergy*, **4**, 75–84.

UN (1991). *World urbanization prospects 1990*. United Nations, New York.
UN (1992). *Energy statistics year book 1990*. United Nations, New York.
UNEP (1994). *Environmental data report 1993–94*. Prepared for UNEP by the GEMS Monitoring and Assessment Research Centre, London, UK, in co-operation with the World Resources Institute, United Nations Environment Program, Washington DC.
UNPF (1992). *The state of world population fund*. United Nations Population Fund, New York.
USDOE (1993). *Electricity from biomass: national biomass power programme five year plan*. US Department of Energy, Washington.
Veena, D. R. (1988). *Rural energy consumption, problems and prospects*. Ashish Publishing House, New Delhi.
Venkataramana, P. (1991). Biogas program in India. *TIDE*, **1**, 1–17.
Venkataramana, P. (1992). *Community biogas system in Methan, Gujarat – a case study*. TERI, New Delhi.
WB (1992). *World development report 1992*. World Bank, Oxford University Press.
WB (1994). *World development report 1994, infrastructure for development*. World Bank, Oxford University Press.
WEC (1993). *Energy for tomorrows world*. World Energy Council, St Martins Press, London.
Williams, R. H. (1994). *Toward an energy industrial renaissance*. Prepared for the European Business School, Schloss Reichartshausen, Oestrich–winkel, Germany.
Williams, R. H. and Larson, E. D. (1993). Advanced gasification-based biomass power generation. In *Renewable energy* (ed. T. B. Johansson, H. Kelly, A. K. N. Reddy, and R. H. Williams), pp. 729–85. Island Press, Washington.
Woods, J. and Hall, D. O. (1994). *Bioenergy for development: technical and environmental dimensions*. FAO Environment and Energy Paper 13, FAO, Rome.
WRI (1992). *World Resources 1992–93*. Oxford University Press, New York.
WRI (1994). *World Resources 1994–95*. Oxford University Press, New York.
Wyman, C. E., Bain, R. L., Hinman, N. D., and Stevens, D. J. (1993). Ethanol and methanol from cellulosic biomass. In *Renewable energy* (ed. T. B. Johansson, H. Kelly, A. K. N. Reddy, and R. H. Williams), pp. 865–923. Island Press, Washington.

Conversion factors

Energy: 1 W (Watt) = 1 J/s (Joule per second)
1 kWh (kilowatt hour) = 3.6 MJ
1 cal (calorie) = 4.2 J

1 t wood = 15 GJ (air dry, 20% moisture)
= 20 GJ (oven dry, 0% moisture)
1 t crop residues = 13 GJ (air dry)
1 t dung cake = 14 GJ (air dry)
1 t sugarcane bagasse = 19 GJ (oven dry)
1 t charcoal = 28 GJ
1 t fuelwood = 1.5 m^3 (approx. depending on species)
1 t ricehusk = 13 GJ

1 t oil equivalent (toe) = 42 GJ
1 million t oil equivalent (Mtoe) = 42 PJ
1 boe (barrel oil equivalent) = 0.159 m^3 (159 litres) = 5.7 GJ
1 t coal equivalent (tce) = 29 GJ
1 million t coal equivalent (Mtce) = 29 PJ

1 t methanol = 20 GJ
1 t ethanol = 27 GJ
1 m^3 biogas = 23 MJ
1 m^3 natural gas = 37 MJ

Metric/Imperial conversions
1 metric tonne = 1000 kg = 0.984 long tons
1 gallon (US) = 0.833 gallons (Imperial) = 3.785 litres
1 hectare (ha) = 2.47 acres

Prefixes
kilo (k) = 10^3
Mega (M) = 10^6
Giga (G) = 10^9
Tera (T) = 10^{12}
Peta (P) = 10^{15}
Exa (E) = 10^{18}

Currency conversion
$US1 = Rs31 (during 1993–95)

Index

energy/carbon balance

buildings/structures
heating/cooling
transport
Machines
food production

---

energy production - the process encompases all above - net carbon and energy balance

integration strategy - approaching the problem from all major angles No point in producing more energy in a better way to use more in another way to negate positive effects.